MW01042743

ROCK SLOPE STABILITY

2ND EDITION

CHARLES A. KLICHE

Published by the
Society for Mining, Metallurgy & Exploration

Society for Mining, Metallurgy & Exploration (SME)
12999 E. Adam Aircraft Circle
Englewood, Colorado, USA 80112
(303) 948-4200 / (800) 763-3132
www.smenet.org

The Society for Mining, Metallurgy & Exploration (SME) is a professional society whose more than 15,000 members represents all professionals serving the minerals industry in more than 100 countries. SME members include engineers, geologists, metallurgists, educators, students and researchers. SME advances the worldwide mining and underground construction community through information exchange and professional development.

ISBN 978-0-87335-369-4
eBook 978-0-87335-370-0

Library of Congress Cataloging-in-Publication Data

Names: Kliche, Charles A., 1951- author.
Title: Rock slope stability / Charles A. Kliche.
Description: Second edition. | Englewood, Colorado : Society for Mining, Metallurgy & Exploration, [2019] | Includes bibliographical references and index.
Identifiers: LCCN 2018017397 (print) | LCCN 2018017699 (ebook) | ISBN 9780873353700 | ISBN 9780873353694
Subjects: LCSH: Rock slopes. | Rock mechanics. | Ground control (Mining) | Stability.
Classification: LCC TA706 (ebook) | LCC TA706 .K62 2019 (print) | DDC 624.1/5132--dc23
LC record available at https://lccn.loc.gov/2018017397

Contents

Preface

The process of assessing a rock slope for stability involves the application of many branches of engineering and natural science. The determination of rock or soil strength parameters requires a working knowledge of rock mechanics and/or soil mechanics. Knowledge of basic hydrological concepts is essential for interpreting the effects of water on a rock slope. An understanding of geology—especially structural geology, mineralogy, petrology, and the geological processes—is extremely important for the rock slope engineer. These areas of knowledge—geology, rock and soil mechanics, and hydrology—along with other basic areas such as statics, engineering mechanics, statistics, and structural engineering are all important elements of rock slope engineering and design.

The computer has become an invaluable tool for the engineer and scientist. With its aide, the geotechnical engineer can make numerous computations or consider many design alternatives in a short period of time. Additionally, modern rock slope instrumentation and monitoring systems use computerized data acquisition and state-of-the-art data transmission. The rock slope engineer must be familiar with computer analysis techniques and some of the current three-dimensional visualization techniques.

The danger, however, with computerized analysis techniques is that engineers who do not have the proper background will not know for certain whether the results are realistic—in other words, "garbage in, garbage out." Therefore, it is imperative that rock slope stability engineers have a solid understanding of the many methodologies in common use for the analysis of the stability of rock slopes. On the basis of experience and a solid engineering background, they should be able to provide and verify the data used in the analysis, verify the results of the investigation, and then make proper recommendations for remedial actions.

The emphasis of this book is on rock slope stability, with sections on geological data collection, geotechnical data collection and analysis, surface water and groundwater effects, kinematic and kinetic stability analysis, rock slope stabilization techniques, and rock slope instrumentation and monitoring. Numerous examples of solutions to typical rock slope engineering problems are given. Additionally, many figures and photographs are included to aid the reader in understanding the various concepts.

During development of the manuscript, the scope was intentionally broadened to include not only mining engineering and geological engineering applications but also civil engineering. Therefore, the text is intended to be a reference and guide for mining and construction engineers responsible for rock slope stability.

Work began on this second edition in 2014. It was envisioned for this edition that, in addition to correcting any errata discovered in the first edition, sections within the book would be expanded to include the following:

- Within Chapter 2, a discussion of three-dimensional imaging techniques for mapping discontinuities and a section on engineering rock mass classification schemes.
- Significantly more on the rotational shear mode of failure. Consequently, Chapter 6 on Kinetic Slope Stability Analysis of Shear Failure was expanded to include examples, and some of the more common techniques for analysis, of rotational shear.
- An expanded section on catch bench design in Chapter 9, Rock Slope Stabilization Techniques.

As the work developed, much was added to, or replaced in, the original chapters to modernize the book as much as possible. The overall goal, however, remained: to provide fundamental knowledge in the field of rock slope stability to those interested in or responsible for rock slope engineering in both construction and mining applications.

Acknowledgments

In the acknowledgments to the first edition, I thanked and acknowledged certain individuals and companies, including the late Richard L. Klimpel, Tad Szwedzicki, Don Berger, Doug Hoy, Mark Keenihan, Russ Kliche, Dave Pierce, Hai-Tien Yu, Jim Bachmann, Howard Grant, Frank Olivieri, Phil Murray, and my loving wife Donna Kliche, Homestake Mining Company, Wharf Resources, Barrick Goldstrike, Newmont Mining, BHP, Modular Mining Systems, and Slope Indicator Company. I thank you again.

To the preceding list I add Scott Durgin, Nichole King, Ekrem Tamkan, Bob McClure, Tamara Wiseman, Brian Wenig, Alvis Lisenbee, F.W. Breithaupt and Sohn, Peabody Energy Company, Cloud Peak Energy, Freeport-McMoRan New Mexico Operations, and RAM Inc.

The first edition was published in 1999. In 1997 our daughter, Alexandra Veturia, to whom the first edition was dedicated, was born. A very substantial portion of the work on the first edition was done by me during the daytime periods and late hours while Alex slept. Alex is now a junior in engineering at my alma mater. How time flies!

The work for the second edition was completed over a period of about the last five years. I started, then stopped, started again, then stopped. It was very difficult staying motivated. However, with the encouragement of family and very special friends, I managed to plod along until this point. I also thank Jane Olivier, SME manager of book publishing, and Diane Serafin, my editor, for bearing with me.

A very special thank you is extended to the following people: Arden Davis, who reviewed and suggested changes to Chapter 3, Groundwater; my old friend, S.A. Gauger, who always provided that extra little bit of encouragement; and the wonderful people I worked with for so very long in the mining engineering program at the South Dakota School of Mines and Technology: Duff Erickson; S.N. Shashikanth; the late Ziggy Hladysz, who is sorely missed by all; and Cindy Hise, the mining program's secretary extraordinaire.

About the Author

Charles A. Kliche is professor emeritus of Mining Engineering and Management at the South Dakota School of Mines and Technology (SDSMT). He has more than 45 years of varied mining experience in industrial minerals (bentonite mining), taconite mining, gold mining, consulting, and mining education. Kliche holds B.S. and M.S. degrees in mining engineering from the SDSMT and a Ph.D. degree in mining engineering from the University of Arizona.

Kliche has worked extensively as an independent mining consultant in the fields of rock slope stability, explosives engineering, and the environmental effects of mining. He has written numerous papers on those subjects as well as this second edition of *Rock Slope Stability* and Chapter 8.3, Slope Stability, in the third edition of the *SME Mining Engineering Handbook*, both published by SME.

Kliche is a registered professional engineer in the states of Minnesota and South Dakota. He is also a member of the National Society of Professional Engineers and the South Dakota Engineering Society.

Kliche is active in SME and the International Society of Explosives Engineers (ISEE), both on the local and national level. He was a founder and first chairman of the Black Hills chapter of the ISEE and also served on its board of directors for many years. Kliche served as chairman of the Black Hills section of SME in 1986, 1999, and, for the third time, from 2006 to 2016. He has served in various capacities for the SME and ISEE national meetings.

CHAPTER 1

Basic Concepts

Because of the discontinuous nature of rock, the design of stable rock slopes is as much an art as it is applied engineering. Experience is as important as the proper utilization of the theories of soil and rock mechanics, structural geology, and hydrology.

Many computerized tools are now available to rock slope design engineers. It is important that these engineers understand the basic theory of rock slope stability (or instability) before they attempt to use many of the computerized methods—especially before they attempt to interpret and apply the results.

SLOPE STABILITY AS AN ENGINEERING ISSUE

The civil engineering field uses mainly soil mechanics principles for slope stability analysis. This field is mostly concerned with slopes cut in loose, granular, or unconsolidated materials. The analysis is used primarily for

- Foundations, buildings, or dam sites;
- Road cuts;
- Cut-and-cover tunneling;
- Irrigation channels;
- Tailings dams; and
- Mine dumps.

The mining engineering field, on the other hand, uses mainly rock mechanics principles to analyze the stability of slopes cut in rocks. Rock mechanics is more complicated than soil mechanics for the following reasons (Brawner and Milligan 1971):

- Rock materials are heterogeneous and usually anisotropic.
- Strength parameters relating to rock masses are infinitely variable and difficult, if not impossible, to determine precisely.

- Generalized models and theories of rock behavior are complex, as are the mathematics involved.
- Field conditions are extremely difficult, and often impossible, to duplicate in the laboratory.
- Field testing is usually complicated and time-consuming and is almost always very expensive.

Rock slope engineering is the application of rock mechanics principles and structural geology principles to the stability of a slope cut in rock. It is a specialized branch of geomechanical engineering. It includes not only kinetic analysis (possible modes of failure) and kinematic analysis (stability of the failure modes) but also probabilistic analysis, methods to stabilize the slope, groundwater analysis, geologic data collection, slope monitoring methods, and so on.

TERMINOLOGY

Before issues of slope stability can be addressed in detail, it is necessary to have a working knowledge of several important introductory terms. (The following terms, as well as additional definitions, may be found in the glossary.*)

To begin with, consider the individual terms in the expression "slope stability." The term *slope* may be defined as any inclined surface cut in natural material or the degree of inclination with respect to horizontal. Slope is usually expressed either as a ratio (such as 1.5:1, which indicates 1.5 units rise per 1 unit of horizontal distance), as a decimal fraction (1.5), as degrees from the horizontal (56.31°), or as a percentage (150%). The term *stability* may be defined as the resistance of a structure, slope, or embankment to failure by sliding or collapsing under normal conditions for which it was designed (e.g., bank stability and slope stability). Hence, *slope stability* may be defined as the resistance of any inclined surface, as the wall of an open pit or cut, to failure by sliding or collapsing.

Rock Strength Parameters

Figure 1.1 depicts important parameters relating to rock strength. The parameters of interest include the following:

- **Internal angle of friction:** The angle (ϕ) at which a body resting on an inclined surface will overcome frictional resistance and begin to slide, as measured between the normal to the surface and the resultant forces acting on the body. This angle is sometimes referred to simply as the friction angle.
- **Cohesion:** A property of like mineral grains, enabling them to cling together in opposition to forces that tend to separate them. Cohesion is that portion of the

* Some definitions have been reprinted with permission from the American Geological Institute (1997).

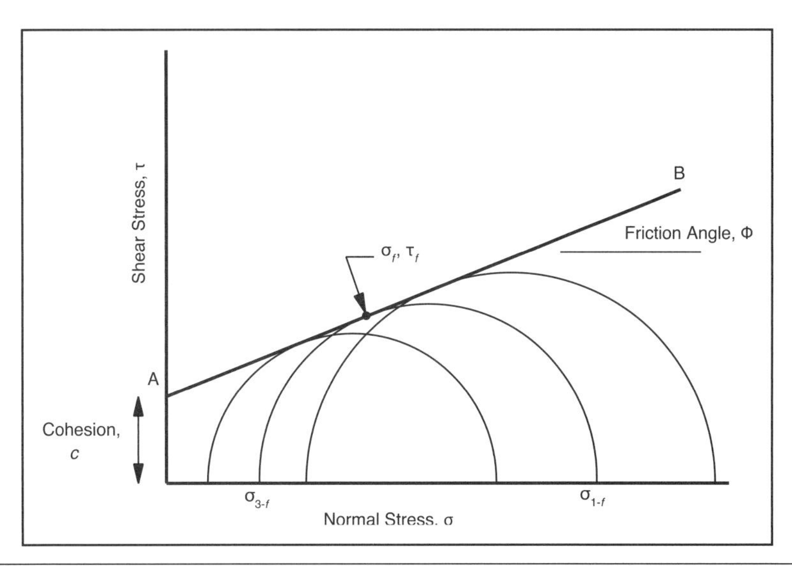

FIGURE 1.1 The Mohr envelope

shear strength, S, or shear stress, τ, indicated by the term c in the Mohr–Coulomb equation (see below).

- **Mohr–Coulomb criterion:** A rock failure criterion that assumes there is a functional relationship between the normal and shear stresses acting on a potential failure surface. The relationship takes the form $\tau = c + \sigma \cdot \tan\phi$, where τ is the shear stress, c is the cohesion, σ is the normal stress, and ϕ is the internal angle of friction. When shear strength, S, is used instead of shear stress, τ, the equation becomes the "shear strength criterion."
- **Mohr envelope:** The envelope of a series of Mohr circles representing stress conditions at failure for a given material. According to Mohr's rupture hypothesis, a failure envelope is the locus of points such that the coordinates represent the combinations of normal and shearing stresses that will cause a given material to fail. The Mohr hypothesis states that when shear failure takes place across a plane, the normal stress (σ) and shear stress (τ) across the plane are related by a functional relationship characteristic of the material (Jaeger and Cook 1979); that is, $|\tau| = f(\sigma)$. (Since the sign of τ affects only the direction of sliding, only the magnitude of τ is important.) The functional relationship can be represented by a curve in the σ-τ plane known as the Mohr envelope, such as curve A-B in Figure 1.1. If we have three principal stresses—σ_1, σ_2, and σ_3—and if $\sigma_2 = \sigma_3$ (as in a triaxial test on a cylindrical specimen), the values of σ and τ can be found by the Mohr construction of Figure 1.1. Failure will not occur if the values of σ and τ are below the curve A-B. However, consider the circle with center $(\sigma_1 + \sigma_3)/2$ and with the magnitude of the maximum principle stress (σ_1) minus the magnitude of the minimum principle stress (σ_3) as its diameter; failure will occur if this circle just touches A-B (Jaeger and Cook 1979). The failure values, σ_f and τ_f, are determined as the tangent point of the circle to the curve A-B,

and the values of σ_1 and σ_3 at failure are denoted as $\sigma_{1\text{-}f}$ and $\sigma_{3\text{-}f}$, respectively. To generate the Mohr envelope A-B, it is often necessary to conduct multiple triaxial strength tests of a given rock type and to plot the results of σ_1 at failure ($\sigma_{1\text{-}f}$) and σ_3 at failure ($\sigma_{3\text{-}f}$) for each test. For soils, the curve is usually straight; for rocks, the curve is usually concave downward as the normal stress increases. At low values of $\sigma_{1\text{-}f}$ and $\sigma_{3\text{-}f}$, as is often the case for rock slope failure, the curve A-B is assumed to be straight.

Slope Configuration

Figure 1.2 shows terms relating to slope configuration:

- *Bench:* A ledge that, in open-pit mines and quarries, forms a single level of operation above which minerals or waste materials are excavated from a contiguous bank or bench face. The mineral or waste is removed in successive layers, each of which is a bench and several of which may be in operation simultaneously in different parts of—and at different elevations in—an open-pit mine or quarry.
- *Bench angle:* The angle of inclination of the bench face, measured from the horizontal.
- *Berm:* A horizontal shelf or ledge built into the embankment or sloping wall of an open pit, quarry, or highway cut to break the continuity of an otherwise long slope and to strengthen the slope's stability or to catch and arrest loose, falling rock.
- *Catch bench:* A berm that is designed to provide a sufficient width to catch loose, fallen rock.
- *Crest:* The top of an excavated slope.
- *Face:* The more or less vertical surface of rock exposed by excavation.
- *Inter-ramp angle:* The slope, or slopes, lying between each ramp or ramp segment that depends on the number of ramps and their widths. Where there are haul roads, working levels, or other wide benches, the overall slope angle will be flatter than the inter-ramp angle, which, in turn, will normally be flatter than the bench angle.
- *Overall slope angle:* The angle measured from the horizontal to the line joining the toe of a wall and the crest of the wall.
- *Toe:* The bottom of a slope or cliff.

Slope Orientation

The following terms relate to slope orientation and are presented in Figure 1.3:

- *Dip angle:* The angle at which a bed, stratum, or vein is inclined from the horizontal, as measured normal to the strike and in the vertical plane. (Note that the inclination of a line, such as a borehole, is more accurately known as the *plunge.*) The term *dip* may also be used as a verb, as in "the vein dips toward the east."
- *Dip direction:* The bearing of the dip of a slope, vein, rock stratum, or borehole, measured normal to the direction of strike.

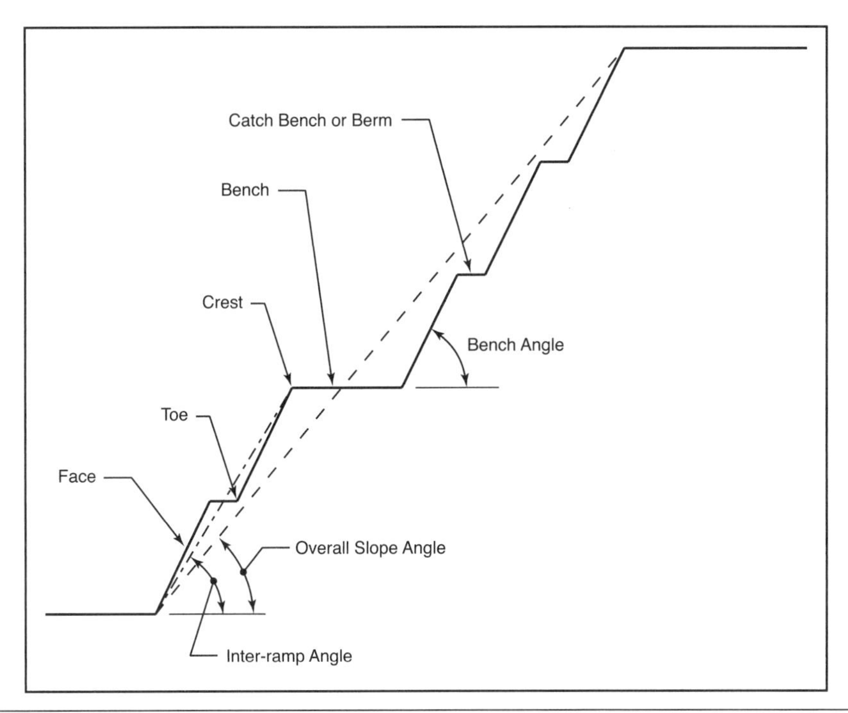

FIGURE 1.2 Highwall slope configuration

- *Strike:* The course or bearing of the outcrop of an inclined bed, vein, joint, or fault plane on a level surface; the direction of a horizontal line perpendicular to the dip direction. (Note that the course or bearing of a line, such as a borehole, is more accurately known as the *trend.*) The term *strike* may also be used as a verb, as in "the vein strikes in a northerly direction."

Rock Mass

Figure 1.4 depicts terms that relate to the rock mass. Shown on the figure are three sets of discontinuities and, for scale effect, a surface cut and an underground opening in the rock mass. The underground opening can be considered to be approximately 10 m^2 (107 ft^2), or 3.16 m × 3.16 m (10.4 ft × 10.4 ft). The circles to the side of the underground opening and on the bench face of the surface cut encompass several zones of transition, from intact rock to the heavily jointed rock mass. The smallest circle encompasses intact rock between the discontinuities; the next circle shows a single discontinuity. The next circles show two discontinuities, then several discontinuities, and finally the rock mass. As the viewing scale expands outward, the rock strength decreases from that of the intact rock to that of the rock mass. The strength parameters for rock slope design purposes, therefore, must be those of the rock mass and not of the intact rock.

Some of the important terms relative to Figure 1.4 are as follows:

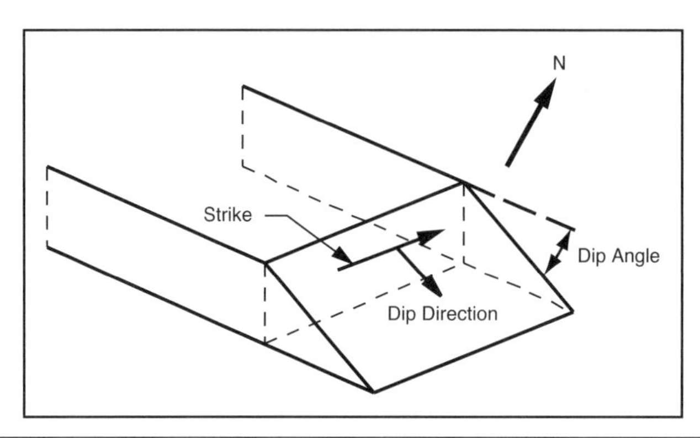

FIGURE 1.3 Orientation of a plane

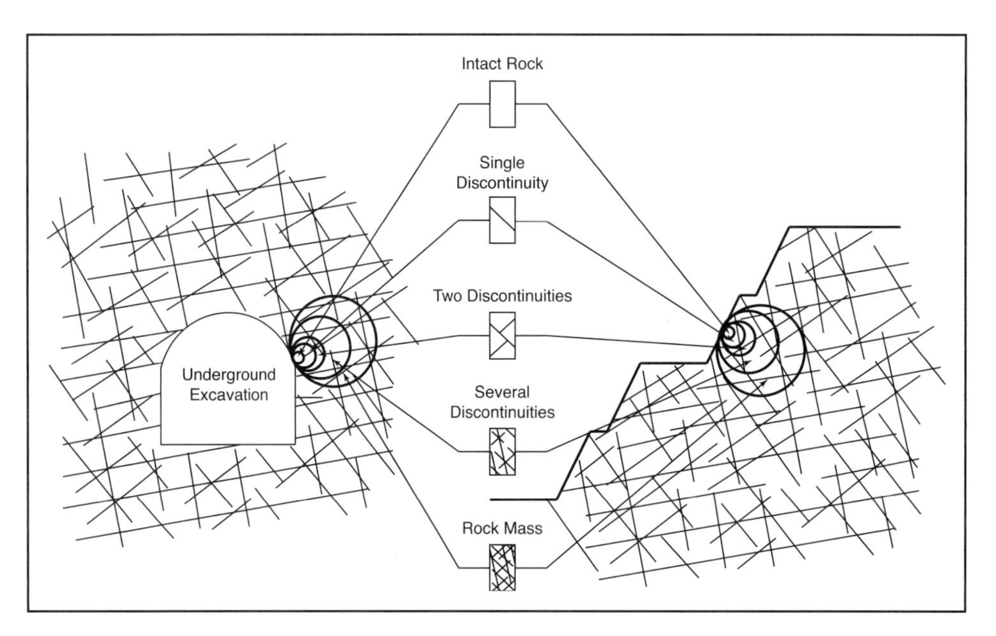

Source: Hoek 2007

FIGURE 1.4 Transition from intact rock to rock mass

- *Intact rock:* The primary unbroken rock as determined from a piece of core cut for compression testing. The term *rock substance* has also been used for the unbroken rock.
- *Discontinuity (or weakness plane):* A structural feature that separates intact rock blocks within a rock mass; a structural weakness plane upon which movement can take place.
- *Rock mass:* The in situ rock made up of the rock substance plus the structural discontinuities.
- *Major structures:* Geologic features such as faults or shear zones along which displacement has occurred and that are large enough to be mapped and located as

Courtesy of Alvis Lisenbee

FIGURE 1.5 Major structural feature (fault)

individual structures. There is actually a continuum between fractures and major structures, but the differentiation is useful for design purposes (Figure 1.5).

- *Minor structures:* These include fractures, joints, bedding planes, foliation planes, and other defects in the rock mass, generally of lesser importance with respect to rock slope instability.

Sectors

Figure 1.6 shows the term *sector*, which corresponds to an area of a cut, slope, or highwall to be analyzed for stability. A sector is the length of wall, pie-slice-shaped portion of wall, or portion of an excavation that can be considered sufficiently homogeneous to allow use of a single set of structural data, strength data, and orientation data. A *subsector* may be used when the orientation of a face or excavation changes within a constant sector.

SLOPE FAILURE CAUSES AND PROCESSES

Slope failure, whether in rock or soil, can be attributed to many causes. However, it is rare that a given failure can be attributed to any single cause. Usually several causes exist simultaneously to eventually trigger the slope failure. These factors can be grouped into two major categories (Varnes 1978): (1) factors that contribute to increased shear stress and (2) factors that contribute to low or reduced shear strength.

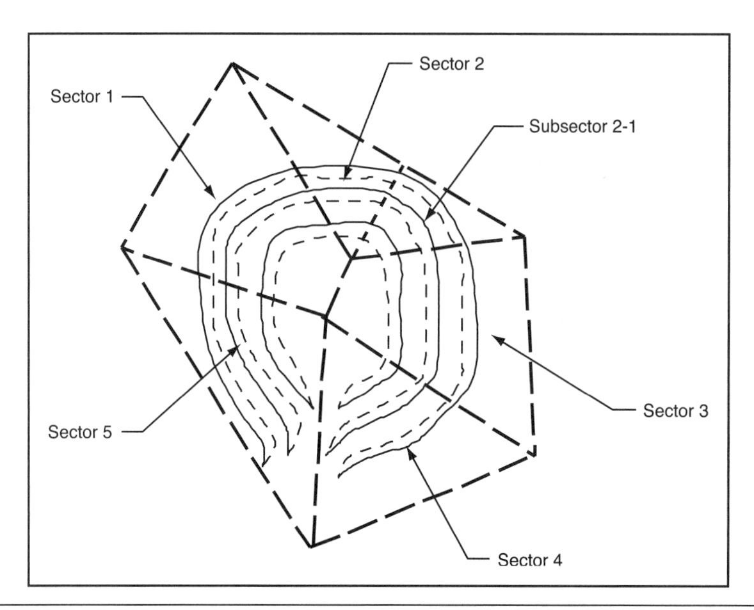

FIGURE 1.6 Pit plan with slope design sectors

Increased Shear Stress

Factors that contribute to increased shear stress include the following:

- **The removal of lateral support.** This is a very common cause of slope instability and may be the result of any of several actions, such as erosion by streams or rivers; wave action on lakes and glaciers; previous rockfall, slide, subsidence, or large-scale faulting that creates new slopes; or the work of humans, as in the creation of mines or quarries, the construction of cuts in rock, the removal of retaining structures, or the alteration of water elevation in lakes and reservoirs. This is a common cause of mine highwall failures due to equipment excavating at the toe of an unstable slope (Figure 1.7).
- **The addition of surcharge to the slope.** Surcharge may be added to a slope by natural actions, such as the weight of rain, hail, snow, or water; accumulation of talus material on top of a landslide; collapse of accumulated volcanic material; and vegetation. Surcharge may also be added to a slope by the action of humans, as in the construction of a fill; the construction of mine waste dumps, ore stockpiles, or leach piles; the weight of buildings, other constructed structures, or trains; and the weight of water from leaking pipelines, sewers, canals, and reservoirs.
- **Transitory earth stresses,** which include vibrations from earthquakes, blasting, heavy machinery, traffic, pile driving, vibratory compactors, and so forth.

  Importantly, the shock wave from an earthquake or other vibration exerts a temporary additional stress on a slope that can cause instability. This has been demonstrated by many landslides triggered by earthquakes (Glass 1982, 2000). This record, though, is misleading with regard to rock slopes, because saturated soil slopes are

Courtesy of South Dakota Mine Safety and Health Administration State Grants

FIGURE 1.7 Removal of lateral support, resulting in a complex slope failure that damaged the cable shovel

subjected to liquefaction, which would result in much greater displacement at lower seismic loading. Thus it is appropriate to include the effect of dynamic stresses in the stability analysis of slopes.

The classic method of including the effect of earthquakes in stability analysis is the pseudo-static approach whereby the maximum site acceleration that could be produced by an earthquake is input into the stability analysis as an additional horizontal driving force ($k_h \cdot W$, where k_h is the horizontal seismic coefficient). The California Department of Mines and Geology stipulates the use of a seismic coefficient, k_h, of 0.15 and a minimum computed pseudo-static factor of safety of 1.0 to 1.2 for analyses of natural, cut, and fill slopes (CDMG 1997). This approach is excessively conservative when applied to pit slopes. A cautionary note is provided stating that the seismic coefficient "k_h" is not equivalent to the peak horizontal ground acceleration value, either probabilistic or deterministic; therefore, peak ground acceleration should not be used as a seismic coefficient in this pseudo-static approach.

- **A slow increase in the overall slope of a region** as a result of tectonic uplift stresses, stress relief, or other natural mechanisms (i.e., regional tilt).
- **The removal of underlying support of the slope.** The support underlying a slope may be decreased or removed by undercutting of banks by rivers, streams, or wave actions; subaerial weathering, wetting and drying, and frost action; subterranean erosion in which soluble material (such as gypsum) is removed and overlying material collapses; mining, quarrying, road construction, and similar actions; loss of strength or failure in underlying material, such as in clays; and the squeezing out of underlying plastic material.
- **Lateral pressure,** most commonly from water in pore spaces, cracks, caverns, or cavities. Other sources of lateral pressure include the freezing of water in cracks, swelling of soils as a result of hydration of clay or anhydrite, and the mobilization of residual stresses.

  It is not just the presence of water, but *water pressure* that is the bad actor. Water *is not* a lubricant, as so many practitioners believe. When groundwater is present above the potential failure surface, an uplift force caused by the water pressure will act on the potential failure plane in a direction opposite that of the failure mass's normal force component. The magnitude of the water force depends on the area of the portion of the potential failure plane below the groundwater surface. Additionally, if near-vertical tension cracks are present that extend below the groundwater table, then a horizontal driving force will develop on the tension cracks because of the water pressure.
- **Volcanic processes,** such as swelling or shrinking of magma chambers.
- **Tectonic activities,** which may alter the stress fields on a very large scale, causing an increase or shift in the direction of geostatic stresses.
- **Processes that created the slope.** These may include creep on the slope or creep in weak strata below the foot of the slope (Figure 1.8).

Low or Reduced Shear Strength

Factors that contribute to low or reduced shear strength include the following:

- **Factors stemming from the initial state or inherent characteristics of the material.** These factors include material composition, texture, and gross structure and slope geometry (i.e., the presence and orientation of discontinuities, slope orientation, the existence of massive beds over weak or plastic materials, and the alternation of permeable beds and weak impermeable beds).
- **Changes in shear strength** due to weathering and other physicochemical reactions. These changes can include softening of fissured clays; physical disintegration of granular rocks due to the action of frost or by thermal expansion and contraction; hydration or dehydration of clay materials (including the absorption of water by clay minerals, which may decrease the cohesion; the swelling—and thus loss of

FIGURE 1.8 Creep in the debris in the zone of depression of an old landslide

cohesion—by montmorillonitic clays; and the consolidation of loess upon saturation); base exchange in clays; migration of water due to electrical potential; drying of clays, which results in cracks; drying of shales, which creates cracks on bedding and shear planes; and removal of cement within discontinuities by solution.

- **Changes in intergranular forces** due to water content and pressure in pores and fractures, which may result from (1) rapid drawdown of a lake or reservoir, (2) rapid changes in the elevation of the water table, (3) rise of the water table in a distant aquifer, and (4) seepage from an artificial source of water.
- **Changes in structure,** which can be caused by remolding clays or clay-like materials upon disturbance, fissuring of shales and preconsolidated clays, and fracturing and loosening of rock slopes due to the release of vertical or lateral restraints upon excavation.
- **Miscellaneous causes,** which can include weakening of a slope due to progressive creep or due to the actions of roots and burrowing animals.

GENERAL MODES OF SLOPE FAILURE IN ROCK MASSES

The four primary modes of slope failure in rock masses are

1. Planar failure,
2. Rotational failure,
3. Wedge failure, and
4. Toppling failure.

Courtesy of Duncan Wyllie

FIGURE 1.9 Rockfalls from steep hillsides frequently disrupt road traffic in the mountainous regions of British Columbia, Canada

FIGURE 1.10 Earth flow landslide

Other modes that are recognized as important under certain conditions include rockfalls (Figure 1.9) or earth falls, rock or earth spread, rock or earth flow (Figure 1.10), buckling of steeply dipping thin beds (i.e., slab failure, as shown in Figure 1.11), step path and step wedge, and the general surface.

Courtesy of Homestake Mining Company

FIGURE 1.11 Slabbing failure mode in the highwall of an open-pit mine reinforced with cable bolts

Planar Failure

In planar failure (Figure 1.12), the mass progresses out or down and out along a more or less planar or gently undulating surface. The movement is commonly controlled structurally by (1) surface weakness, such as faults, joints, bedding planes, and variations in shear strength between layers of bedded deposits; or (2) the contact between firm bedrock and overlying weathered rock.

In order for the likelihood of failure to exist, the following conditions must be met:

- The strike of the plane of weakness must be within ±20° of the strike of the crest of the slope.
- The toe of the failure plane must daylight between the toe and the crest of the slope. (The term *daylight* is a slang word that has common usage in the field of slope stability. For a plane, vein, discontinuity, etc., it means "to intersect the face of the excavation above the toe level.")
- The dip of the failure plane must be less than the dip of the slope face, and the internal angle of friction for the discontinuity must be less than the dip of the discontinuity (Hoek and Bray 1981).

Rotational Failure

The most common examples of rotational failures are little-deformed slumps, which are slides along a surface of rupture that is curved concavely upward. In slumps, the movement is more or less rotational about an axis that is parallel to the slope (Figures 1.10 and 1.13).

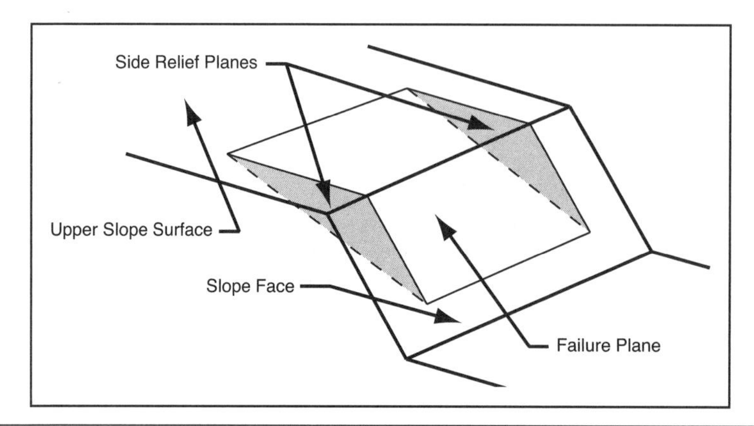

FIGURE 1.12 Planar failure mode

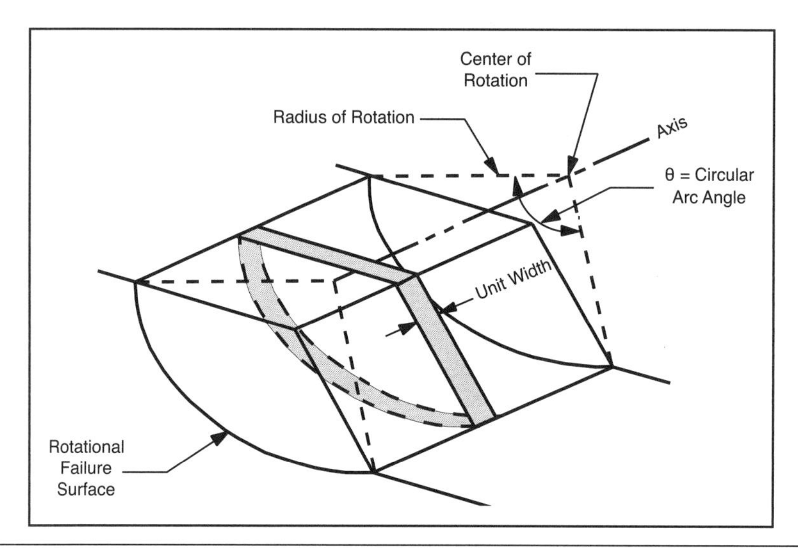

FIGURE 1.13 Rotational failure mode

In the head area, the movement may be almost wholly downward, forming a near-vertical scarp, and have little apparent rotation; however, the top surface of the slide commonly tilts backward away from the preexisting slope face, thus indicating rotation. A purely circular failure surface on a rotational failure is quite rare because frequently the shape of the failure surface is controlled by the presence of preexisting discontinuities, such as faults, joints, bedding, shear zones, and so forth. The influence of such discontinuities must be considered when a slope stability analysis of rotational failure is being conducted. Rotational failures occur most frequently in homogeneous materials, such as constructed embankments, fills, and highly fractured or jointed rock slopes.

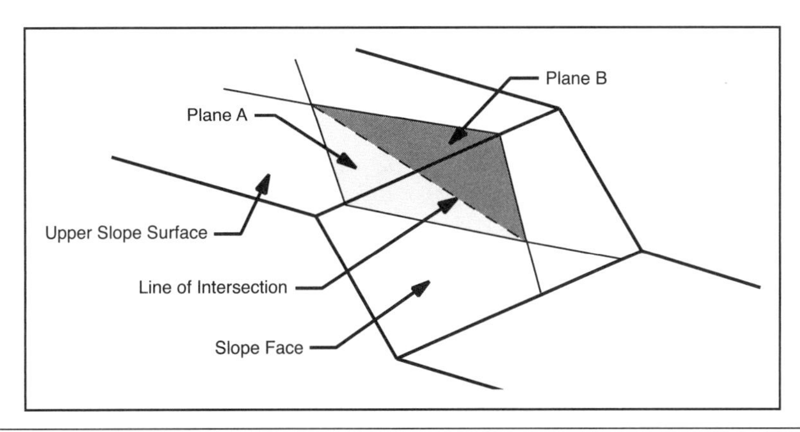

FIGURE 1.14 Wedge failure mode

Wedge Failure

The possibility of wedge failure exists where two discontinuities strike obliquely across the slope face and their line of intersection daylights in the slope face (Figure 1.14). The wedge of rock resting on these discontinuities will slide down the line of intersection provided that (1) the inclination of the line of intersection is significantly greater than the angle of internal friction along the discontinuities, and (2) the plunge of the line of intersection daylights between the toe and the crest of the slope.

Toppling Failure

Toppling failure occurs when the weight vector of a block of rock resting on an inclined plane falls outside the base of the block. This type of failure may occur in undercutting beds (Figure 1.15).

MECHANICAL APPROACHES TO STABILITY ANALYSIS

Numerous approaches to the analysis of slope stability problems have been used, including static equilibrium methods, probabilistic methods, finite element and finite difference procedures, back analysis, the "keyblock" concept, and stochastic medium theory. The most common method employed is the simple limiting equilibrium technique to evaluate the sensitivity of possible failure conditions to slope geometry and rock mass parameters (Piteau and Martin 1982). More detailed limiting equilibrium analyses, finite element analyses, or statistical and/or probabilistic analyses are employed for cases where the slope stability is sensitive to the failure mechanisms and/or to the operating parameters. Limiting equilibrium techniques are best suited for stability analysis of benches or overall slopes where the failure mechanism can readily be defined and strength parameters are known or can be estimated. On the other hand, more advanced techniques are required for analyses involving the consideration of

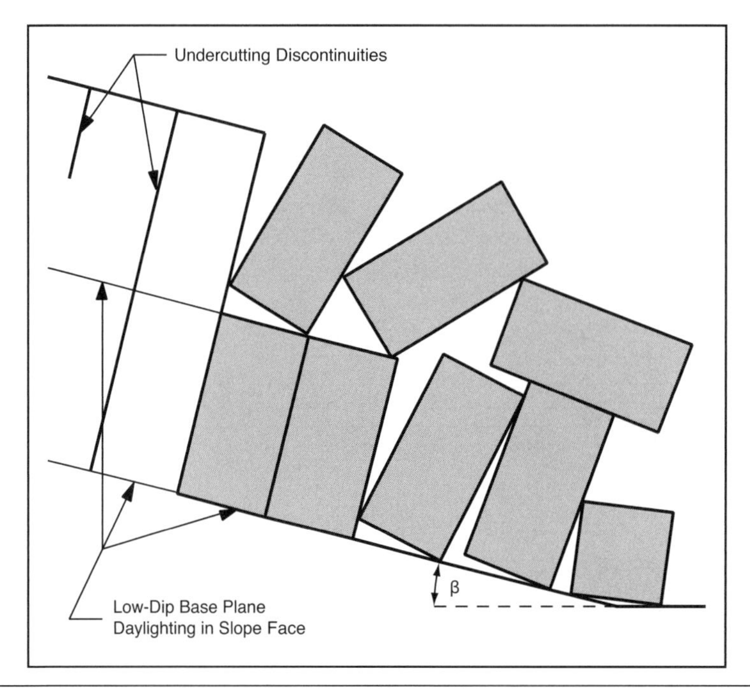

FIGURE 1.15 Toppling failure mode

multiple failure modes within a slope, complex slope geometry, complex or multiple structural geometry (i.e., modeling of the distributions of discontinuity orientations), complex hydrologic conditions, and/or variable rock strength parameters.

This section provides an introduction to two of these types of mechanical stability analysis methods: the limiting equilibrium approach and probabilistic analysis. An example of using the latter technique is included as well.

The Limiting Equilibrium Concept

At limiting equilibrium, by definition, all points are on the verge of failure. At this point in time, the driving forces (or stresses, moments) just equal the resisting forces (or stresses, moments), and the factor of safety equals unity. Where the resisting forces of a slope are greater than the driving forces, the factor of safety is greater than unity and the slope is stable; when the resisting forces are less than the driving forces, the slope is unstable.

Planar failure. The simplest model applied to planar failure is that of a block resting on an inclined plane at limiting equilibrium (Figure 1.16). The following equations describe the forces acting on the block:

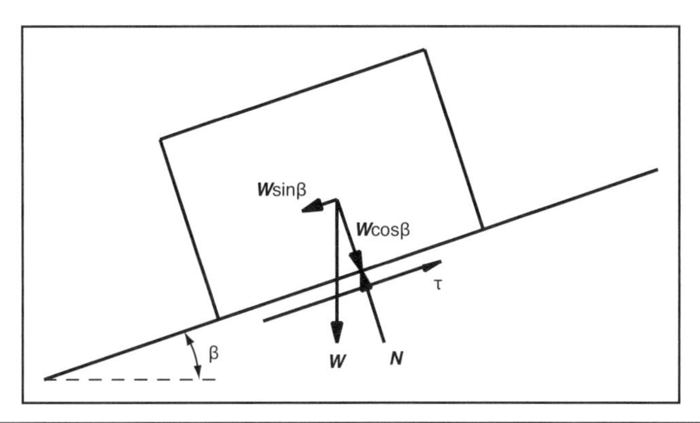

FIGURE 1.16 Block on an inclined plane at limiting equilibrium

$$\tau = c + \sigma \cdot \tan\phi$$

$$\sigma = \frac{N}{A} = \frac{W\cos\beta}{A}$$

$$\tau = c + \frac{W\cos\beta}{A}\tan\phi$$

$$\text{shear force} = \tau A = \text{resisting force} = cA + W\cos\beta \cdot \tan\phi$$

$$\text{driving force} = W\sin\beta$$

where

τ = shear stress along the failure plane
c = cohesion along the failure plane
σ = normal stress on the failure plane
ϕ = angle of internal friction for the failure plane
N = magnitude of the normal force across the failure plane
A = area of the base of the plane
W = weight of the failure mass
β = dip angle of the failure plane

Equating the driving forces and the resisting forces at limiting equilibrium, we get an equation for the factor of safety, FS:

$$\text{FS} = \frac{\text{resisting forces}}{\text{driving forces}} = \frac{cA + W\cos\beta\tan\phi}{W\sin\beta} \quad \text{(EQ 1.1)}$$

Rotational failure. For rotational failure, the model generally applied is one of many methods of slices (i.e., the Bishop method, Taylor method, the Morgenstern–Price method, Janbu method, the ordinary method of slices, etc.). These models generally differ only in the assumptions needed to make the equations statically determinate.

Rotational failure may occur in slopes composed of soil or soil-like material (i.e., overburden repositories, waste dumps, fills, tailings dams, etc.). These slopes are generally composed of individual particles of soil or rock that are not interlocked (so that a reasonably continuous failure surface may form) and that are very small compared with the size of the slope (Piteau and Martin 1982). For failure to occur, the total shear stress along the failure surface must be equal to or greater than the shear strength along the surface.

The strength of the material is generally S given by the Mohr–Coulomb failure criterion:

$$S = c' + (\sigma - u)\tan\phi'$$

where

S = shear strength

c' = effective cohesion of the material

σ = normal stress on the sliding surface

u = pore water pressure on the sliding surface

ϕ' = effective friction angle of the material

A detailed discussion of the many methods available for determining the factor of safety versus failure for rotational failure is beyond the scope of this book. The reader is referred to excellent dialogues of this subject by Perloff and Baron (1976), Smith (1982), Schuster and Krizek (1978), Hoek and Bray (1981), Brawner and Milligan (1971, 1972), and Brawner (1982). The first three of these references discuss rotational failure in soil slopes, whereas the last four are concerned primarily with failures in rock slopes.

Consider a generalized failure surface such as that shown in Figure 1.17. This figure shows a slope and a failure surface of circular-segment shape. The slope may be composed of several different materials that may vary in both the horizontal and vertical directions, along with pore pressure that may also vary with position. To investigate the stability conditions at some point along the failure surface, the slope has been divided into a number of thin, vertical slices of width Δx. A typical slice, the ith slice, is shown in Figure 1.17 along with the forces acting on it: the weight of the slice, $\boldsymbol{W}_i$; the horizontal forces, $\boldsymbol{E}_i$ and $\boldsymbol{E}_{i+1}$; the shear forces, $\boldsymbol{X}_i$ and $\boldsymbol{X}_{i+1}$; the normal force on the bottom of the slice, $\boldsymbol{N}_i$; and the resisting shear force on the bottom of the slice, $\boldsymbol{S}_i$. $\boldsymbol{N}_i$ acts at a position $\boldsymbol{Z}_i$ with respect to x; $\boldsymbol{E}_i$ and $\boldsymbol{E}_{i+1}$ act at positions Y_i and Y_{i+1} with respect to x and $x + \Delta x$, respectively.

The total number of unknowns for n slices is $6n - 3$ (Figure 1.17). However, the factor of safety is one additional unknown, bringing the total to $6n - 2$. If, through laboratory testing, we determine a value for S, which is assumed constant for all n slices, then the total number of unknowns becomes $5n - 2$. If we are attempting to solve the problem in the x-y plane (utilizing a unit width of slice through the assumed failure as shown in Figure 1.12), and if we assume the slope is in equilibrium, then the total number of equations available is $3n$ (Figure 1.17).

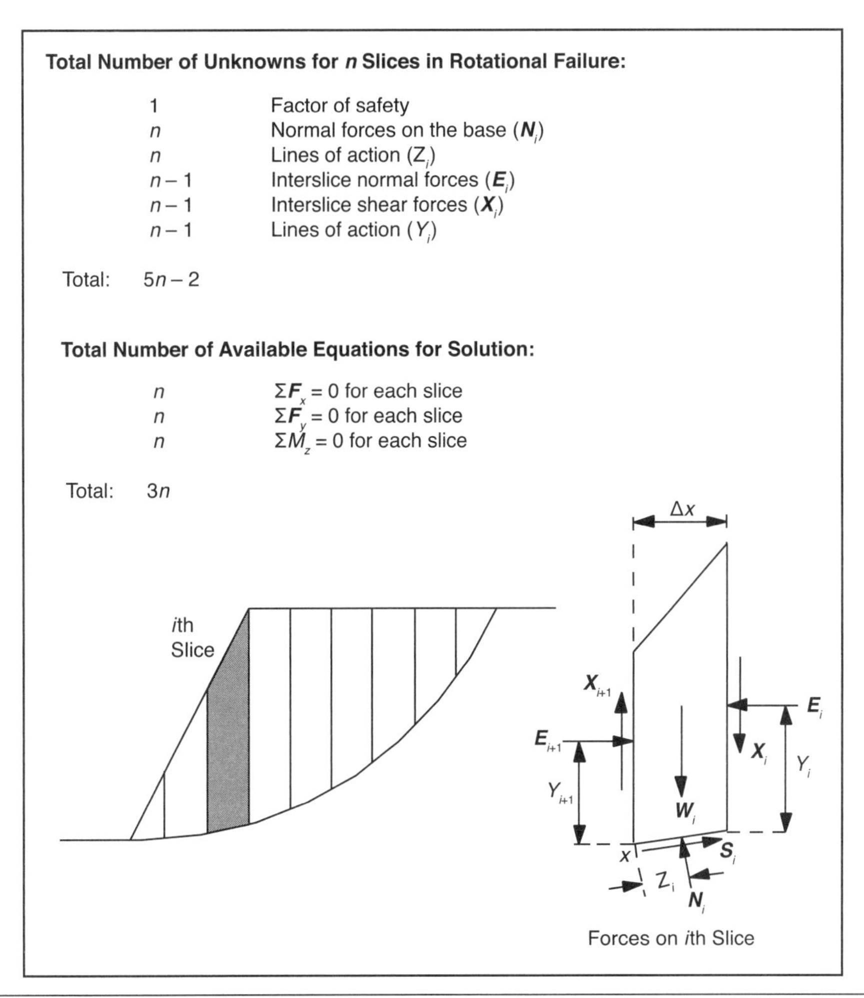

FIGURE 1.17 Statics for the method of slices for rotational failure

Wedge failure. The basic mechanics of wedge failure are very simple; however, because of the large number of variables involved, the mathematical treatment of the mechanics can become very complex unless a very strict sequence is adhered to in the development of the equation. For the simplest case—friction only—the following equations apply:

$$\text{resisting forces} = N_A \tan\phi_A + N_B \tan\phi_B$$

$$\text{driving forces} = W \sin\beta_i$$

$$\text{F.S.} = \frac{N_A \tan\phi_A + N_B \tan\phi_B}{W \sin\beta_i}$$

where

N_A = magnitude of normal force on plane A

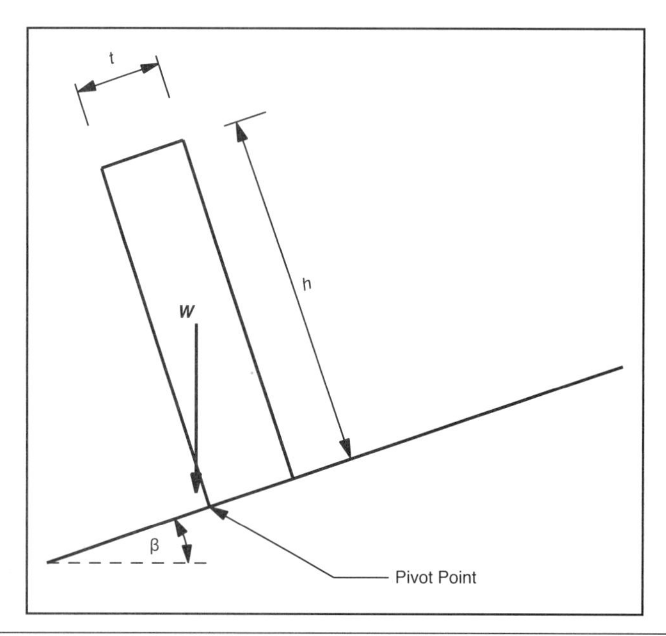

FIGURE 1.18 Potential toppling failure when W is outside the pivot point

N_B = magnitude of normal force on plane B
ϕ_A = friction angle on plane A
ϕ_B = friction angle on plane B
β_i = plunge of the line of intersection of planes A and B

Toppling failure. For toppling analysis, the sum of the moments causing toppling of a block (i.e., the horizontal weight component of the block and the sum of the driving forces from adjacent blocks behind the block under consideration) is compared to the sum of the moments resisting toppling (i.e., the vertical weight component of the block and resisting forces from adjacent blocks in front of the block under consideration). If the driving moments exceed the resisting moments, then toppling can occur.

A measurement of the potential for toppling is the ratio of the thickness to the height, t/h (called the slenderness ratio), for the block. This ratio regulates the location of the resultant force due to the weight of the block with respect to a pivot point at the lowest corner of the block (Figure 1.18). Whenever $t/h < \tan\beta$ (where β is the dip of the failure plane), the resultant force occurs outside the toe of the block, and an overturning force develops about the pivot point that could lead to toppling.

The Probabilistic Approach

The probabilistic approach takes into account the uncertainty of variables such as cohesion, friction angle, dip, strike, and joint length (Coates 1977). Some of the more important definitions relating to probabilistic analysis include the following (Miller and Freund 1985):

- *Target population:* The entire group of data from which representative samples are to be taken (e.g., dip values for all the joints in a set).
- *Sample population:* The group of data from which actual samples are taken—which may or may not be equivalent to the target population (e.g., the dip values of joints available for measurement on the face of the benches may not be representative of all joints in the wall rock).
- *Random sample:* A sample taken in such a way that there is an equal chance of every member of the target population being selected or observed.
- *Biased sample:* A sample taken in a manner resulting in a greater possibility of some members being selected or observed than others (e.g., a set of dip values obtained from a drill core will be biased against dips parallel to the hole).

Another issue of concern in probabilistic analysis is, of course, the term *probability* itself. Probability is concerned with events that individually are not predictable but that in large numbers are predictable. It is the relative expected frequency of occurrence of a given event in an infinitely large population of events. A probability distribution depicts the relationship between the relative likelihood of occurrence of an event and the numerical value associated with the event (e.g., if the event is the dip of joints in a set, the probability distribution may be as shown in Figure 1.19).

Three important axioms associated with probabilities are as follows:

1. The probability, P, associated with any discrete event among all possible outcomes must be between 0 and 1.
2. The sum of all probabilities defined in the probability density function (PDF) must add up to 1.
3. If A and B are mutually exclusive events, the probabilities associated with A and B must be additive; that is, $P(\mathrm{A} \cup \mathrm{B}) = P(\mathrm{A}) + P(\mathrm{B})$.

A probability density function, $f(x)$, when integrated between any two constants a and b, gives the probability that the corresponding random variable takes on a value between these two limits. The first of the preceding axioms states that the probability must be between 0 and 1, whereas the second axiom states that the sum of the probabilities within a given PDF must equal 1.

The Normal Distribution

A common type of probability density—the normal probability density (usually referred to simply as the *normal distribution*)—is a mathematical function that takes the form of the familiar bell-shaped curve. The equation of the normal probability density is

$$f\left(x;\mu,\sigma^2\right)=\frac{1}{\sqrt{2\pi}\sigma}e^{-\frac{1}{2}\left(\frac{x-\mu}{\sigma}\right)^2} \quad -\infty<x<\infty$$

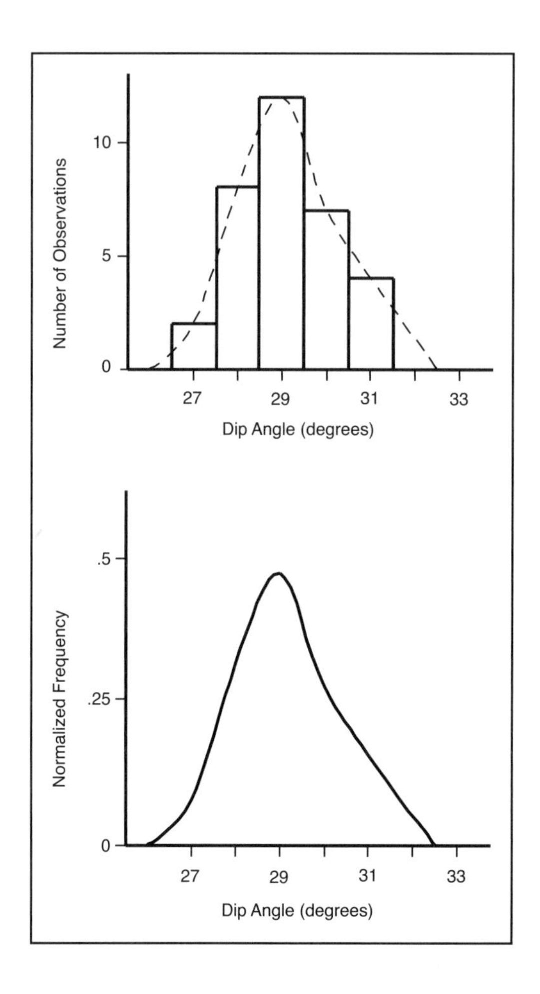

Source: Coates 1981

FIGURE 1.19 Histogram and probability density function of joints in a set

Since the normal probability density cannot be integrated in closed form between every pair of limits *a* and *b*, probabilities relating to normal distributions are usually obtained from a special table (Figure 1.20). The table pertains to the standard normal distribution, namely, the normal distribution with a mean (μ) of 0 and a standard deviation (σ) of 1; its entries are the values of the following equation:

$$F(z)=\frac{1}{\sqrt{2\pi}}\int_{-\infty}^{z} e^{-\frac{1}{2}t^2}\,dt \qquad \text{(EQ 1.2)}$$

for $z = 0.00, 0.01, 0.02, \ldots, 3.49$, and also for $z = 4.00$, $z = 5.00$, and $z = 6.00$ (Miller and Freund 1985). In other words, we cannot directly solve Equation 1.2 for the value of z by integration, so we must use special tables that give us close approximate solutions for the value of z given $F(z)$. For experimental data, such as a series of dip angle measurements, and

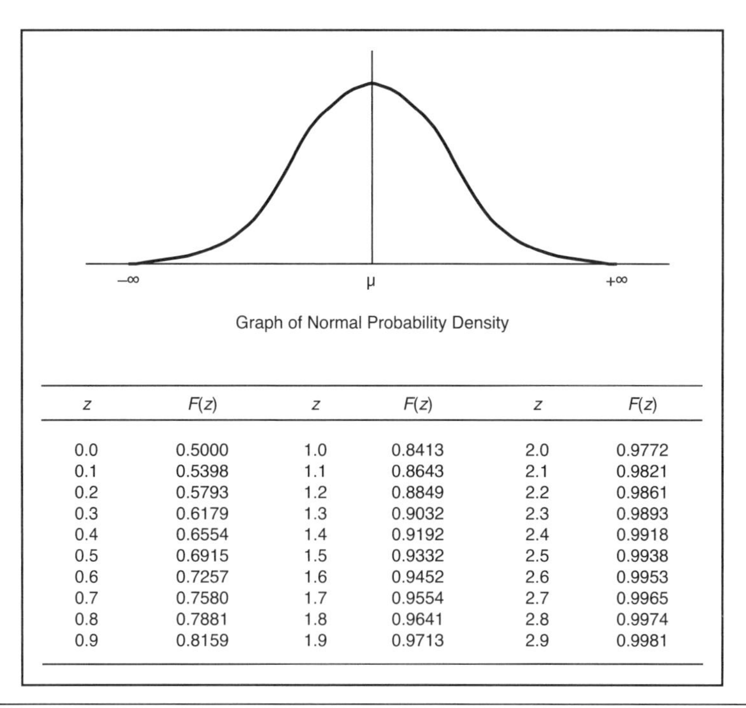

z	F(z)	z	F(z)	z	F(z)
0.0	0.5000	1.0	0.8413	2.0	0.9772
0.1	0.5398	1.1	0.8643	2.1	0.9821
0.2	0.5793	1.2	0.8849	2.2	0.9861
0.3	0.6179	1.3	0.9032	2.3	0.9893
0.4	0.6554	1.4	0.9192	2.4	0.9918
0.5	0.6915	1.5	0.9332	2.5	0.9938
0.6	0.7257	1.6	0.9452	2.6	0.9953
0.7	0.7580	1.7	0.9554	2.7	0.9965
0.8	0.7881	1.8	0.9641	2.8	0.9974
0.9	0.8159	1.9	0.9713	2.9	0.9981

FIGURE 1.20 Standardized normal distribution

assuming that the variations follow closely a normal distribution, the curve is defined by the mean ($\overline{x}$), and the variance (s^2), or standard deviation (s), according to the following:

- The *sample mean* or *expected value* or *first moment* indicates the center of gravity of a probability distribution. Assuming that there are n individual test values x_i, the mean $\overline{x}$ is given by

$$\overline{x} = \frac{\sum_{i=1}^{n} x_i}{n}$$

- The *sample variance*, s^2, or the *second moment about the mean* of a distribution is defined as the mean of the square of the difference between the value of x_i and the mean value, $\overline{x}$:

$$s^2 = \frac{\sum_{i=1}^{n} \left(x_i - \overline{x}\right)^2}{n-1}$$

- The *standard deviation* is given by the positive square root of the variance, s^2. In the case of the commonly used *normal distribution*, about 68.3% of the data will fall within ±1 standard deviation, and approximately 95.4% will fall within ±2 standard deviations.

$$s = \sqrt{s^2}$$

- In probability theory and statistics, the *coefficient of variation* (CV) is a normalized measure of dispersion of a probability distribution. It is defined as the ratio of the standard deviation, s, to the mean, $\bar{x}$:

$$\mathrm{CV} = \frac{s}{\bar{x}}$$

where

x_i = individual sample values
n = total number of samples

In the preceding equations, note that the mean and standard deviation of the x terms were written as x and s, respectively, not as μ and σ. The reason is that general practice in applied statistics is to use Latin letters to denote descriptions of actual data and Greek letters to denote descriptions of theoretical distributions. (Note also that had we divided by n instead of $n - 1$ in the equation for the variance, the resulting formula could have been used for the standard deviation of the distribution of a random variable that assumes the values of x_i with equal probabilities of $1/n$; its standard deviation would be denoted by σ [Miller and Freund 1985].)

The mean describes the center, median, or average of the data for 1, 2, and 3 standard deviations on either side of the mean, 68.27%, 95.45%, and 99.73% of the data are included, respectively.

The table in Figure 1.20 can be applied to find values of $F(z)$, given z, from a normally distributed sample population of observed values, where $F(z)$ is the probability of an occurrence. To use the table for the normal distribution in connection with a random variable that has the value x and a normal distribution with the mean μ and variance σ^2, we refer to the corresponding standardized random variable, z, which has the value

$$z = \frac{x - \mu}{\sigma}$$

or

$$x = \mu + \sigma z$$

and to the standard normal distribution. This equation uses the mean and standard deviation of our observed data (μ and σ), along with a value (x) within the range of the observed data, to calculate the standard score (z). The standard score (z) is then entered into the table in Figure 1.20 to determine the probability of occurrence of x, that is, $F(z)$. The example given later in this chapter provides more detail on how a standard normal distribution is used to find values for a nonstandard normal distribution.

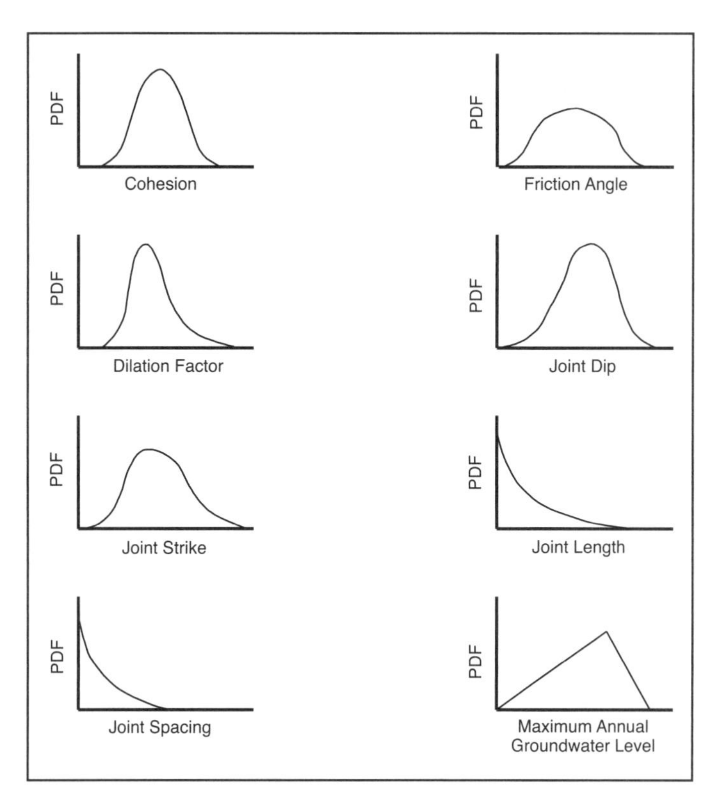

Source: Coates 1981

FIGURE 1.21 Typical probability density functions for variables affecting slope stability

Typical PDFs for some of the more important variables affecting slope stability are shown in Figure 1.21. Plotted on the abscissa (x-axis) is the range of measured values (cohesion, friction angle, strike, etc.); plotted on the ordinate (y-axis) is a ratio representing the normalized number of occurrences within each class interval (i.e., number of occurrences of a dip measurement within the range of 30° to 32° divided by the total number of dip measurements). This normalization procedure results in the sum of the area under the PDF equaling 1.0. Because the sum under the PDF curve equals 1.0, we can now determine the probability of occurrence, P(x), of any value x within the range of measured values plotted along the abscissa. If we sum the normalization ratios from left to right along the abscissa for each class interval, we obtain the cumulative probability distribution that has an asymptote of 1.0. Figure 1.22 is a cumulative distribution of the dip angles of Figure 1.19.

Probabilistic Analysis of the Plane Shear Failure Mode*

The probability of plane shear failure is a combination of two probabilities: (1) the probability that the failure plane exists, P_e, and (2) the probability that sliding on the failure plane will occur, P_s. The equation for the probability of plane shear failure (PF) is

* This section draws heavily from Marek and Savely (1978).

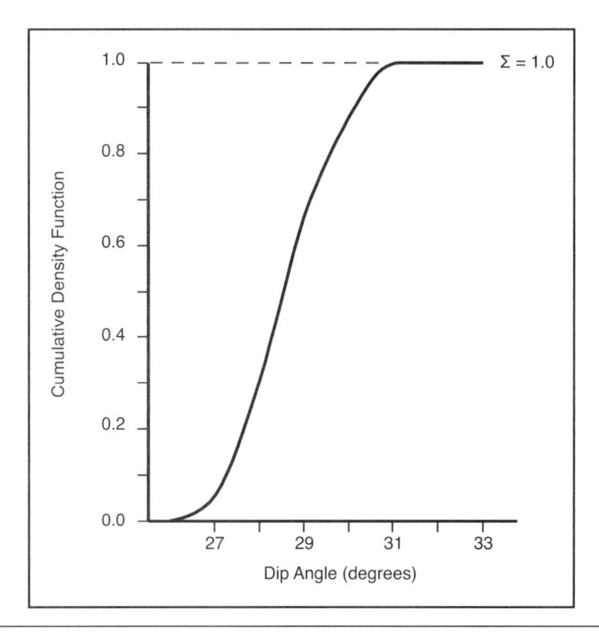

FIGURE 1.22 Cumulative probability distribution for dip angles

$$PF = \int P_e P_s \, dx$$

where x represents the joint orientation. This is the joint probability that a fracture is present and that sliding occurs, summed over all possible joint orientations. In the plane shear analysis, many orientations are not considered because their probabilities of sliding are zero. Only those joint sets with strikes that parallel the slope face and dips that could be daylighted are considered. In reality, the integral is approximated by

$$PF = \sum_i P_{e_i} P_{s_i}$$

where the summation i is taken over a discrete set of dips with a range between zero and the slope face angle. This equation is valid for any failure mode.

For the plane shear condition, the probability of existence is the combined probability of two occurrences:

$$P_{e_i} = P_{d_i} P_{l_i}$$

That is, the probability of existence is the combined probability that the fracture has a certain dip attitude (P_d) and that the fracture is long enough to reach from the toe to the top surface of the slope at that particular dip. The probability of length (P_l) is determined directly from the cumulative length distribution (see Figure 1.22). For any dip in the joint set, a certain fracture length is required to reach from the toe to the top surface of the slope.

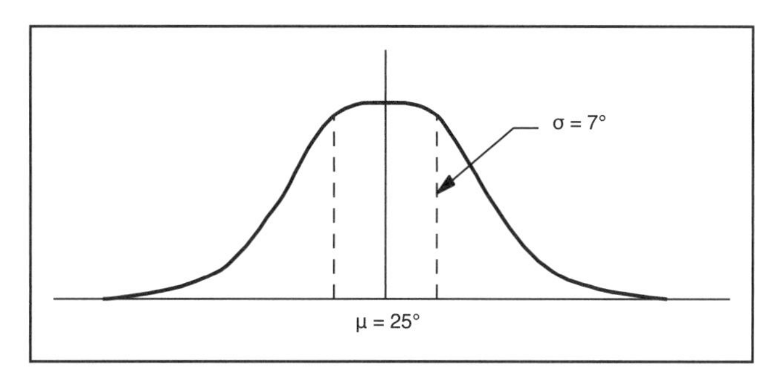

FIGURE 1.23 Distribution of friction angles for probabilistic method example

The probability of meeting or exceeding this length is determined directly from the probability distribution for fracture length. The probability of failure thus becomes

$$PF = \sum_i P_{d_i} P_{l_i} P_{s_i}$$

summed over a selected range of dips.

Probabilistic method example. As a simple example of the probabilistic method, assume that a discontinuity strikes parallel to the slope face and dips at a constant 23°. For simplicity, assume also that cohesion is equal to zero. The applicable equation for the factor of safety of the rock wedge—i.e., Equation 1.1 for planar failure—reduces to

$$FS = \frac{\tan\phi}{\tan\beta}$$

where

ϕ = the friction angle, which is normally distributed with a mean value, μ, of 25° and a standard deviation, σ, of 7° (Figure 1.23)

β = 23°, the dip of the discontinuity

If we generate a table of random numbers between zero and 1, as shown in Table 1.1, and let the random numbers be equal to $F(z)$, or the probability of occurrence on the standardized normal distribution (see Figure 1.20), then we can determine the corresponding z-value for each random number. The z-value is determined by interpolation between values of $F(z)$ in the table within Figure 1.20. Note that the table in Figure 1.20 includes only values of $F(z)$ from 0.5000 to 0.9981 (1.0, that is). This is only the area under the right side of the standardized normal distribution. For probabilities, $F(z)$, less than 0.5000, then $z = 1 - F(z)$; that is, z has a negative sign (see Table 1.1). Using the z-value, we can then calculate the x-value (friction angle, ϕ) from the equation

$$x = \mu + \sigma z$$

TABLE 1.1 Random numbers, *z*-values, friction angles (ϕ), and associated factors of safety for example problem

Random Number, Equated to $F(z)$	Corresponding z-value	Corresponding $x = \phi$	Resulting Value of FS
0.516048	0.0401	25.2807	1.0128
0.497778	-0.0056	24.9608	0.9982
0.943212	1.5826	36.0782	1.6100
0.898150	1.2709	33.8963	1.4408
0.567629	0.1703	26.1921	1.0549
0.60527	0.2670	26.8690	1.0865
0.272314	-0.6059	20.7587	0.8129
0.354349	-0.3737	22.3841	0.8832
0.789751	0.8057	30.6399	1.2703
0.447875	-0.1311	24.0823	0.9585
0.507101	0.0178	25.1246	1.0057
0.884756	1.1992	33.3944	1.4137
0.259576	-0.6451	20.4843	0.8011
0.143982	-1.0627	17.5611	0.6787
0.801948	0.8487	30.9409	1.2855
0.011362	-2.2780	9.0540	0.3417
0.841238	0.9997	31.9979	1.3399
0.979903	2.0521	39.3647	1.7593
0.214619	-0.7906	19.4658	0.7580
0.147134	-1.0490	17.6570	0.6826
0.022191	-2.0100	10.9300	0.4141
0.737278	0.6305	29.4135	1.2090
0.325784	-0.4517	21.8381	0.8594
0.328797	-0.4433	21.8969	0.8620
0.896458	1.2614	33.8298	1.4372
0.964635	1.8067	37.6469	1.6543
0.965152	1.8136	37.6952	1.6572
0.127559	-1.1378	17.0354	0.6571
0.994595	2.5495	42.8465	1.9891
0.320719	-0.4658	21.7394	0.8551
0.024061	-1.9768	11.1624	0.4232
0.343429	-0.4032	22.1776	0.8742
0.278295	-0.5903	20.8679	0.8175
0.622958	0.3133	27.1931	1.1018
0.677264	0.4602	28.2214	1.1509
0.288918	-0.5565	21.1045	0.8277
0.553400	0.1343	25.9401	1.0432
0.152844	-1.0244	17.8292	0.6897

Table continues next page

TABLE 1.1 Random numbers, *z*-values, friction angles (ϕ), and associated factors of safety for example problem (continued)

Random Number, Equated to *F*(*z*)	Corresponding *z*-value	Corresponding *x* = ϕ	Resulting Value of FS
0.233209	−0.7284	19.9012	0.7764
0.064456	−1.5187	14.3691	0.5494
0.942048	1.5723	36.0061	1.5584
0.804880	0.8592	31.0144	1.2893
0.597861	0.2478	26.7346	1.0802
0.760960	0.7095	29.9665	1.2365
0.648361	0.3810	27.6670	1.1243
0.921845	1.4176	34.9232	1.4973
0.077356	−1.4232	15.0376	0.5761
0.762841	0.7156	32.7156	1.3776
0.298600	−0.5285	21.3005	0.8361
0.353741	−0.3753	22.3729	0.8827
0.166698	−0.9672	18.2296	0.7063
0.601258	0.2566	26.7962	1.0831
0.979675	2.0475	39.3325	1.7573
0.001911	−2.8889	4.7777	0.1792
0.894796	1.2522	33.7654	1.4337
0.164106	−0.9776	18.1568	0.7033
0.819464	0.9133	31.3931	1.3087
0.252786	−0.6657	20.3401	0.7950
0.175954	−0.9309	18.4837	0.7169
0.435390	−0.1626	23.8618	0.9486

For the numerous values of ϕ and the constant value of β, we can then calculate the corresponding factor of safety, as tabulated in Table 1.1.

Upon calculating the factor of safety for the 60 random numbers generated, we can next develop a frequency distribution for various class intervals of the factor of safety (Figure 1.24A). If we normalize the frequency distribution by dividing the number of occurrences in each class by the total number of occurrences (i.e., 60), then we develop a PDF for the factor of safety (Figure 1.24B). We can then use this PDF to determine the probability of, for example, the factor of safety being less than or equal to 1.0. For the example, we get the following results:

Mean value of FS, μ: 1.035557

Standard deviation of FS, σ: 0.382114

$P(FS \le 1.0)$: 0.4629, or 46.3%

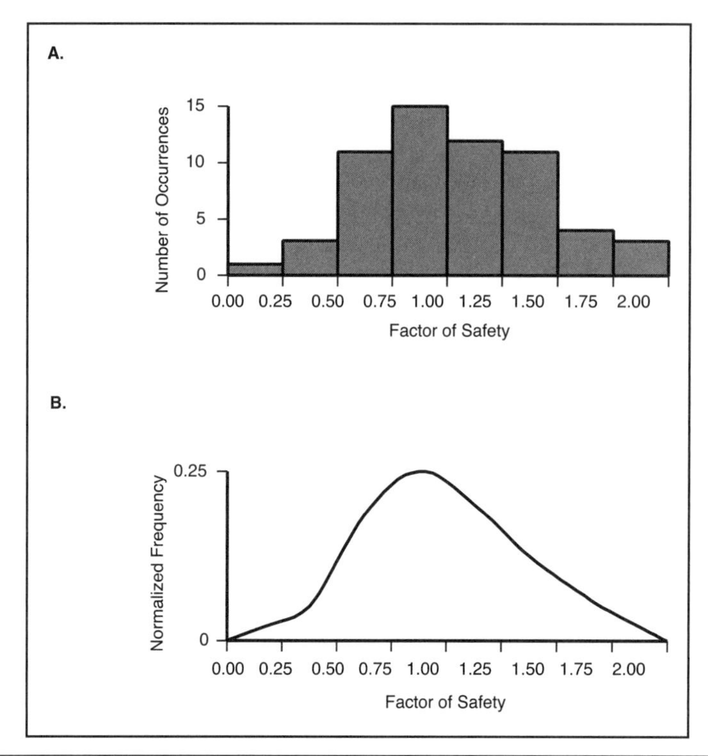

FIGURE 1.24 (A) Frequency distribution for factor of safety, and (B) probability density function for factor of safety

This result, then, implies that the slope is marginally stable (i.e., the FS is calculated to be greater than 1.0 for 53.7% of the iterations), and it also has a 46.3% probability of eventually failing.

Numerical Methods

Rapid advances in computer technology over the past five decades have pushed the finite element method (FEM) and other numerical analysis approaches to the forefront of geotechnical analysis. Numerical models are computer programs that attempt to represent the mechanical response of a rock mass subjected to a set of initial conditions such as in situ stresses and water levels, boundary conditions, and induced changes (e.g., slope excavations). The result of a numerical model simulation is typically either equilibrium or collapse. If the result obtained is equilibrium, then the resultant stresses and displacements at any point in the rock mass can be compared to measured values. If the result obtained is collapse, then the predicted mode of failure is demonstrated.

The rock mass is divided into elements (or zones) for numerical models. Each element is assigned a material model and properties. The material models are idealized stress–strain relations that describe how the material behaves. The simplest model is a linear elastic one, which uses the elastic properties—Young's modulus, Poisson's ratio—of the material.

Elastic-plastic models use strength parameters to limit the shear stress that an element (or zone) may sustain.

The zones may be tied together, termed a *continuum model*, or separated by discontinuities, termed a *discontinuum model*. Discontinuum models allow slip and separation at explicitly located surfaces within the model.

In the mid-1970s, techniques for applying the FEM to slope stability analysis started appearing in geotechnical literature (Rocscience 2004). For slopes, the factor of safety is often defined as the ratio of the actual shear strength to the minimum shear strength required to prevent failure, or the factor by which soil shear strength must be reduced to bring a slope to the verge of failure (Duncan 1996). The methodology used in many of the numerical modeling techniques (finite element or finite difference) to compute the factor of safety is to reduce the shear strength until collapse occurs. This is now commonly referred to as the *shear strength reduction (SSR) method*. The slope's factor of safety is then the ratio of the actual strength to the calculated reduced shear strength at failure.

To perform slope stability analysis with the SSR technique, simulations are run for a series of increasing trial factors of safety (F). For Mohr–Coulomb material, shear strength reduced by a factor (of safety), F, can be determined from the following equation:

$$\frac{\tau}{F}=\frac{c}{F}+\frac{\tan\phi}{F}$$

This equation can be written as

$$\frac{\tau}{F}=c_{\text{trial}}+\tan\phi_{\text{trial}}$$

Thus, the reduced Mohr–Coulomb shear strength parameters become

$$c_{\text{trial}}=\left(\frac{1}{F}\right)c$$

$$\phi_{\text{trial}}=\arctan\left(\frac{1}{F}\right)\tan\phi$$

If multiple materials and/or joints are present, the reduction is made simultaneously for all materials. For Mohr–Coulomb materials, the steps for systematically searching for the critical factor of safety value, F, which brings a previously stable slope to a state of limiting equilibrium, are as follows (Rocscience 2004):

- *Step 1:* Develop a model (finite element, finite difference) of a slope, using the deformation and strength properties established for the slope materials.

  Compute the model and record the maximum total deformation of the slope.

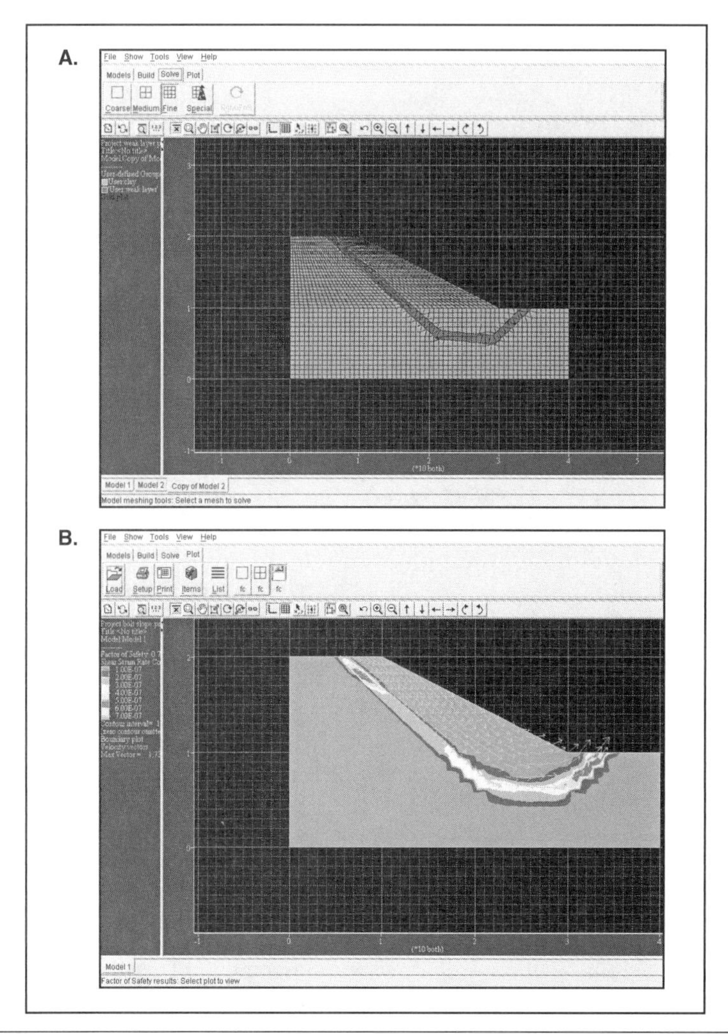

FIGURE 1.25 The shear strength reduction method: (A) grid mesh model; (B) movement vectors and failure surface

- *Step 2:* Increase the value of F by some incremental amount and calculate factored Mohr–Coulomb material parameters as previously described. Enter the new strength properties into the slope model and recompute. Record the maximum total deformation.
- *Step 3:* Repeat step 2 using systematic increments of F until the model does not converge to a solution (i.e., continue to reduce material strength until the slope fails). At failure, the factor of safety equals the trial factor of safety (i.e., FS = F).

Lorig and Verona (2004) state that the SSR method has two main advantages over the limit equilibrium technique for slope stability analysis. Firstly, the critical slide surface is found automatically, and it is therefore not necessary to specify the shape of the surface (e.g., circular, curvilinear, spiral) in advance. Secondly, numerical methods automatically satisfy translational and rotational equilibrium, whereas not all limit equilibrium methods do so. Therefore, the SSR technique will normally determine a factor of safety equal to or slightly less than the one obtained by limit equilibrium.

Another advantage of the SSR approach is its elimination of arbitrary assumptions regarding the locations and inclinations of the inter-slice forces. Similarly, the method can automatically monitor development of failure zones, from localized areas all the way to total slope failure. This is especially important given the large, deep open-pit mines of today. If the correct input material property parameters are used, then the SSR method can predict expected deformations at the stress levels found in slopes (Rocscience 2004).

Because of the finite element method's ability to compute deformations and quantities such as bending moments, it can to be used to design the support members for a slope. For example, in a slope stabilized with piles, the FEM is capable of predicting axial loads, bending moments, and deformations of the piles, making it possible for engineers to select the proper material and dimensions of the support members to obtain the desired safety factor.

The SSR finite element technique can model construction procedures and sequences (i.e., loading paths), which is a particularly important aspect for some embankments and excavations. Drawing on the robustness of the FEM, the technique performs very well under a wide range of conditions. And finally, it can be more readily applied to three-dimensional slope modeling than limit equilibrium or probabilistic methods.

Figure 1.25A and 1.25B show a slope analyzed by the SSR method with a finite difference computer code.

REFERENCES

American Geological Institute. 1997. *Dictionary of Mining, Mineral and Related Terms*, 2nd ed. Alexandria, VA: AGI.

Brawner, C.O. 1982. *Stability in Surface Mining*, Vol. 3. New York: SME-AIME.

Brawner, C.O., and V. Milligan. 1971. *Stability in Open Pit Mining*. New York: SME-AIME.

Brawner, C.O., and V. Milligan. 1972. *Geotechnical Practice for Stability in Open Pit Mining*. New York: SME-AIME.

CDMG (California Division of Mines and Geology). 1997. *Guidelines for Evaluating and Mitigating Seismic Hazards in California*. Special Publication 117. Sacramento: Department of Conservation, Division of Mines and Geology. http://gmw.consrv.ca.gov/shmp/webdocs/sp117.pdf

Coates, D.F. 1977. Design. In *Pit Slope Manual*. Report 77-5. Edited by D.F. Coates. Ottawa, ON: Canada Centre for Mineral and Energy Technology.

Coates, D.F. 1981. *Rock Mechanics Principles*. Monograph 874. Ottawa, ON: Canada Centre for Mineral and Energy Technology.

Duncan, J.M. 1996. State of the art: Limit equilibrium and finite-element analysis of slopes. *Journal of Geotechnical Engineering*, 122(7):577–596.

Glass, C.E. 1982. Influence of earthquakes on rock slope stability. In *Proceedings, 3rd International Conference on Stability in Surface Mining*. New York: SME-AIME. pp. 89–112.

Glass, C.E. 2000. The influence of seismic events on slope stability. In *Slope Stability in Surface Mining*. Edited by W.A. Hustrulid, M.K. McCarter, and D.J.A. Van Zyl. Littleton, CO: SME.

Hoek, E. 2007. *Practical Rock Engineering*. Toronto, ON: Rocscience. www.rocscience.com/hoek/corner/Practical_Rock_Engineering.pdf

Hoek, E., and J.W. Bray. 1981. *Rock Slope Engineering*. London: Institution of Mining and Metallurgy.

Jaeger, J.C., and N.G.W. Cook. 1979. *Fundamentals of Rock Mechanics*, 3rd ed. London: Chapman and Hall.

Lorig, L., and P. Verona. 2004. Chapter 10, Numerical analysis. In *Rock Slope Engineering Civil and Mining*. Edited by Duncan C. Wylie and Christopher W. Mah. London: Spon Press.

Marek, J.M., and J.P. Savely. 1978. Probabilistic analysis of the plane shear failure mode. In *Proceedings, 19th U.S. Symposium on Rock Mechanics, University of Nevada—Reno*, Vol. 2. Reno, NV: Conferences and Institutes, Extended Programs and Continuing Education, University of Nevada.

Miller, I., and J.E. Freund. 1985. *Probability and Statistics for Engineers*. Englewood Cliffs, NJ: Prentice-Hall.

Perloff, W.H., and W. Baron. 1976. *Soil Mechanics Principles and Applications*. New York: John Wiley and Sons.

Piteau, D.R., and D.C. Martin. 1982. Mechanics of rock slope failure. In *Stability in Surface Mining*, Vol. 3. Edited by C.O. Brawner. New York: SME-AIME.

Rocscience. 2004. A new era in slope stability analysis: Shear strength reduction finite element technique. *RocNews*, www.rocscience.com/library/rocnews/summer2004/StrengthReduction.pdf

Schuster, R.L., and R.J. Krizek, eds. 1978. *Landslides, Analysis and Control*. Special Report 176. Washington, DC: Transportation Research Board, Commission on Sociotechnical Systems, National Research Council, National Academy of Sciences.

Smith, G.N. 1982. *Elements of Soil Mechanics for Civil and Mining Engineers*, 5th ed. New York: Granada Publishing.

Varnes, D.J. 1978. Slope movement types and processes. In *Landslides, Analysis and Control*. Edited by R.L. Schuster and R.L. Krizek. Special Report 176. Washington, DC: Transportation Research Board, Commission on Sociotechnical Systems, National Research Council, National Academy of Sciences.

CHAPTER 2

Rock Mass Properties

In the assessment of the slope stability of an excavation cut in rock, it is important to accurately determine many of the rock mass engineering properties. These properties consist of (1) the physical and mechanical properties of intact blocks of rock and (2) the properties of discontinuities—such as joints, faults, foliation, shear zones, and bedding planes—that bound the individual blocks. Excavations within a rock mass are affected by the shear strength of discontinuities within the mass. If a rock mass contains unfavorably oriented discontinuities, its strength for engineering purposes may be greatly reduced. The discontinuities may intersect and form potentially unstable wedges, or they may dip and strike so as to form potential planes of weakness. Rock mass behavior is also influenced by rock joint traits such as surface roughness, weathering, and the presence of infilling. The surface roughness of a rock joint controls the joint's shear strength and dilatancy. The strength of the whole mass of rock (the intact blocks plus the contained discontinuities) must be considered in the design of structures, such as underground openings or rock slopes (Goodman 1976).

ENGINEERING PROPERTIES OF DISCONTINUITIES

The two most important rock mass factors that control slope stability are (1) discontinuity shear strength and (2) the location and orientation of the discontinuities with respect to the slope. Other important factors include the rock material type, discontinuity type, discontinuity (joint) intensity, discontinuity surface roughness, the amount of attachments across the discontinuity, and the type and nature of filling material (if any) between the two planes of the discontinuity.

Type of Rock

Generally, in geological terms, rocks are grouped by origin as igneous, sedimentary, or metamorphic and are classed according to petrographic characteristics, which include their mineral content, texture, and fabric:

TABLE 2.1 Common examples of the three main categories of rock origin

Igneous	Sedimentary	Metamorphic
Granite	Sandstone	Marble
Basalt	Limestone	Schist
Diorite	Shale	Slate
Gabbro	Mudstone	Gneiss
Syenite	Dolomite	Quartzite
Rhyolite	Gypsum	Amphibolite
Porphyry	Coal	Hornfels
Monzonite	Hematite	Serpentine

- Igneous rocks are formed by crystallization of masses of molten rock originating from below the earth's surface.
- Sedimentary rocks are formed from sediments that have been transported and deposited, sometimes as chemical precipitates, or from the remains of plants and animals that have been lithified under the tremendous heat and pressure of overlying sediments or by chemical reactions.
- Metamorphic rocks are those derived from preexisting rocks by mineralogical, chemical, and/or structural changes, essentially in the solid state, in response to marked changes in temperature, pressure, shearing stress, and chemical environment, generally at depth in the earth's crust.

Rock can be either fresh or decomposed, and it can be either intact or non-intact. Freshly broken rock can usually be distinguished by clean, angular, often sharp surfaces, whereas decomposed rock may be angular or smooth but is often discolored and softer than fresh rock. Intact (or in-place) rock may be fresh or weathered but has not moved from its original position. Non-intact rock has been moved from its original position by some force, of either natural or human origin. Table 2.1 lists some of the more common types of igneous, sedimentary, and metamorphic rocks.

Discontinuity Type

The term *discontinuity* refers to any of several different types of defects in the rock fabric (Table 2.2). Discontinuities, representing weakness planes in the mass, control the engineering properties by dividing the mass into blocks separated by fractures such as faults, joints, foliations, and bedding.

Joints are the most common defect in rock masses; faults are the most serious. All discontinuities have the following physical properties:

- Orientation
- Spacing
- Width of opening

TABLE 2.2 Types of discontinuities and their characteristics

Type of Discontinuity	Definition	Characteristics
Fracture	A separation in the rock mass, a break	Signifies joints, faults, slickensides, foliations, and cleavage.
Joint	A fracture along which essentially no displacement has occurred	Most common defect encountered. Present in most formations in some geometric pattern related to rock type and stress field. Open joints allow free movement of water, increasing decomposition rate of the mass. Tight joints resist weathering and the mass decomposes uniformly.
Fault	A fracture along which displacement has occurred due to tectonic activity	A fault zone usually consists of crushed and sheared rock through which water can move relatively freely, increasing weathering. Waterlogged zones of crushed rock are a cause of running ground in tunnels.
Slickenside	A preexisting failure surface from faulting, landslides, expansion	A shiny, polished surface with striations. Often the weakest element in a mass, since strength is often near residual.
Foliation planes	A continuous foliation surface that results from orientation of mineral grains during metamorphism	Can be present as open joints or merely orientations without openings. Strength and deformation relate to the orientation of applied stress to the foliations.
Foliation shear	A shear zone resulting from folding or stress relief	Thin zones of gouge and crushed rock occur along the weaker layers in metamorphic rock.
Cleavage	A stress fracture from folding	Found primarily in shales and slates; usually very closely spaced.
Bedding planes	Contacts between sedimentary rocks	Often are zones containing weak materials such as lignite or montmorillonite clays.
Myonite	An intensely sheared zone	Strong laminations; original mineral constituents and fabric crushed and pulverized.
Cavities	Openings in soluble rocks resulting from groundwater movement, or in igneous rocks from gas pockets	In limestone, these range from caverns to tubes. In ryolite and other igneous rocks, these range from voids of various sizes to tubes.

Adapted from Hunt 1984

- Intensity
- Roughness

They can be filled with some material and display the strength parameters of the infilling material (cohesion and friction) along their surfaces.

Geometric Properties

In relation to any stability problem in rock, of all the properties of the joint set, its orientation with respect to the slope face is the most important. Patterns of jointing are known to be preferentially developed over reasonably large areas; that is, attributes of the joints—such as dip, strike (or dip direction), and joint spacing—often do not vary markedly within the rock mass over large geologic areas. Therefore, the degree of stability of slopes within the same pit may vary since some of the slopes will be oriented adversely with respect to the discontinuity orientations and others will not.

There are two basic conventions for recording discontinuity orientation: the geologic convention and the rock slope engineering convention (Figure 2.1). In the geologic convention, the orientation of a plane in space (i.e., a bedding plane, joint, fault, shear zone, etc.) is recorded by measuring the plane's strike and dip. The strike is measured as the direction, with respect to north, of a horizontal line on the plane. The dip is the inclination angle of the plane measured normal to the strike direction. However, because there are two possible directions of dip for each strike line (see Figure 2.1), a direction indicating the survey quadrant is given with the dip. With the rock slope engineering convention, the dip direction and the dip angle are measured. The dip direction is measured as an azimuth angle (0° to 360°, where north = 0° and 360°, east = 90°, south = 180°, and west = 270°) of the direction of the dip from north, and the dip is measured as the inclination angle of the plane without any direction reference.

A geologic compass is often used to determine the dip/strike or dip direction/dip of the discontinuity. Two types of geologic compasses may be used: one that measures the bearing angle or one that measures the azimuth angle. Bearings are directions measured either from north or south as the reference direction. With reference to the top portion of Figure 2.1 and to Figure 2.2, a bearing in quadrant 1 would have a north-east direction and would be denoted as NαE, where α is the measured deflection angle from north. A bearing in quadrant 4 would have a north-west direction and would be denoted as NαW. Angles measured in quadrants 2 and 3 are referenced from south. A bearing in quadrant 2 would have a south-east direction, and a bearing in quadrant 3 would have a south-west direction, denoted as SαE and SαW, respectively. Azimuths, as mentioned previously, are direction angles measured from north as the reference direction. In Figure 2.1, the azimuth angle from north is denoted as α'. Frequently, the geologic compass user must convert from bearing to azimuth or azimuth to bearing. The technique for doing so is illustrated in Figure 2.2.

Discontinuity Surface Roughness

Rock slope engineers generally recognize two orders of joint wall roughness (Patton 1966) that affect either the movement characteristics or strength properties of the joints. (Some practitioners recognize three, though this text follows the two-order convention.) Figure 2.3 is a sectional view (side view) of a rock surface showing first-order surface asperities on the bottom portion of the figure and second-order surface asperities on the top. Waviness of a mean joint surface is regarded as a major, or first-order, type of asperity. These asperities are considered to be of such dimensions that they are unlikely to shear off; for practical purposes, they appear as undulations of the joint surface. Waviness is considered to modify the apparent angle of dip (β) but not the joint frictional properties (ϕ and

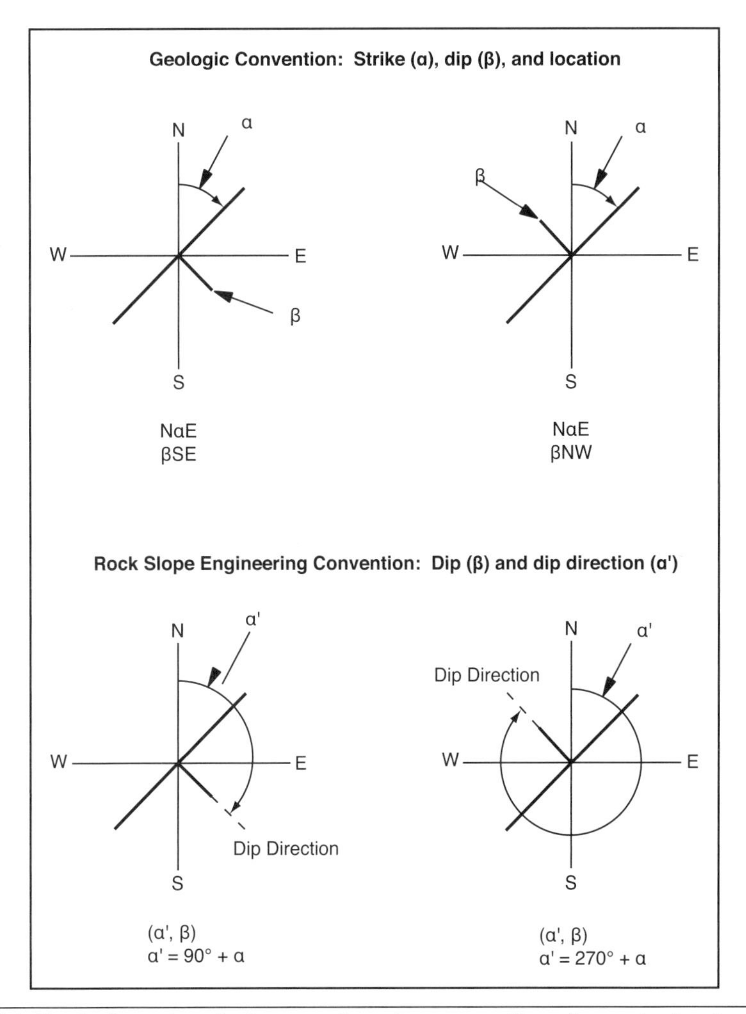

FIGURE 2.1 Geologic and rock slope engineering conventions for rock structure orientation

cohesion, *c*), and in the stability analysis, effects of waviness are considered as influencing the direction of shear movement of the sliding block.

Minor, or second-order, asperities are designated as roughness. Asperities of roughness are considered to be sufficiently small that they are likely to be sheared off during movement along the joint plane (Piteau 1970). Increased roughness of the joint walls results in an increased friction angle along the joint. However, the effects of first- or second-order asperities have been found to be greatly reduced by gouge or other infilling materials. Figure 2.4 shows a rock face with second-order undulations of roughness.

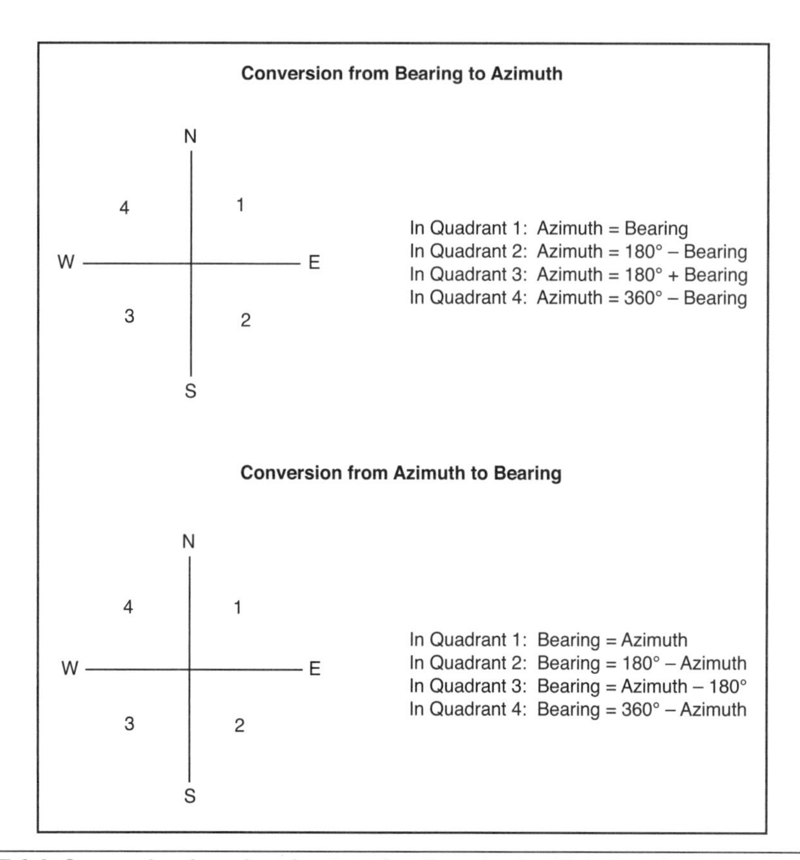

FIGURE 2.2 Conversion from bearing to azimuth and azimuth to bearing

Openness and Filling Material

Two important properties of a discontinuity that must be considered are (1) its openness and (2) the presence and type of filling material. A closed or cemented discontinuity will have a shear strength approaching that of the intact rock (or the cementing material). An open discontinuity, or one filled with soft filling material such as clay gouge, has a low shear strength.

The effect of gouge on the strength properties of a discontinuity is of great importance. The resistance to sliding along a discontinuity plane can be either increased or decreased depending on the nature (type and hardness) and thickness of the gouge and character of the discontinuity walls. If the gouge is sufficiently thick, the discontinuity walls will not touch and the strength properties of the discontinuity will be those of the gouge. There are four possible cases with respect to gouge thickness and asperity size, as shown in Figure 2.5.

1. The sliding plane passes entirely through the gouge (case 1); the shear strength is dependent only on the gouge material, and no modification is considered for roughness.

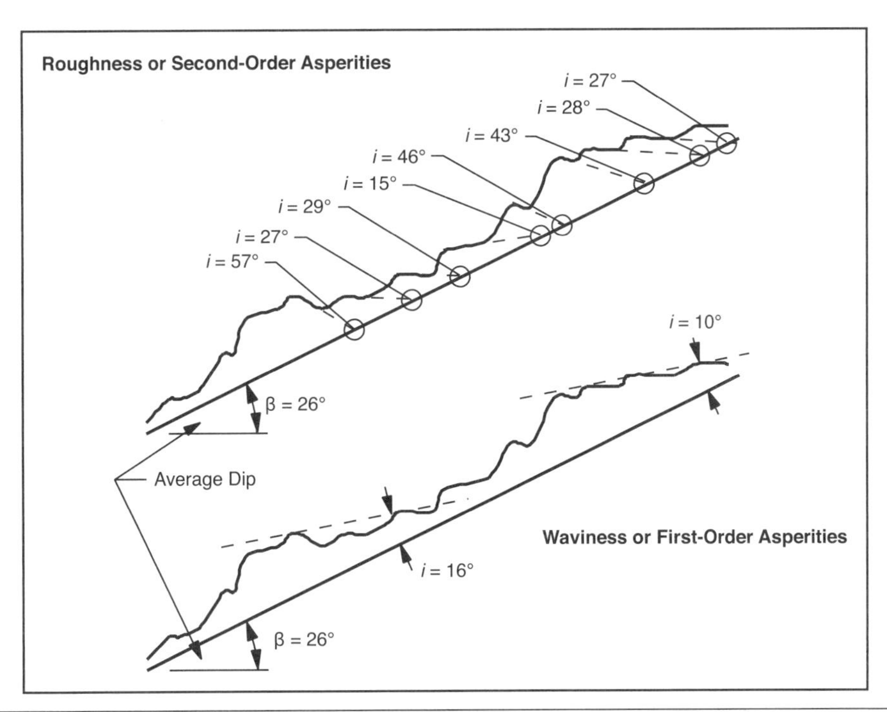

FIGURE 2.3 First- and second-order discontinuity wall roughness

FIGURE 2.4 Rock face showing second-order asperities of roughness

2. The sliding plane passes partly through gouge and partly through discontinuity wall rock (case 2); the shear strength will be more complex, being made up of contributions of both gouge and wall rock.
3. Gouge is present but very thin (case 3); gouge is considered only as a modification of the friction angle.

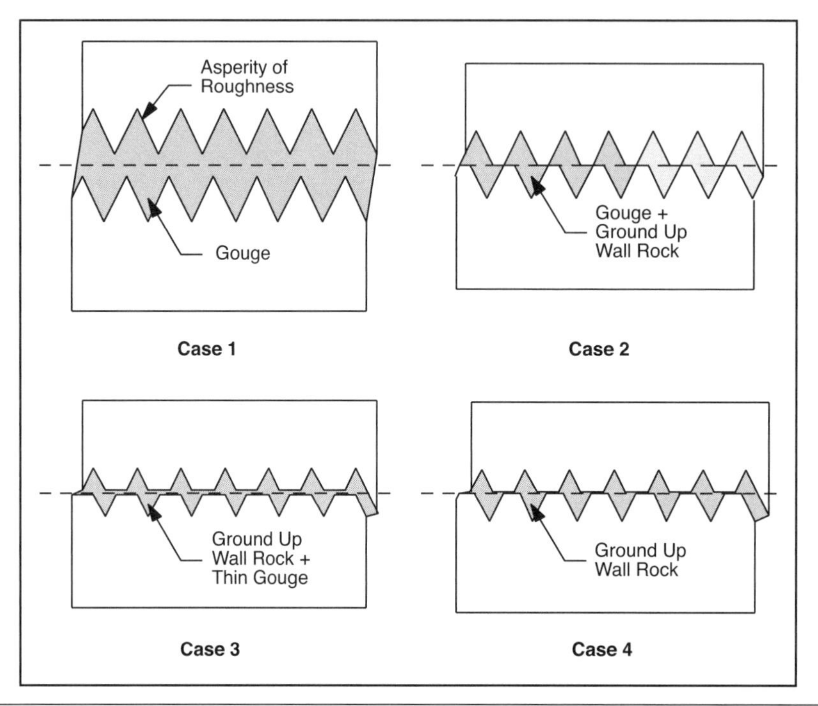

FIGURE 2.5 The effect of gouge on discontinuity strength

4. No gouge is present and the sliding plane passes entirely through joint wall rock (case 4); here the strength is dependent only on the properties of the wall rock.

SHEAR STRENGTH OF DISCONTINUITIES

As mentioned earlier, the most important factor to be considered in analyzing the stability of a rock slope is the geometry of the rock mass behind the slope face (Hoek and Bray 1977). The next most important factor is the shear strength of the potential failure surface. The shear strength is controlled by many factors, such as the rock–rock frictional resistance, cohesion along the failure surface, rock mass density, joint continuity, and surface roughness.

Impact of Surface Roughness on Shear Strength

John (1965) considered that limiting stress ratios (ratios of the normal stress, σ_n, to the shearing stress, τ) could be determined based on the parameters of joints and confining pressures when a jointed rock fails by sliding along one set of joints. At low normal stresses (low stress ratios), the shearing stresses along a joint with relatively smooth asperities produce a tendency for one block to ride up onto and over the asperities of another. As the asperities are mounted on the upside, the two sides of the discontinuity separate, producing a swelling or increase in bulk volume. This mechanism is called *dilatancy*. As movement takes place on the downside of the discontinuity, the block goes back down the asperities and the bulk volume decreases. At high normal stresses (high stress ratios), shearing tends to occur through the asperities as sliding progresses.

Patton (1966) investigated the shearing strength along irregular joint planes. For his study, a series of shear tests were performed on the horizontal surfaces of specimens cast from plaster that contained irregularly shaped asperities. These specimens were tested at several normal stresses to obtain maximum shear strengths. It was found that if displacement continued after the initial failure, then a residual shearing resistance could be obtained. The peak and residual shear strengths were plotted on a Mohr diagram (Figure 2.6). The peak strength envelope is shown by line OAB, and the residual shear strength envelope is shown by line OC. The line OA was obtained at low normal stresses and can be represented by the following expression:

$$S = c + \sigma_n \tan(\phi_r + i) \qquad \text{(EQ 2.1)}$$

where

S = shear strength along the rock joint
c = cohesion of the rock joint
σ_n = normal stress acting on the surface of the rock discontinuity
ϕ_r = residual angle of frictional sliding resistance of the rock surface
i = asperity angle of the failure surface

The line AB represents failure at high normal stresses where failure takes place through the base of the asperities. The vertical distance between the two lines (OAB and OC) represents the amount of shearing resistance that is lost with displacement. Figure 2.6 also shows the appearance of cohesion (or, more correctly, apparent cohesion) at the point where shearing of the asperities begins (point A). The angle of peak shearing resistance, $\phi_r + i$, can be obtained by measuring the angle of frictional sliding of the planar surface, ϕ_r, and the asperity angle, i. The angle of frictional sliding of the planar surface is approximately equal to the residual friction angle (the remaining friction angle along the discontinuity after shearing along the discontinuity has occurred). Therefore, the average asperity angle, i, can be derived by subtracting ϕ_r from $\phi_r + i$. The previous expression for S offers a method of interpreting the results of laboratory and field tests on jointed rocks, but its practical application requires an evaluation of the asperity angle in the field.

Patton (1966) demonstrated the significance of this relationship by measuring the average value of the asperity angle from photographs of bedding planes in rock slopes in the Rocky Mountains. He concluded that, to obtain a reasonable relationship between his field observations and the sum of the asperity angle and the angle of frictional sliding of the planar surface ($\phi_r + i$), it was necessary to measure only the first-order roughness of the surface. Second-order roughness did not significantly affect the frictional resistance of the planar surface.

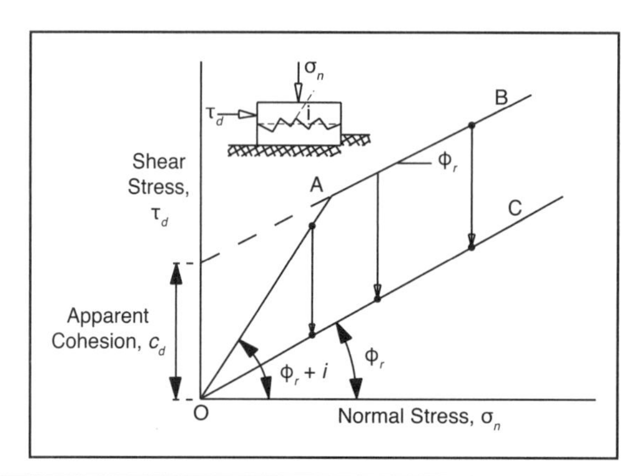

FIGURE 2.6 Peak and residual failure envelopes for multiple inclined surfaces

In a later study, Barton (1973) showed that Patton's results were related to the normal stress acting on the bedding planes that Patton measured. As the normal stress increased, the effect of the roughness angle on shear strength decreased. Barton concluded that, at low normal stresses, the second-order roughness projections play an important role in shear strength of discontinuities but that the role is reduced at high normal stresses.

Figure 2.7 shows a method for determining the average waviness of a discontinuity surface. The interlimb angle (ILA) of waviness can be measured by placing a 60-cm (24-in.) clinorule on the rock surface at the trough of the waviness and measuring the angle in the direction of potential sliding. For example, in the case of plane shear, the ILA would be measured in the down-dip direction of the discontinuities. Subtracting the ILA from 180° and dividing by 2 gives the waviness angle. A plot of the histogram of waviness [(180° – ILA)/2] versus frequency of occurrences of each waviness angle gives a mean value for the waviness (Figure 2.6), which describes the expected dilatancy (d_0) for that particular surface (Herget 1977). Note that d_0 is equivalent to the term *i* used in the earlier discussion.

Farmer (1983) conducted shear box tests along a single discontinuity—having frictional resistance but no cohesion—at low normal stress. He reported that the shear-displacement curves for tests on smooth and rough discontinuities in the same rock and at the same constant normal stress were similar to the hypothetical curves published by Roberds and Einstein (1978). These hypothetical curves are shown in Figure 2.7. Both Farmer's actual tests and the hypothetical curves of Roberds and Einstein show a change from poorly controlled, strongly dilatant, strain-softening behavior through a brittle-ductile transition to stable, mildly dilatant, strain-hardening behavior (Farmer 1983).

The hypothetical model of Figure 2.8 shows four cases of shear testing a discontinuity with a rough surface. At low normal stress—case A in Figure 2.8—shear resistance rose with displacement to a peak shear stress level. During the test there was dilation as the asperities that make up the surface roughness were mounted. The dilation (or opening of the discontinuity) is shown in the bottom portion of Figure 2.8. As movement took place on the

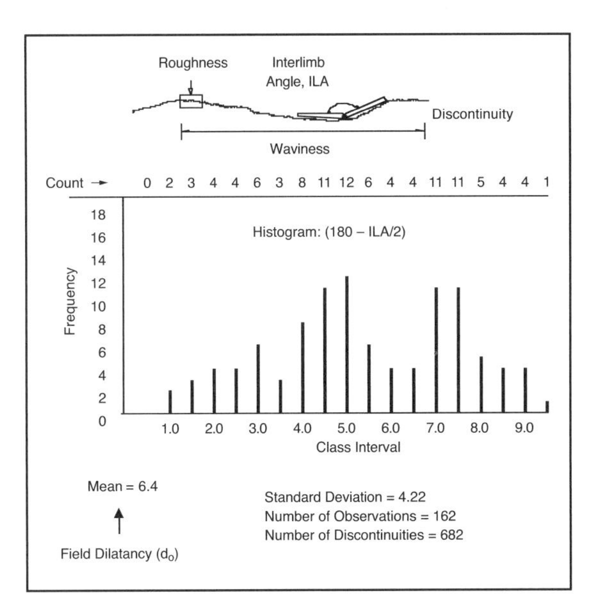

Adapted from Herget 1977

FIGURE 2.7 Principle of and output from waviness measurements

downside of the discontinuities, the shear resistance fell to a residual level, then rose again as the next series of asperities were mounted. When the normal stress was increased—and to a lesser extent when the roughness of the discontinuity was reduced—that pattern of behavior changed. Initially the asperities were still mounted, but the smaller asperities were sheared at high strains, which gave the type of modified strain-softening behavior shown for case B in Figure 2.8. This tendency of shearing the asperities and strain softening continued with increased normal stress until a typical strain-softening curve, as in case C, was obtained. Ultimately, at very high normal stresses, fracturing through the discontinuity occurred immediately and there was no mounting of the asperities, which led to the strain-hardening type of behavior as in case D.

Joint Roughness Coefficient

Barton (1973) proposed a different approach to the problem of predicting the shear strength of rough joints. On the basis of tests and observations carried out on artificially produced rough "joints" in material used to model studies of slope behavior, Barton and Choubey (1977) derived the following empirical equation:

$$S = \sigma_n \tan\left[\phi_b + \mathrm{JRC} \cdot \log_{10}\left(\frac{\mathrm{JCS}}{\sigma_n}\right)\right] \qquad \text{(EQ 2.2)}$$

where

S = shear strength along the rock joint

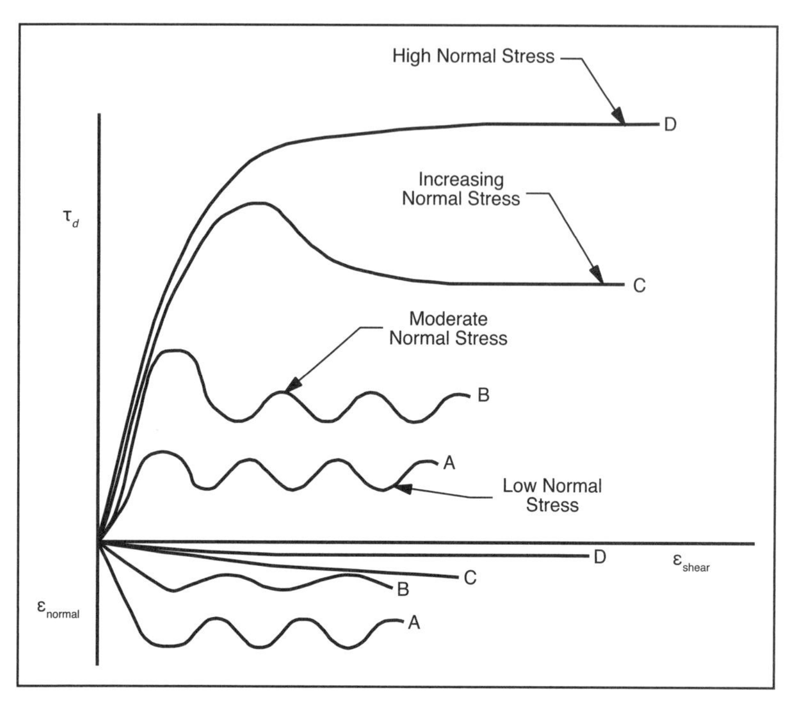

Source: Roberds and Einstein 1978

FIGURE 2.8 Discontinuity shear resistance (τ_d) against shear strain (ε_{shear}), as well as normal strain (ε_{normal}) against shear strain, during testing along a typical discontinuity

σ_n = normal stress acting on the surface of the rock joint

ϕ_b = angle of frictional sliding of the planar surface

JRC = joint roughness coefficient

JCS = joint compressive strength

The JRC varies from 0 for perfectly smooth rock to 20 in very rough rock (Farmer 1983). The rock JCS is equal to the uniaxial compressive strength of the rock if the joint is unweathered; smoothly walled joints are less affected by the value of JCS because failure of asperities plays a less important role. The smoother the walls of the joints, the more significant the role played by the mineralogy of the rock mass.

The JRC can be empirically determined by comparing profiles of rock surface roughness with 10 standardized rock roughness profiles (Figure 2.9). This method of roughness determination has been supported by the International Society for Rock Mechanics (Brown 1981).

Tse and Cruden (1979) pointed out that fairly small errors in estimating a JRC could produce serious errors in estimating the peak shear strength from Equation 2.2. This is especially true if the ratio of JCS to σ_n is large. They recommended a numerical method of

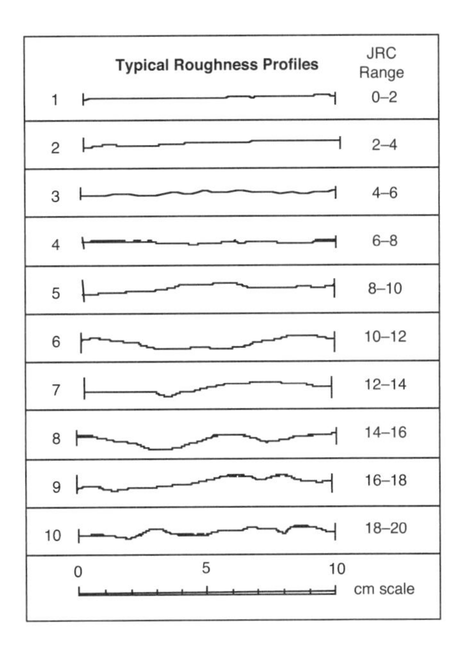

Source: Brown 1981

FIGURE 2.9 Rock joint roughness profiles showing the typical range of JRC values associated with each

checking the value of JRC based on the surface slope of the asperities in a profile of the rock joint surface. Previously, Barton and Choubey (1977) had suggested that the tilt and push tests provided a more reliable means of estimating the JRC than did comparison with typical profiles. Barton and Bandis (1980) supported the use of such tests, particularly in heavily jointed rock masses.

In a tilt test, two contiguous blocks are extracted from an exposure, and the upper is laid upon the lower in exactly the same position as it was in the rock mass. Both are then tilted, and the angle at which sliding occurs is recorded (Figure 2.10). Tilt tests can also be conducted on jointed core to measure the friction angle of joints intersecting drill core. The value of JRC is back-calculated directly from the tilt test by rearrangement of the Mohr–Coulomb peak strength equation (Barton 1982):

$$JRC = \frac{\alpha^{\circ} - \phi_r}{\log_{10}\left(JCS/\sigma'_{no}\right)} \qquad \text{(EQ 2.3)}$$

where

α° = tilt angle when sliding occurs = $\arctan(\tau/\sigma'_{no}) = \phi'$

ϕ_r = residual angle of frictional sliding resistance of the rock surface

σ'_{no} = effective normal stress acting across the joint when sliding occurs

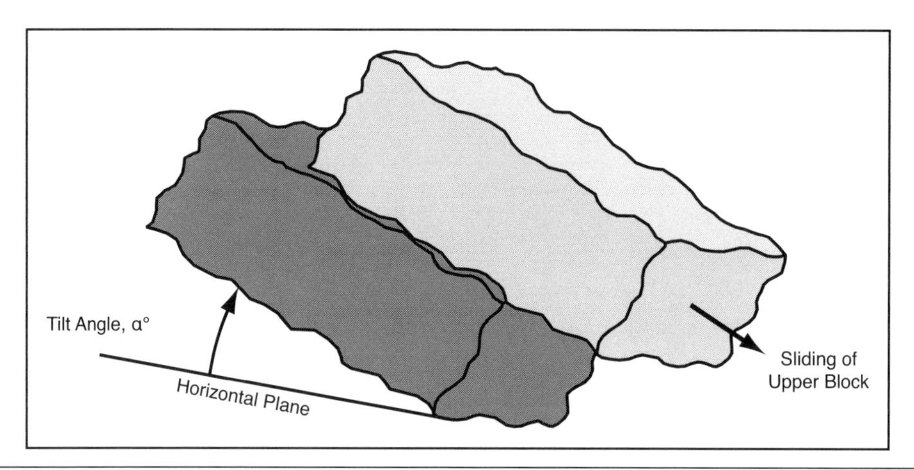

FIGURE 2.10 The tilt test using two contiguous blocks of rock extracted from the same location to measure the friction angle and back-calculate the JRC

For example, if $\alpha° = 70°$, $\phi_r = 22°$, JCS = 125 MPa, and $\sigma'_{no} = 0.0025$ MPa, then

$$JRC = (70° - 22°)/4.70 = 10.21 \approx 10$$

Barton and Bandis (1980) found that JRC is a constant only for a fixed joint length. Generally, longer profiles of the same joint have lower JRC values. Barton and Bandis suggested that the rate for full development of peak strength along a joint seems to be dependent on the distance the joint has to be displaced for the asperities to be brought into contact. This distance increases with increasing joint length. Consequently, during testing, longer samples tend to give lower values of peak shear strength. The curves of shear force versus displacement change with increasing scale; that is, for longer joint lengths, behavior along a joint during shearing changes from brittle to plastic and the shear stiffness is reduced. Barton and Bandis (1980) attributed the changing shape of the curves to the progressive damage that occurs to larger and larger asperities as the scale is increased. They suggested that the appropriate test size appears to be the natural block size.

Brown (1981) suggested that the peak friction angle can be estimated by using the following equation:

$$\phi_{\text{peak}} = JRC \cdot \log_{10}\left(\frac{JCS}{\sigma_n}\right) + \phi_{\text{residual}} \qquad \text{(EQ 2.4)}$$

where

ϕ_{peak} = peak friction angle
ϕ_{residual} = residual friction angle
σ_n = normal stress acting on the surface of the rock joint

This method begins by comparing measured roughness profiles with the standardized roughness profiles shown in Figure 2.9. Then the discontinuity walls are tested with a Schmidt hammer, or another method of rapid testing, to estimate the JCS and the residual friction angle. The residual friction angle is estimated by using an empirical relationship between Schmidt rebound hardness and residual friction angle.

Krahn and Morgenstern (1979) carried out a series of direct shear tests on natural and artificially produced discontinuities in limestone to demonstrate how the residual shearing resistance was influenced by surface roughness. They found that, except at low normal stresses, there was essentially no difference between peak and residual shearing resistance for smooth and flat artificial surfaces. All the natural discontinuities, however, showed a significant drop from peak to residual strength. The wide variation in the shearing resistance was mostly attributable to the surface roughness of the discontinuities. Krahn and Morgenstern (1979) concluded that the ultimate frictional resistance of jointed, hard, unweathered rock depends on the initial surface roughness along the joint and on the type of surface alteration that occurs during shearing. Each residual shearing resistance reflects the characteristic roughness of the particular type of discontinuity. To determine the residual frictional resistance of a rock mass, those authors suggested that the differing geologic histories of the various discontinuities must be recognized and that the sampling and testing program should proceed accordingly.

The Generalized Hoek–Brown Strength Criterion

Hoek and Brown (1980a, 1980b) proposed a method for obtaining estimates of the strength of jointed rock masses as a means to estimate rock mass strength by scaling the relationship derived according to the geologic conditions present. The criterion was developed based on Hoek's (1968) experiences with brittle rock failure and his use of a parabolic Mohr envelope derived from earlier work by Griffith (1920, 1924) to define the relationship between shear and normal stress at fracture initiation. By associating fracture initiation with fracture propagation and rock failure, Hoek and Brown (1980a, 1980b) proceeded through trial and error to fit a variety of parabolic curves to triaxial test data to derive their criterion. Accordingly, the Hoek–Brown criterion is empirical with no fundamental relationship between the constants included in the criterion and any physical characteristics of the rock (Eberhardt 2012).

Initially, Hoek and Brown (1980a, 1980b) sought to link the empirical criterion to geological observations by means of one of the available rock mass classification schemes, and, for this purpose, they chose the rock mass rating proposed by Bieniawski (1976).

The generalized Hoek–Brown strength criterion is expressed in terms of the major and minor principal stresses, and is as follows:

$$\sigma'_1 = \sigma'_3 + \sigma_{ci}\left(m_b \frac{\sigma'_3}{\sigma_{ci}} + s\right)^a \qquad \text{(EQ 2.5)}$$

where

σ'_1 and σ'_3 = maximum and minimum effective stresses at failure
σ_{ci} = uniaxial compressive strength of the intact rock pieces
m_b = value of the Hoek–Brown constant m for the rock mass
s and a = constants that depend on the rock mass characteristics

For the intact pieces that make up the rock mass, $s = 1$ and $m = m_i$, then Equation 2.5 simplifies to (Hoek 2006):

$$\sigma'_1 = \sigma'_3 + \sigma_{ci}\left(m_i \frac{\sigma'_3}{\sigma_{ci}} + 1\right)^{0.5} \tag{EQ 2.6}$$

The relationship between the principal stresses at failure for a given rock is defined by two constants, the uniaxial compressive strength, σ_{ci}, and a constant, m_i. Wherever possible, the values of these constants should be determined by statistical analysis of the results of a set of triaxial tests on carefully prepared core samples.

Note that the range of minor principal stress (σ_3') values over which these tests are carried out is critical in determining reliable values for the two constants. In deriving the original values of σ_{ci} and m_i, Hoek and Brown (1980a) used a range of $0 < \sigma_3' < 0.5\sigma_{ci}$ and, to be consistent, it is essential that the same range be used in any laboratory triaxial tests on intact rock specimens. At least five well-spaced data points should be included in the analysis.

Laboratory tests should be carried out at moisture contents as close as possible to those that occur in the field. Many rocks show a significant strength decrease with increasing moisture content, and tests on samples, which have been left to dry in a core shed for several months, can give a misleading impression of the intact rock strength. Once the five or more triaxial test results have been obtained, they can be analyzed to determine the uniaxial compressive strength σ_{ci} and the Hoek–Brown constant m_i as described by Hoek and Brown (1980a). In this analysis, Equation 2.6 is rewritten as follows (keeping s in general form instead of setting $s = 1$, as in the case for intact rock):

$$\sigma'_1 = \sigma'_3 + \left(\sigma_{ci} m_i \sigma'_3 + s\sigma_{ci}^2\right)^{0.5} \tag{EQ 2.7}$$

$$\left(\sigma'_1 - \sigma'_3\right)^2 = \sigma_{ci} m_i \sigma'_3 + s\sigma_{ci}^2 \tag{EQ 2.8}$$

or

$$y = m \cdot \sigma_{ci} \cdot x + s\sigma_{ci}^2 \tag{EQ 2.9}$$

where $y = (\sigma_1' - \sigma_3')^2$ and $x = \sigma_3'$.

For n specimens, the uniaxial compressive strength, σ_{ci}, the constant m_i, and the coefficient of determination, r^2, can be calculated by the statistical regression equations for a straight line, where x and y are as previously defined:

$$\sigma_{ci}^2 = \frac{\sum y}{n} - \left[\frac{\sum xy - \left(\sum x \sum y / n\right)}{\sum x^2 - \left(\left(\sum x\right)^2 / n\right)}\right]\frac{\sum x}{n}$$

$$m_i = \frac{1}{\sigma_{ci}}\left[\frac{\sum xy - \left(\sum x \sum y / n\right)}{\sum x^2 - \left(\left(\sum x\right)^2 / n\right)}\right]$$

$$r^2 = \frac{\left[\sum xy - \left(\sum x \sum y / n\right)\right]^2}{\left[\sum x^2 - \left(\sum x\right)^2 / n\right]\left[\sum y^2 - \left(\sum y\right)^2 / n\right]}$$

The Mohr envelope, relating normal and shear stresses, can be determined by the method proposed by Hoek and Brown (1980a). In this approach, Equation 2.5 is used to generate a series of triaxial test values, simulating full-scale field tests, and a statistical curve fitting process is used to derive an equivalent Mohr envelope defined by the following equation:

$$\tau = A \cdot \sigma_{ci}\left(\frac{\sigma'_n - \sigma_{tm}}{\sigma_{ci}}\right)^B \qquad \text{(EQ 2.10)}$$

where

A and B = material constants

σ'_n = normal effective stress

σ_{tm} = "tensile" strength of the rock

Figure 2.11 shows a comparison of the linear Mohr–Coulomb envelope to the Hoek–Brown curvilinear failure envelope (Eberhardt 2012).

The Hoek–Brown failure criterion, which assumes isotropic rock and rock-mass behavior, should only be applied to those rock masses in which there are a sufficient number of closely spaced discontinuities, with similar surface characteristics, that isotropic behavior involving failure on multiple discontinuities can be assumed (Hoek and Karzulovic 2001). When the structure being analyzed is large and the block size is small in comparison, the rock can be treated as a Hoek–Brown material. On the other hand, when the block size is of the same order as that of the structure being analyzed, or when one of the discontinuity sets is significantly weaker than the others, then the Hoek–Brown criteria should not be used. In these cases, the stability of the structure should be analyzed by considering failure

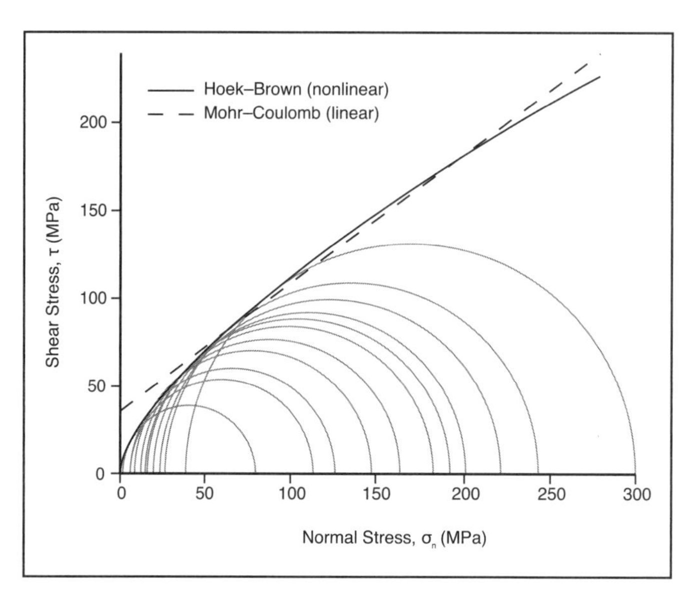

Source: Eberhardt 2012

FIGURE 2.11 A comparison of the linear Mohr–Coulomb and nonlinear Hoek–Brown failure envelopes plotted against triaxial test data for intact rock

mechanisms involving the sliding or rotation of blocks and wedges defined by intersecting structural features (Hoek and Karzulovic 2001).

The Direct Shear Test

To obtain shear strength parameters for use in rock slope design, some sort of testing is often required. This testing may take the form of a simple field test or of a very sophisticated laboratory or an in situ test in which all the characteristics of the in situ rock discontinuity are reproduced as accurately as possible (Hoek and Bray 1977).

The direct shear machine (Figure 2.12) is a useful tool to determine the strength parameters of cohesion, c, and friction angle, ϕ, for a discontinuity. The machine uses a split shear box configuration (Figure 2.13) to force shearing along a predetermined surface. A simple discussion of the test procedure follows.

Suppose that a number of samples of rock have been obtained, each of which has been cored or cut from the same block of rock that contains a through-going, cemented discontinuity. One of the samples containing the discontinuity is subjected to a normal stress, σ_1, and a slowly increasing shear stress, τ_1, causing a displacement along the surface of the discontinuity, Δl, until failure of the discontinuity occurs (Figure 2.14A). Plotting the shear stress level for the first test, at a constant normal stress level versus measured displacement, results in the type of curve shown in Figure 2.14B. At very small displacements, the specimen behaves elastically and the shear stress increases linearly with displacement. As the

FIGURE 2.12 Direct shear machine

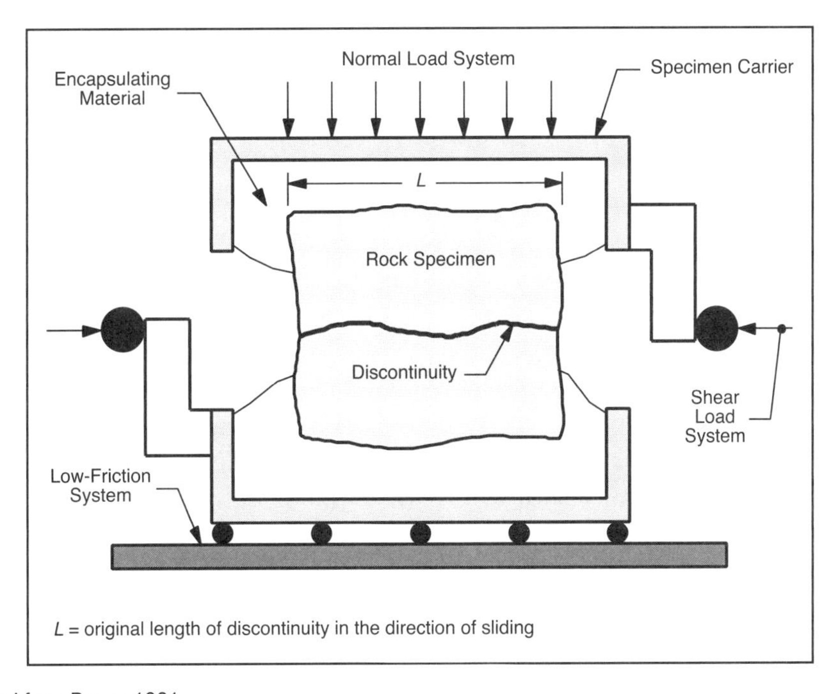

Adapted from Brown 1981

FIGURE 2.13 Suggested arrangement for a laboratory direct shear test on a single discontinuity

forces resisting movement are overcome, the curve becomes nonlinear and then reaches a peak value, τ_p. Thereafter, the shear stress required to cause further shear displacement drops rapidly and then levels out at a constant residual value, τ_r.

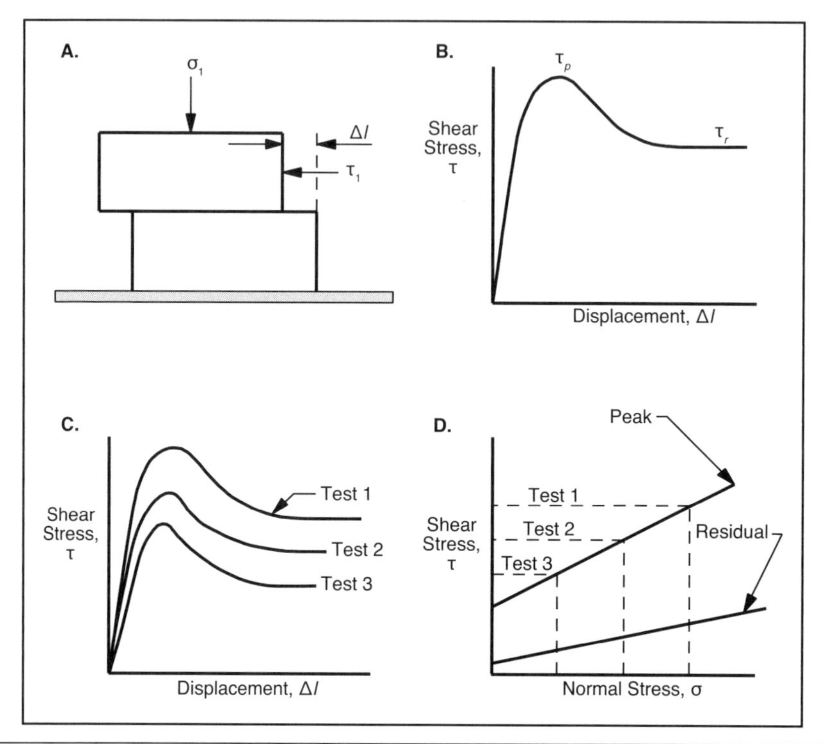

FIGURE 2.14 Direct shear test

If the remaining samples are tested in the same manner—but at different levels of constant normal stress, σ_n—a series of shear stress–displacement curves will be generated (Figure 2.14C). If the peak shear strength values are plotted versus normal stress for each of the tests as shown in Figure 2.14D, then the Mohr–Coulomb peak shear strength envelope is developed for that specimen. (NOTE: The x- and y-axes of the plot should be the same scale.) Similarly, a plot of τ_r versus σ_n on the same (or different) graph develops the Mohr–Coulomb residual shear strength envelope (Figure 2.14D).

The shear envelopes will be approximately linear, and the values of c_p and c_r can be determined from the y-intercept of the peak and residual envelopes. Likewise, the values of ϕ_p and ϕ_r can be determined from the slopes of the two lines. The peak and residual shear strengths, then, are defined by the following equations:

$$S_p = c_p + \sigma \tan\phi_p$$

$$S_r = c_r + \sigma \tan\phi_r$$

where p and r denote peak and residual strength values, respectively.

The value of residual cohesion, c_r, is generally zero, or very nearly so, indicating that the cohesive strength of the cementing material in the discontinuity has been lost after failure.

Also, the residual friction angle, ϕ_r, is usually lower than the peak friction angle, ϕ_p, indicating grinding of surface asperities of roughness along the discontinuity into powder.

GEOLOGIC DATA COLLECTION

Whether one is conducting a preliminary slope stability assessment or a detailed stability study, the collection of appropriate geologic data is an important step in the process of determining rock slope stability. Information that should be collected includes survey data, rock attributes, discontinuity attributes, discontinuity orientation, structure type, water conditions, filling material attributes, and so on.

Preliminary Slope Stability Investigation

The first phase in the slope stability investigation is the preliminary investigation. This phase may very well involve simply compiling available information from previous studies and records. It should include the following details:

- A study of the material type (soils, rock, or both)
- Mapping of structural geology (bedding planes, faults, joints, folds, etc.)
- Gathering of available hydrological data

From the preliminary investigation, the following can be determined:

- Possible modes of slope failure
- Location of zones susceptible to failure
- Prediction for the overall slope angle
- Plans for the feasibility study
- Possible water problems

Detailed Slope Stability Study

The detailed slope stability study can be divided into two parts: (1) field work and (2) office and laboratory work.

Field work. The field work is the data collection phase of the detailed slope stability investigation. It should include the following:

- Compiling and editing topographic maps
- Detailed geologic mapping (1:1,000 to 1:500):
  - Map formation boundaries on about a 1:1,000 scale.
  - Map structural features within the boundaries. Use a geological compass to map the structures. Put the mapped structures on the survey map.

- Ascertaining the location of groundwater discharge points in the study area. Determining discharge amounts (monitor at least twice per year for at least two years).
- Collecting rock samples for laboratory testing
- Establishing a core-drilling program (core recovery should be at least 90%), if needed
- Drilling, and setting up instrumentation for, any needed groundwater-monitoring wells
- Conducting any necessary geophysical studies

Office and laboratory work. This stage is the data reduction and interpretation phase of the detailed slope stability study. It should include the following:

- Development of systematic working cross sections: radial for open pit, axial for long pits or road cuts
- Discontinuity evaluations
- Laboratory testing of samples collected from the field:
  - Testing of intact rock specimens (uniaxial, triaxial, maybe direct shear)
  - Testing of discontinuity surface samples (direct shear)
- Evaluation of the stability of proposed slopes by sector
- Determination of the seismicity of the region
- Factor-of-safety analysis sector by sector, based on available data and sound engineering judgment. If the factor of safety along a cross section is below the acceptable level, then take one or more of these steps: (1) accept the risk of failure; (2) flatten the slope; (3) use artificial support; and/or (4) drain the slope, if appropriate.

Procedure for Geologic Data Collection

The following general procedure is suggested for geologic data collection prior to, or during, a slope stability investigation:

1. Determine the boundaries between geologic materials with different properties:
   - Weathering (color changes, hammer impact, etc.)
   - Differences in sedimentation (grain size distribution)
   - Differences in joint intensity (high versus less)
2. Determine structural features:
   - Folds
   - Faults and faulting systems
   - Bedding in sedimentary rocks
   - Schistocity and cleavage in metamorphic rocks
   - Discontinuities (joints, shear zones, bedding planes, etc.)

3. Map discontinuities (Figure 2.15 shows a typical discontinuity survey data sheet):
   - Location (coordinates, generally along a line survey)
   - Material type (host rock)
   - Type of discontinuity (note whether it is a fault, shear zone, bedding plane, etc.)
   - Orientation of the discontinuity (dip and dip direction, or dip and strike)
   - Persistence or the continuity of the discontinuity (The coefficient of continuity, K_c, is the ratio of the sum of the unattached area of the discontinuity to the whole area of the surface; $K_c = 1$ for major features, such as faults or shear zones, and $K_c < 1$ for minor structural features.)
   - Joint intensity (number of joints per unit distance normal to the strike of the set) and degree of separation
   - Openness or closed nature. If the discontinuity is open, look for the presence of filling material such as gouge or transported material; if filling is present, look at the type (clay, granular material, crystalline material, or veins); measure the thickness of the filling.
   - Waviness and roughness (first and second order of asperities)
4. Conduct a sampling program:
   - Obtain samples of intact rock, both weathered and unjointed.
   - Obtain samples of the discontinuity surface for shear testing.
   - If filling material is present, obtain representative samples—soil mechanics tests may be needed to determine strength parameters.
5. Determine groundwater conditions:
   - Locate springs or discharge points.
   - Determine the permeability of the rock and the joints (generally more important along the joints—drawdown or recovery tests may be required).
   - Measure the discharge of springs or seeps at least twice per year.

HEMISPHERICAL PROJECTION TECHNIQUES

Hemispherical projection techniques offer a graphical method for analyzing three-dimensional (3-D) problems involving planes, lines, and points in a convenient and easily interpreted two-dimensional (2-D) form. The method is often referred to as *stereographic projection*, which literally means the projection of solid or 3-D drawings (Priest 1985). Hemispherical projection methods are frequently used in rock mechanics studies for analyzing planar discontinuities such as faults, bedding planes, shear planes, and joints. The techniques are valuable in rock mechanics studies given that they present data in the form of a graphical representation that can be inspected visually rather than numerically.

Figure 2.16 illustrates the basic concept behind hemispherical projections, where the hemisphere is represented by a plane, the plane is represented by a line, and a line is represented

DISCONTINUITY SURVEY DATA SHEET

GENERAL INFORMATION:

Date: ____/____/____ Site: ____________________

Rock Material Information ________________________________

____

Rock Mass Information ________________________________

____

Comments ________________________________

____

Locality Type: ________

1. Natural exposure
2. Construction excavation
3. Mine bench
4. Tunnel
5. Other (specify)

LINE SURVEY INFORMATION:

Line Number	Line Trend	Line Plunge	Initial Northing	Initial Easting	Initial Elevation

DISCONTINUITY SURVEY DATA:

Line Number	Line Distance	Dip Direction	Dip	Structure Type	Continuity	Filling Type	Waviness	Surface Rough-ness	Rock Type	Rock Hard-ness	Water

Structure Type	Continuity	Filling Type	Waviness	Surface Roughness	Rock Hardness	Water
0. Fault zone	1 = High	1. Clean	Length	1. Polished	1 = soft	yes or no
1. Fault	5 = Low	2. Surface stain	(in meters)	2. Slickensided	10 = Extremely hard	
2. Joint		3. Noncohesive		3. Smooth		
3. Cleavage		4. Cohesive		4. Rough		
4. Schistosity		5. Cemented		5. Defined ridges		
5. Shear		6. Calcite		6. Small slope		
6. Fissure		7. Chlorite, talc		7. Very rough		
7. Tension crack		8. Others				
8. Foliation						
9. Bedding						

Adapted from Herget 1977

FIGURE 2.15 Example discontinuity survey data sheet

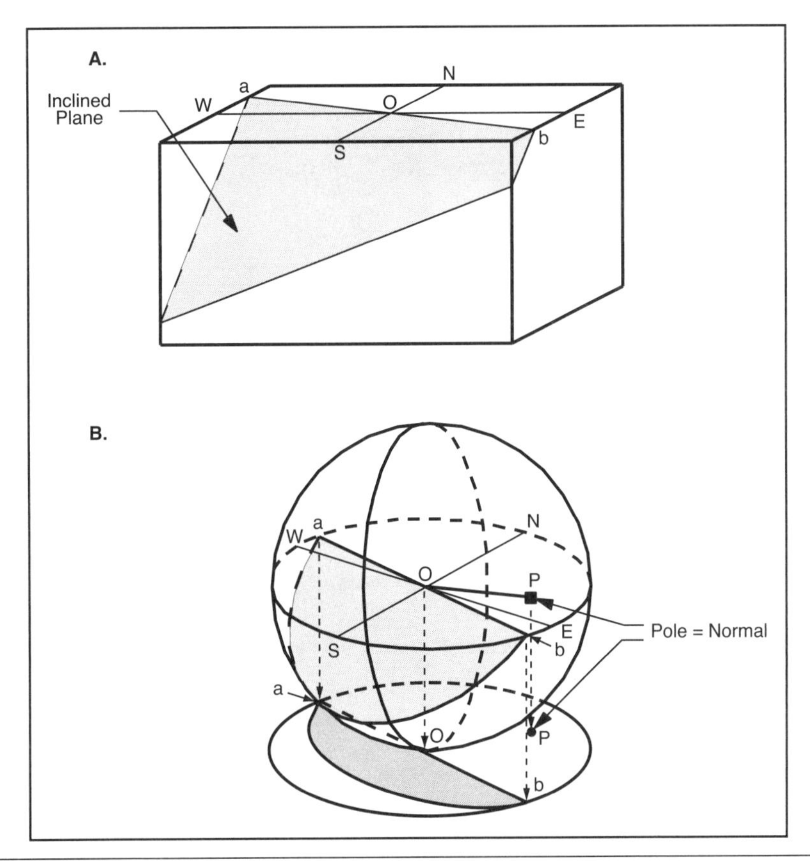

FIGURE 2.16 Basic concept behind hemispherical projections: (A) inclined plane; (B) stereographic representation of inclined plane

by a point. Suppose a plane with orientation of dip direction and dip (α', β) exists in space (Figure 2.16A). A horizontal line on the plane has the trace aOb on Figure 2.16B. The direction of this trace aOb is the strike of the inclined plane. If we encompass the plane in a reference sphere with center at O, which is free to move in space but is not free to rotate in any direction, then the plane will intersect the sphere in both the upper and the lower hemispheres (see Figure 2.16B for the lower-hemisphere intersection). If we consider only the lower hemisphere, which is common in rock mechanics and structural geology work, and project the lower hemisphere of the sphere and the trace of the inclined plane onto a flat surface, then we get the stereographic projection of the *great circle* representing the plane on the lower hemisphere. If a line normal to the plane is constructed at O and projected until it pierces the lower hemisphere, then a point P is defined. Point P will have an inclination of ($90° - \beta$) from O in the opposite direction of α'. If P is projected onto the flat surface, then the pole, P, of the plane is defined.

Consider a set of planes similar to the one in Figure 2.16A, with dip directions of 90° and 180° (i.e., planes with a strike direction of N-S or a dip direction of E-W) and dips varying in 10° increments from 0° to 90°. If we now project their intersections with the sphere in space onto the flat surface as in Figure 2.14B, then we get a series of great circles oriented with dip direction E and W and dips ranging from 0° to 90°. If we define *true dip* as the dip angle, in degrees, measured in the dip direction, then the true dip of each of these great circles is measured inward along the E-W line. We can define another term, *pitch angle*, as the deflection angle, from 0° to 90° toward N or S, of *apparent dip* from true dip. Apparent dip is always less than true dip.

With the preceding terms and definitions in mind, we can now generate a meridional stereographic projection, also known as a stereonet:

1. Construct the loci of planes with E-W dip direction (N-S strike) and various dip angles, as in Figure 2.17A. Dip angles of 10° increments are used in Figure 2.17A.
2. Project radial lines of apparent dip from the center of the stereonet to each of the great circles representing planes, as in Figure 2.17B. Pitch angles of 10° are used in Figure 2.17B.
3. Project systematic (10°) radial lines in the planes with systematic dip angles, as in Figure 2.17C. That is, for each great circle representing each plane, connect all the points of 10°, 20°, 30°, and so forth, pitch angles to form the small circles on the stereonet. Figure 2.17C shows construction of the small circles at 30° intervals from the E-W line.
4. Construct all great circles and small circles at 2° intervals to generate the stereonet shown later in Figure 2.19.

Consider the location of the pole of the plane in Figure 2.16B. If the plane aOb is dipping at 0°, then the pole is located directly below O in the top portion of Figure 2.16B; if the plane is dipping at 90°, then the pole is located at the outside perimeter (equator) of the reference sphere. As the dip of the plane increases from 0° to 90°, the location of the pole on the lower-hemisphere projection (bottom portion of Figure 2.16B) moves outward from the center to the perimeter. If the plane with strike line a-b and known dip direction/dip (α', β) is rotated around the centerline of the reference sphere, O, then the normal to the plane will pierce the surface of the sphere and form a circle around O with radius β (Figure 2.18). If this technique is employed for all planes dipping in the same direction as the plane aOb in Figure 2.16, with dips varying in 2° intervals from 0° to 90°, then we generate the *polar stereographic net* as shown later in Figure 2.21. Since the normal, or pole, of a plane is located at ($\alpha' \pm 180°$), the polar stereonet will have N-S and E-W reversed. Poles of planes can be plotted directly on the polar net.

A great many different types of hemispherical projections are available, and the choice of the proper one depends on the particular problem being analyzed. These projection methods are extremely useful in rock mechanics studies, such as analyzing discontinuous rock blocks both underground and on the surface, as in rock slope stability analyses. Structural geologists frequently use hemispherical projections for analyzing rock structure interactions and borehole discontinuity data; these projections have also been used for analyzing force vectors (Priest 1985).

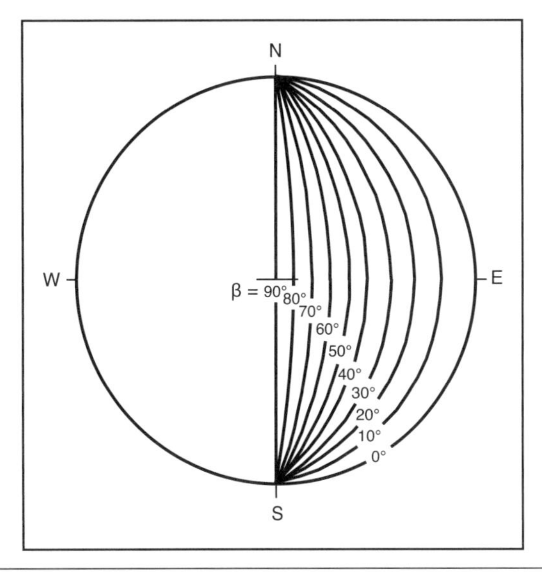

FIGURE 2.17A Loci of planes with N-S strike and various dip angles

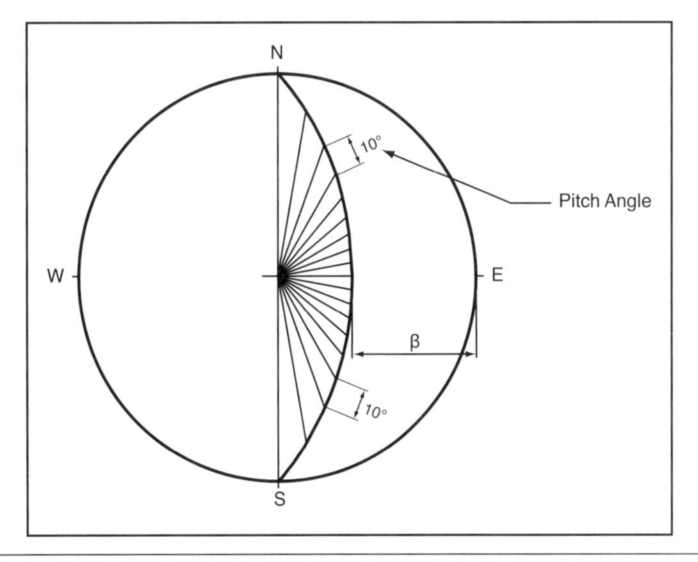

FIGURE 2.17B Projection of radial lines in a plane with β dip angle

Some of the more common types of stereonets in use include the following (Ragan 1973; Goodman 1980):

- The equal-angle stereonet, also known as the meridional stereonet or the Wulff stereonet (Figure 2.19)
- The equatorial equal-area stereonet, also known as the Schmidt stereonet (Figure 2.20)

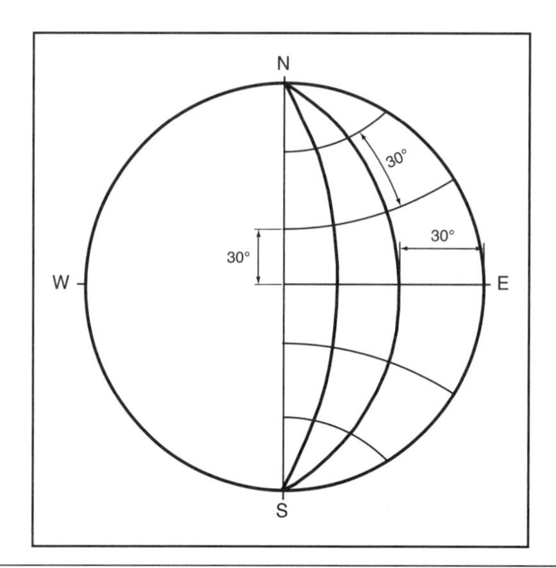

FIGURE 2.17C Projection of small circles in planes with systematic dip angles

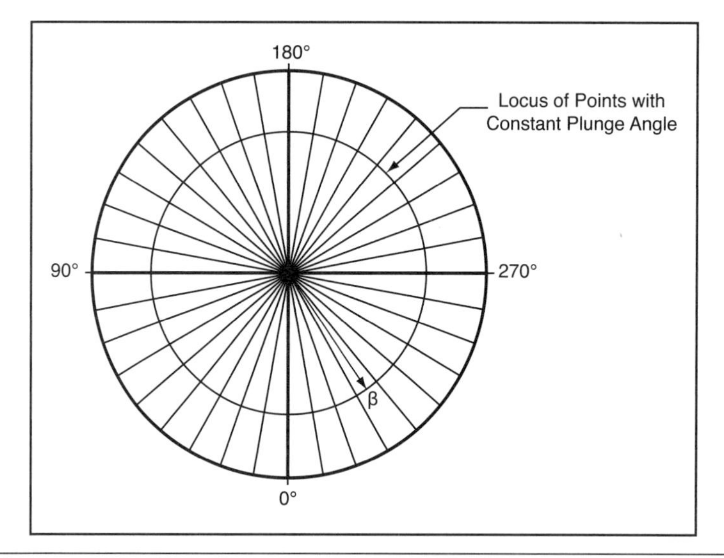

FIGURE 2.18 Locus of points of constant plunge angle, ß, around the center of the reference sphere

- The polar equal-angle net (Figure 2.21)
- The polar equal-area net (Figure 2.22)
- Counting nets such as the Kalsbeek net (Figure 2.23) or the Denness net

The polar equal-angle net is used in conjunction with the equal-angle stereonet for the plotting of the normals (poles) to the discontinuities; the polar equal-area net is used in

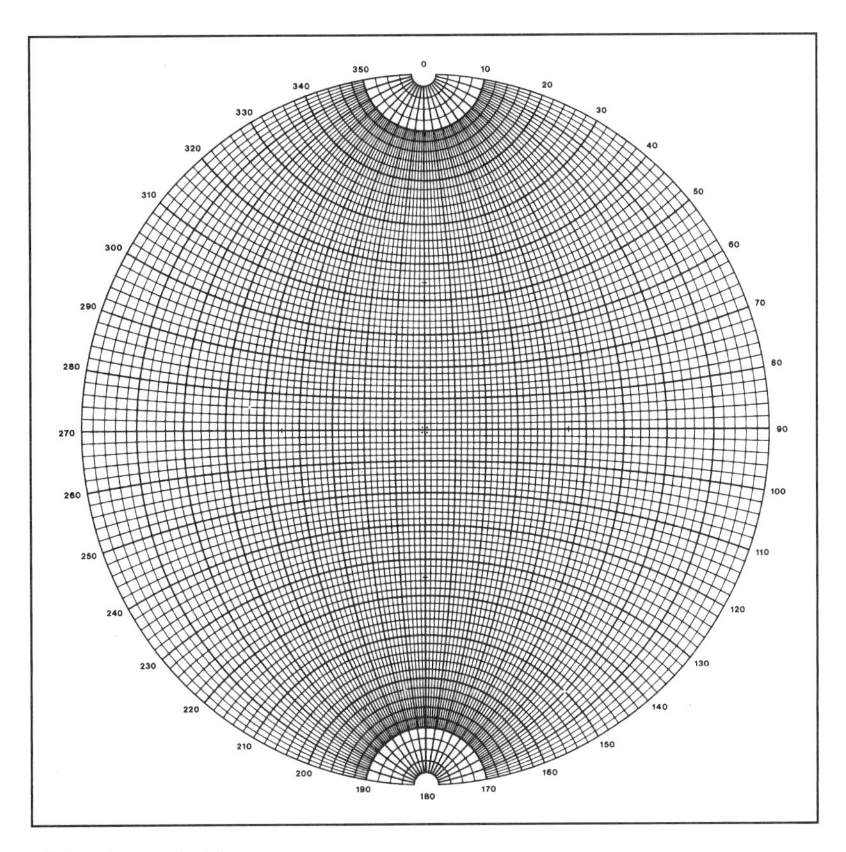

Courtesy of Christopher St. John

FIGURE 2.19 Equatorial equal-angle stereonet

conjunction with the equal-area stereonet for the plotting of poles; and counting nets are used with both types of polar nets to determine clusters or concentrations of orientations.

The basis for all projection techniques is the imaginary reference sphere of radius R, positioned with its center at the center of the area of projection. Consider a line oriented with trend α and plunge β, and positioned so that it passes through the center of our reference sphere. If this line is extended, it will pierce the perimeter of the reference sphere at two points: P on the lower hemisphere and Q on the upper hemisphere. If we further consider only the point on the lower hemisphere, P, it can be projected onto the horizontal plane by several methods. Two of these methods are the equal-angle projection and the equal-area projection (Priest 1985). Figures 2.24 and 2.25 show a sectional view (vertical plane) and plan view (horizontal plane) of the reference sphere for the equal-angle projection and the equal-area projection, respectively, illustrating the geometry for the position of P on both types of projections.

For the equal-angle projection (Figure 2.24), the given line of trend α and downward plunge β will intersect the lower reference sphere at point P'. If a straight line is drawn

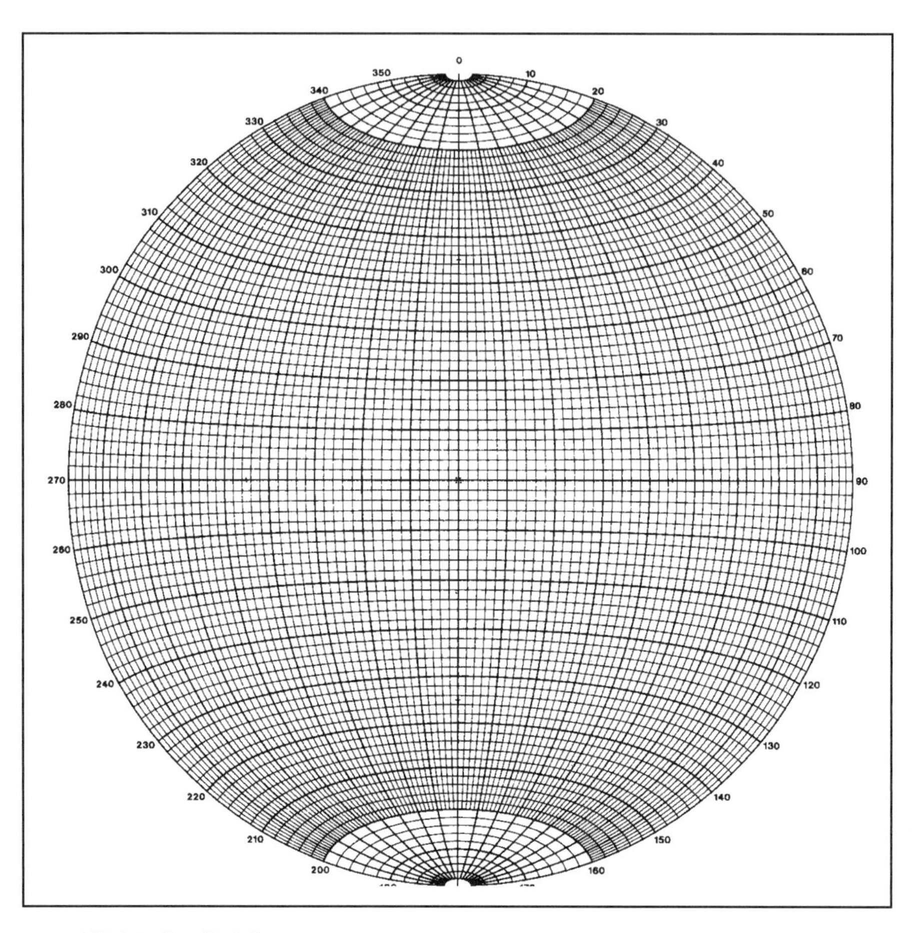

Courtesy of Christopher St. John

FIGURE 2.20 Equatorial equal-area stereonet

from P' to a zenith point, Z, which is at a distance R vertically above the center point, O, then the line will intersect the plane of projection (horizontal plane) at P. For this projection, the relationship between r, the radial distance of point P from O, and β is given by (Priest 1985)

$$r = R\tan(\beta/2)$$

The trace of a great circle (Priest 1985) of dip direction α_c and dip β_c on the equal-angle projection will have a radius R_c given by (Figure 2.24):

$$R_c = R/\cos\beta_c$$

and will have a center point of

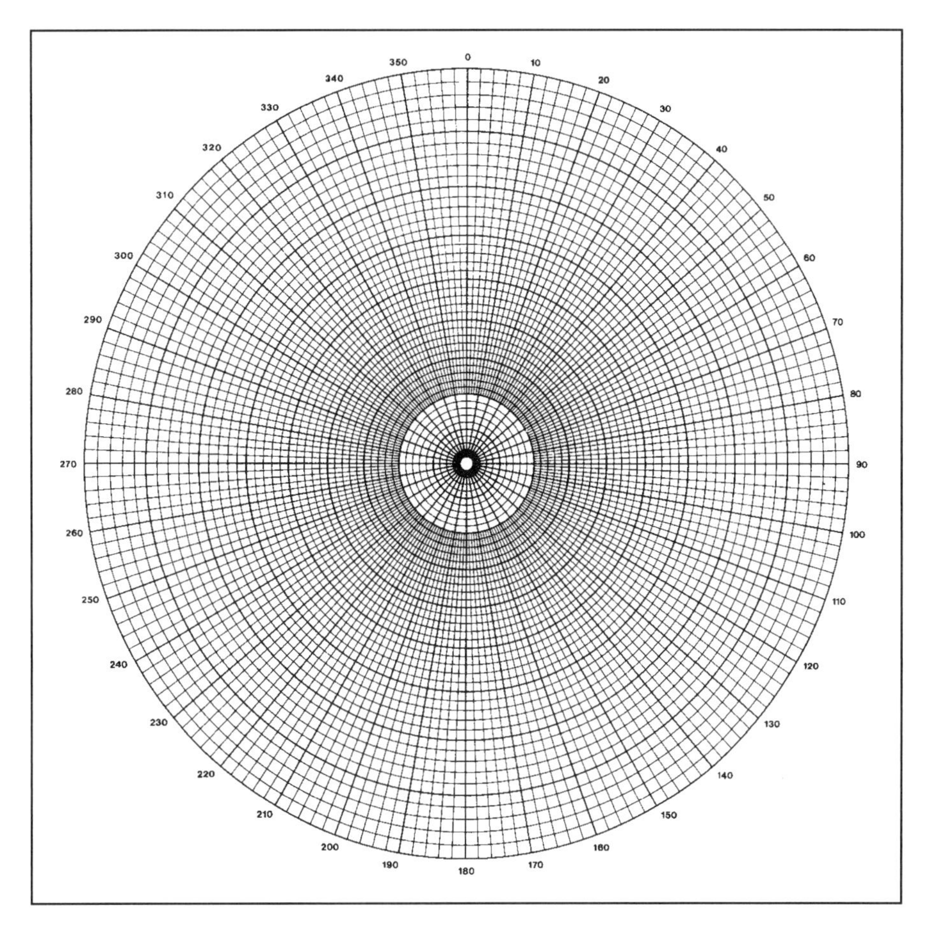

Courtesy of Christopher St. John

FIGURE 2.21 Polar equal-angle stereonet

$$r_c = R \tan \beta_c$$

where

r_c = horizontal length from point O in the opposite direction of α_c

R_c = radius of the great circle of dip direction α_c and dip β_c

For the equal-area projection (Figure 2.25), the given line of trend α and downward plunge β will again intersect the lower reference sphere at point P'. This point is projected by swinging it in a vertical plane through a circular arc centered at B, located at distance R vertically below O, to point P''. P'' is projected to P', where the arc intersects the lower reference hemisphere, then in a straight-line extension of the cord P'' – P' to P on the plane of projection (horizontal plane). For this projection, the relationship between r, the radial distance of point P from O, and β is given by (Priest 1985):

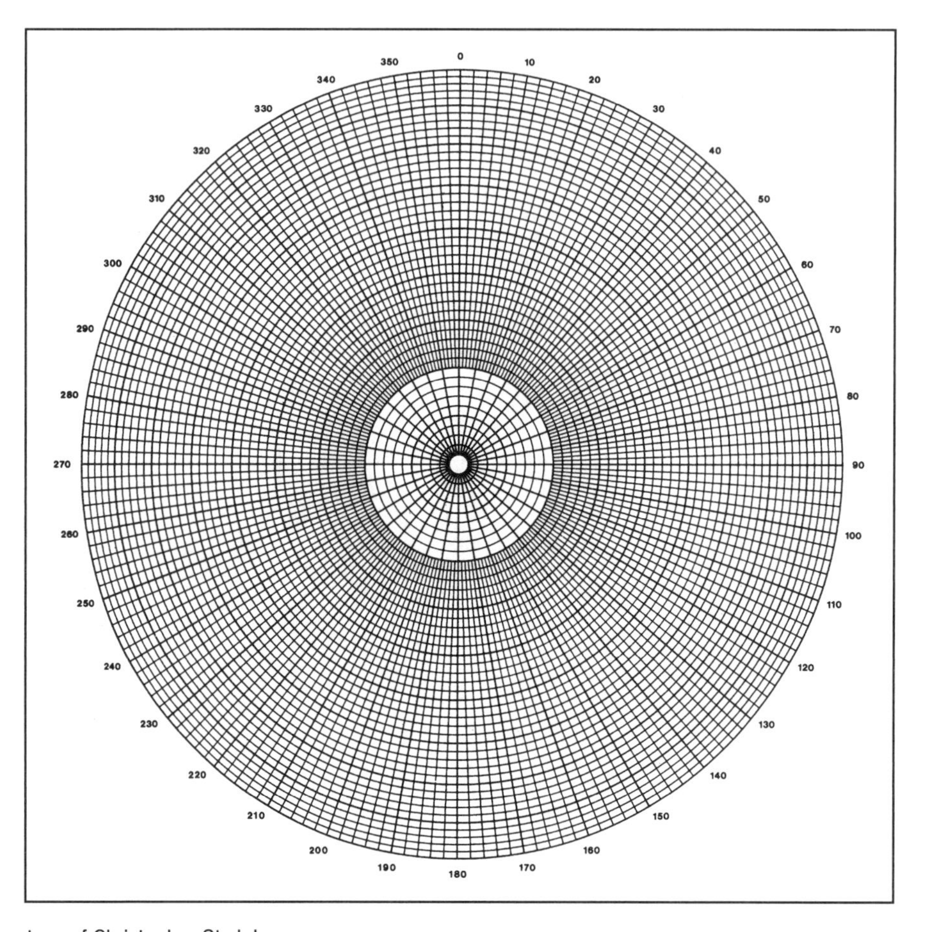

Courtesy of Christopher St. John

FIGURE 2.22 Polar equal-area stereonet

$$r = \sqrt{2R\sin(\beta/2)}$$

On an equal-area construction, the great circle projects as part of a fourth-order curve.

Mapping Visible Structure

The geological compass. Mapping of visible structural features that are found on outcrops or excavated faces is a slow and tedious process; unfortunately, there are few alternatives to the traditional techniques used by geologists. Historically, the most important tool for use in mapping has been the geological compass (Figure 2.26).

The geological compass, or pocket transit, has been in use since 1896. It is not just a compass; it combines the principles of a surveyor's compass, a prismatic compass, a clinometer, a hand level, and a plumb. It is used to determine azimuth angles or compass bearings (and

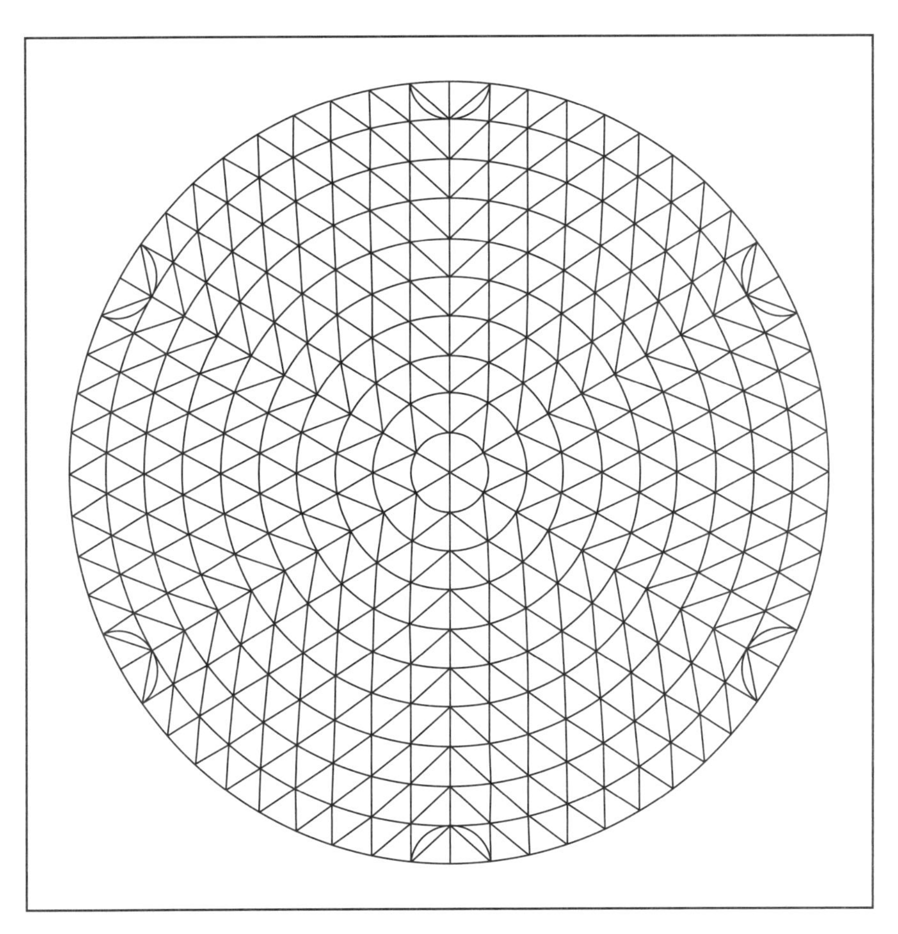

Courtesy of Christopher St. John

FIGURE 2.23 Kalsbeek counting net

thus to determine horizontal angles); to measure vertical angles, percent of grade, or slope; to run levels; and to measure the inclination of objects.

The azimuth or bearing (depending on the type of compass) is read directly on the compass circle, with the compass needle acting as the pointer. The direct reading method is considered to be the most accurate way to obtain an angle with a magnetic compass.

Inclination (i.e., dip) is taken by laying the compass on its side in the open position with the large sight pointing up the slope. The long bubble is then centered by using the lever at the back of the case. The inclination or slope can then be read in degrees or percentage on the vertical angle scale.

Before a geological compass (or any magnetic compass) is used at a location for mapping, the declination should be set. The declination is the angle, in degrees, between the magnetic North Pole and the true North Pole. The amount of declination, as well as whether

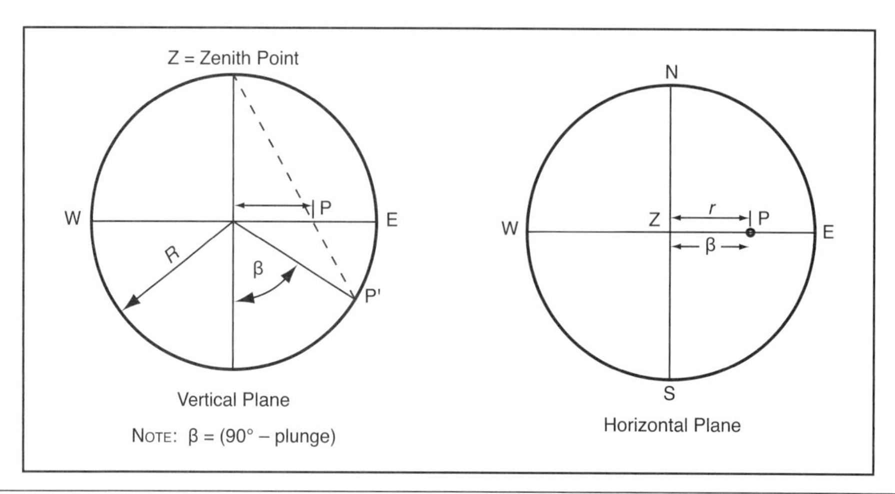

FIGURE 2.24 Geometry for the position of P on the equal-angle projection

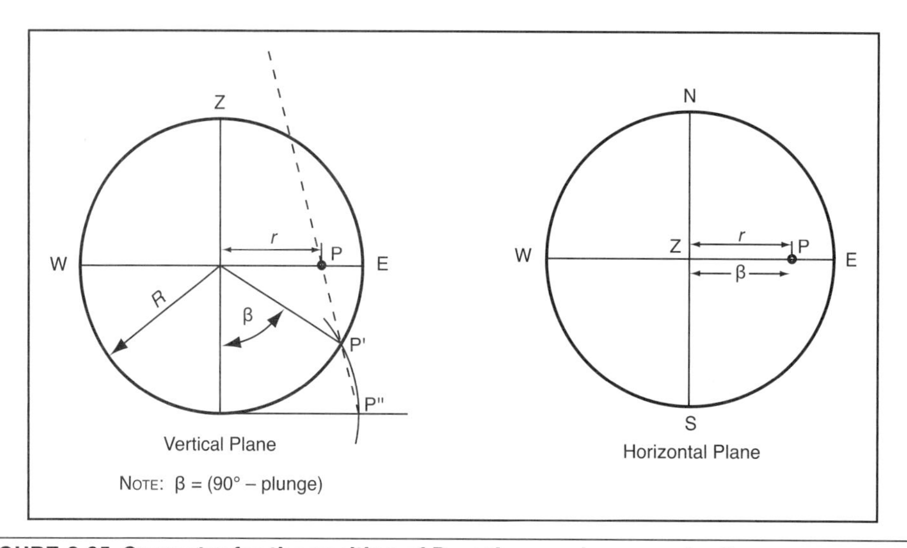

FIGURE 2.25 Geometry for the position of P on the equal-area projection

the angle is measured east or west, depends on the compass user's location on the globe. If the declination is not set, a consistent error equal to the amount and direction of the declination will be introduced into all the directional measurements.

The Clar-type compass. The Clar-type stratum compass (Figure 2.27A) is another portable handheld instrument for geological, geophysical, mining, tectonic, and rock mechanical applications for measuring the azimuth of dip and angle of dip in a single operation (F.W. Breithaupt and Sohn 2015). A modern digital read-out version of the instrument is also available (Figure 2.27B).

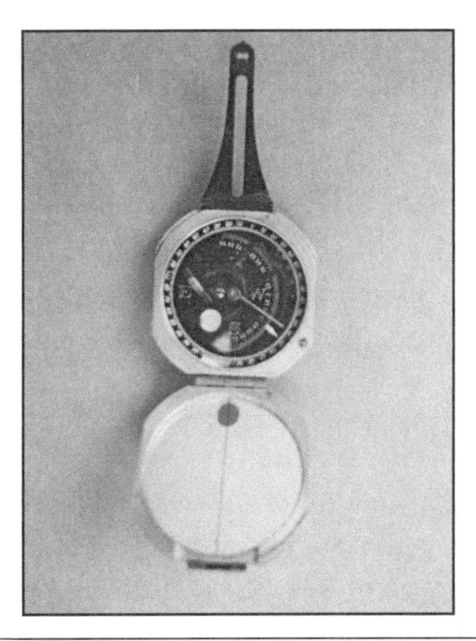

FIGURE 2.26 The geological compass

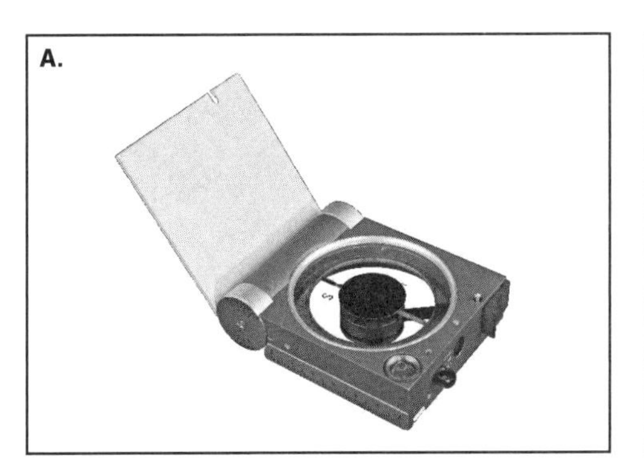

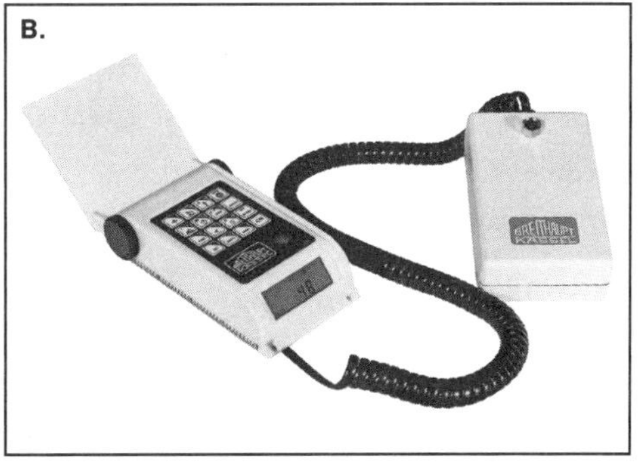

Courtesy of F.W. Breithaupt and Sohn, Kassel, Germany

FIGURE 2.27 Clar-type stratum compass: (A) standard model; (B) digital model

The compass is solidly built with a sturdy housing and cover of corrosion-resistant aluminum alloy. The large circular opening in the compass housing contains the graduated circle along with the magnetic needle and the damping mechanism. At its base, the housing is enclosed by a bottom plate of transparent material, allowing overhead measurements. At its top, the housing is encased with a glass cover. On the left-hand lateral face of the compass housing is a graduated ruler.

The cover serves as the dip angle measuring plate. It pivots on an axis that is parallel to the plane of the graduated circle. One of the trunnions fixed on the cover bears the vertical circle on which the angle of dip can be measured. The vertical circle is graduated in 5° increments, and can be reliably read to an accuracy of 1°. This vertical graduation on the cover hinge allows for simultaneous measurements or the dip and dip angle without rearranging the compass.

Cellular phone applications. Numerous new applications (apps) have been developed by users for the use of cell phones as geological compasses and data loggers. Field mappers of geological structure can download one of several apps for either the iPhone or Android operating system to use as a geological structure mapping tool. Several of these apps are able to capture plane or line data (dip and strike or dip direction; plunge and trend), record the data into a project record database, display the orientation of the mapped structure on a stereonet, and provide other pertinent information such as global coordinates of the mapping point (latitude, longitude, and elevation) in conjunction with displaying a map of the operator's current location.

The advantage of these systems is that they use the world Global Positioning System satellite network to determine the location of the mapped point to at least 4.9 m (16 ft) user accuracy without using any receiver augmentation system (GPS.gov 2017). *User accuracy* refers to how close the device's calculated position is from the truth, expressed as a radius.

3-D Imaging

The method for undertaking geological interpretation has come a long way from the old sectional interpretation on paper. Three-dimensional modeling, 3-D imaging, and terrestrial laser scanning, among others, are recently developed techniques that can be applied to the mapping of geological structure. Additionally, airborne magnetics and aerial photography can provide a wealth of information on structure at a large scale.

Digital photogrammetry is the science of determining the 3-D locations of objects from 2-D digital photographs (Booth and Meyer 2013). It uses digital photographs of a scene captured from two or more different locations (i.e., "stereo pairs") to perceive depth and dimension of space. By locating common points in the groups of two or more images, modern photogrammetric software can automatically calculate the 3-D location of a large number of points in the scene and use the information to generate a digital terrain model (DTM). When coupled with the known (surveyed) location of a small number of "control points" in the scene (and/or known camera locations), the DTM can be georeferenced and the user is able to digitize and measure features that can be seen in at least two images to an accuracy of, or greater than, 1:10,000 of the size of the area covered by a single image (Booth and Meyer 2013). The location of structures and other areas of interest can be directly included into the geological model. Small-scale information such as joint infill and joint roughness cannot be discerned from these images; however, this can be mitigated by correlating joint set data from mapping with drill-hole data, or by carrying out field observations in areas that can be accessed (Rees and Graaf 2013).

Some specific applications for the quarrying, mining, and civil geotechnical fields include

- Remote geological and geotechnical mapping;
- Remote mapping and surveying of inaccessible and unsafe areas, as in underground excavation mapping;

- Aerial topographic surveys, using either conventional manned aircraft or unmanned aerial vehicles (drones);
- Slope stability and slope failure assessment and monitoring;
- Volumetric surveys and measurements (of excavations, stockpiles, and tailings facilities, for example);
- Thickness measurements of shotcrete and other liner material layers; and
- Convergence monitoring.

Once the 3-D image from digital photographs is captured using calibrated cameras, geometric measurements can be determined using the appropriate software. Software tools allow for the measurement of basic features, such as point locations, areas and volumes, the dip angle and dip direction of fracture surfaces and surfaces, the determination of joint sets, and the determination of joint set spacing (3GSM GmbH 2015).

An example of a 3-D image of a quarry bench face illustrating the process from 3-D point measurements through mapping of geological structure, hemispherical plots and the determination of joint spacing, structural measurements, and stereographic plots of dip direction and dip are shown in Figures 2.28–2.32.

Terrestrial laser scanning, a ground-based LiDAR (light detection and ranging, which is a remote sensing technology that measures distance by illuminating a target with a laser and analyzing the reflected light) technique, allows the rapid acquisition of 3-D data to describe rock slopes and, at high resolutions, has several applications that extend from classical mapping: the construction of digital surface models, landslide forensics, monitoring deformation and rock slope displacement, and precursory to slope failure (Afana et al. 2013).

The newest laser imaging technology allows detailed structural information to be made available in near real time, improving accuracy and safety of mining operations. Commercial packages, such as the Maptek I-Site laser scanner (Maptek 2015) can collect a high-resolution panoramic image (Figure 2.33) at the same time as laser scanning, using all of the pixel information to render onto a 3-D point cloud. From this, geologists can digitize boundaries, structures, and other features directly onto a 3-D image in true local grid coordinates without any interpolation from 2-D to 3-D.

Using high-resolution data, a detailed 3-D model can be easily generated from the point cloud for geotechnical interrogation of fault planes or joint sets. The 3-D computer-aided design drawings can be exported and combined with other geological information to create a geotechnical database for the site.

Structural and kinematic analysis can be conducted on point cloud data collected from working surfaces to provide up-to-date detail on joint planes, faults, and geotechnical hazards (Figures 2.34 and 2.35). This data can also be used to update the geological model that guides the excavation process.

Courtesy of RAM Inc., Powell, Ohio

FIGURE 2.28 (A) A 3-D image of a bench face in a quarry. It combines a dense set of 3-D point measurements (B) plus the photographs. Note that the photographs perfectly align with the point cloud.

Courtesy of RAM Inc., Powell, Ohio

FIGURE 2.29 (A) Depth coloring of a 3-D image and (B) mapped geological structures

Repetitive surveying of the shape and position of a wall can identify geotechnical stability hazards by analyzing deformation and creep characteristics. In an open-cut environment, this may mean balancing batter (berm) angles with geotechnical stability. Steeper batter (berm) angles can add significant value to a resource by reducing overburden removal and enabling access to a greater proportion of an ore body.

Additionally, structural joints mapped from laser scan data can be represented on stereonets, along with joint persistence and spacing. Kinematic analysis can be performed directly on laser scan data. These data can be used by drill-and-blast engineers in blast designs to optimize the initiation direction for minimizing in situ wall damage. Laser scan data can be modeled to optimize pattern spacing and burden to achieve the desired fragmentation, as well as to guide charge design for energy distribution and propagation.

High-resolution laser scan data contains detailed information about geotechnical structures that can be used to build an accurate geotechnical database. Analysis of these data

A.

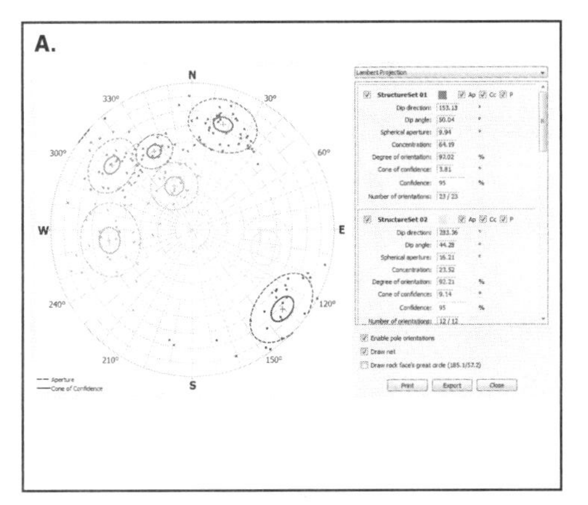

B.

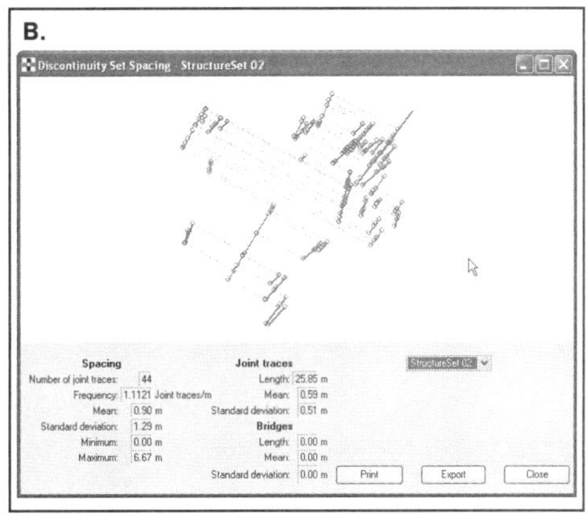

Courtesy of RAM Inc., Powell, Ohio

FIGURE 2.30 (A) Hemispherical plot and (B) determination of spacing

A.

B.

Courtesy of RAM Inc., Powell, Ohio

FIGURE 2.31 (A) Unordered set of structural measurements and (B) the resulting plot from automatic clustering

over time is useful in defining structural boundaries between primary and secondary mineralization, allowing geologists to accurately update and validate the geological model as the working face advances.

With the advent and application of digital data acquisition technology in recent years, so much geotechnical data can be generated using this technology that it makes data management and reporting a high-priority specialization. Too much information to someone who does not require all the tables, spreadsheets, charts, and maps that can be generated by the various software packages can inhibit rather than aide the decision-making process.

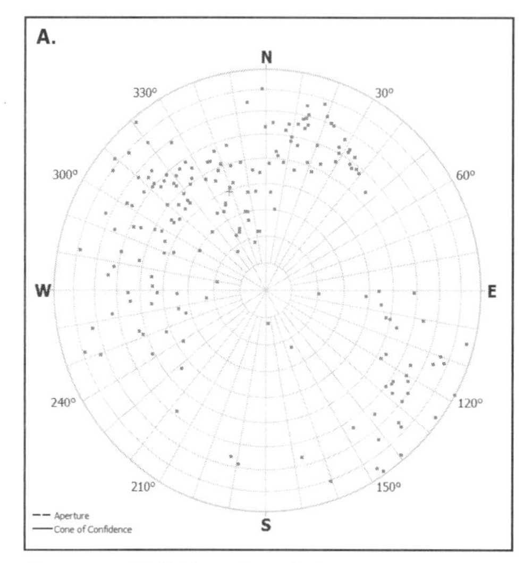

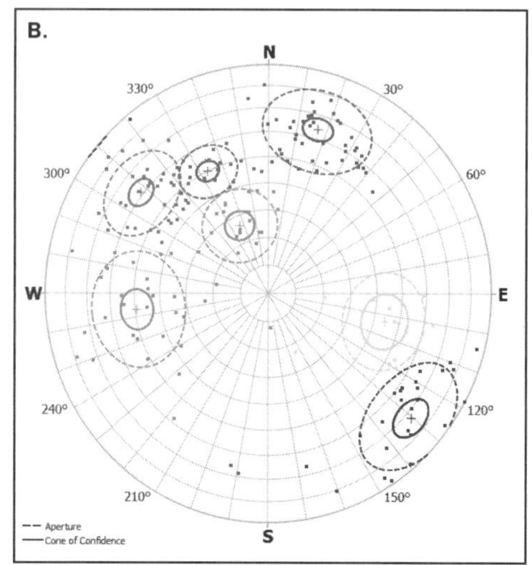

Courtesy of RAM Inc., Powell, Ohio

FIGURE 2.32 Stereonet of (A) unordered structural measurements and (B) after clustering

Courtesy of Maptek USA, Denver, Colorado

FIGURE 2.33 Whole-pit scan—scanning the pit walls with laser imaging technology captures the detail required for geotechnical analysis. The long dashes outline the area of scan in Figure 2.34; the short dashes outline the area of Figure 2.35.

Plotting Geologic Structure on Stereonets

This section discusses how the main features of geologic structure on a stereonet—great circles, poles, and lines—are plotted.

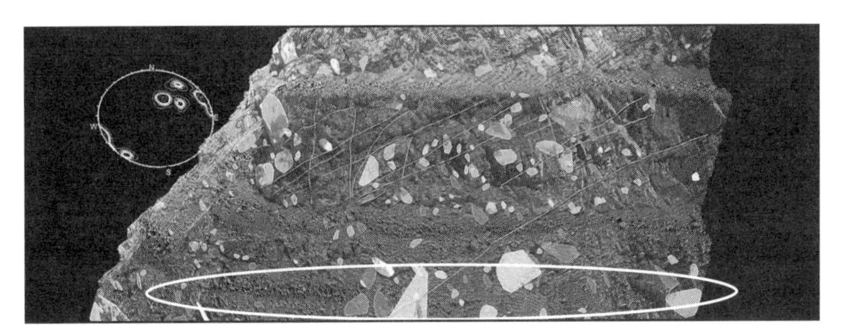

Courtesy of Maptek USA, Denver, Colorado

FIGURE 2.34 Wall stereonet—kinematic analysis helps identify the type and cause of failures

Courtesy of Maptek USA, Denver, Colorado

FIGURE 2.35 Highwall digitizing—geological boundaries can be traced directly onto high-resolution imagery captured by laser scanning

Great circles. Great circles of planar structures, such as discontinuities and faults, may be plotted on stereonets if the mapped orientation data of dip direction and dip are available. Either the equal-angle or equal-area equatorial net may be used.

To start the procedure, an overlay of transparent tracing paper or polyester film should be prepared. A tack should be positioned through the center point of the stereonet, pointing upward, and the center point of the tracing paper should then be pressed down over the point of the tack. The outside perimeter of the stereonet should be traced on the overlay (or drawn with a compass), and the north, south, east, and west (N, S, E, and W) directions should be marked. Either an equal-area or equal-angle stereonet can be used.

The steps for plotting the great circle representing the orientation of a structure with dip direction of 62° and dip of 51° (Figure 2.36) are as follows, as shown in Figure 2.37:

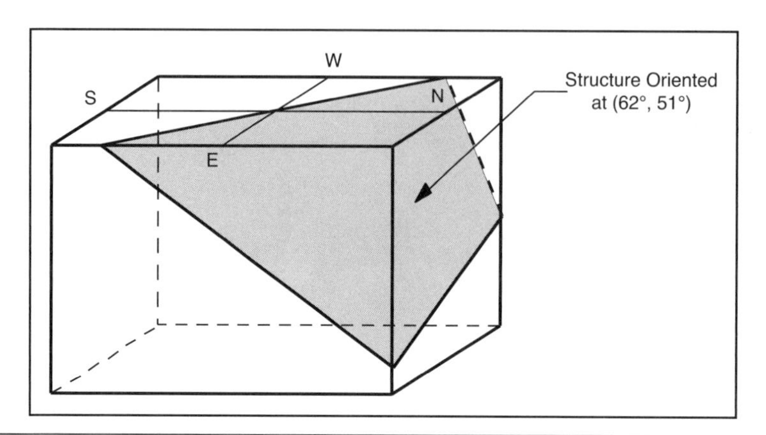

FIGURE 2.36 Geologic structure with dip direction of 62° and dip of 51°

1. With the overlay aligned so that the N, S, E, and W directions correspond with their proper axes on the stereonet, begin by marking the dip direction amount, in degrees, clockwise from north along the traced perimeter.
2. Rotate the overlay until the dip direction mark lies on the E-W axis line. Use *either* the E direction or the W direction, whichever is the most convenient.
3. With the overlay rotated, count inward from the dip direction mark along the E-W axis line the dip amount. Trace the great circle representing this dip amount.
4. Rotate the overlay back to its original position with the N, S, E, and W marks aligned with their corresponding axes on the stereonet. The great circle representing the structure is now drawn with its proper orientation on the lower hemispherical projection.

Additional great circles can be drawn on the overlay in the same manner. However, a large number of mapped discontinuity orientations can be more easily interpreted if a polar plot is used.

Poles. A pole represents a line normal to the structure, as shown in Figure 2.16B. Remember, lines plot on a stereonet as points. The pole will have a location on the equatorial stereonet 180° from the dip direction of the structure and 90° from the dip of the structure. Poles can be plotted either on an equatorial net or on a polar net. It is much more convenient to plot large amounts of orientation data as poles on a polar net.

To begin the procedure, prepare an overlay in the same manner as for the great circle plots. This is done so that the points representing the poles are not plotted on the polar net, which could get cluttered when it is time to contour the points to find the predominant orientation. *A very important point to note for direct polar plotting of discontinuity orientations is that the major axes are rotated 180°.* That is, N is marked at the *bottom* of the overlay, S is marked at the *top* of the overlay, E is marked on the *left* side of the overlay, and W is marked on the *right* side of the overlay, because the poles and not the dip vectors are being plotted.

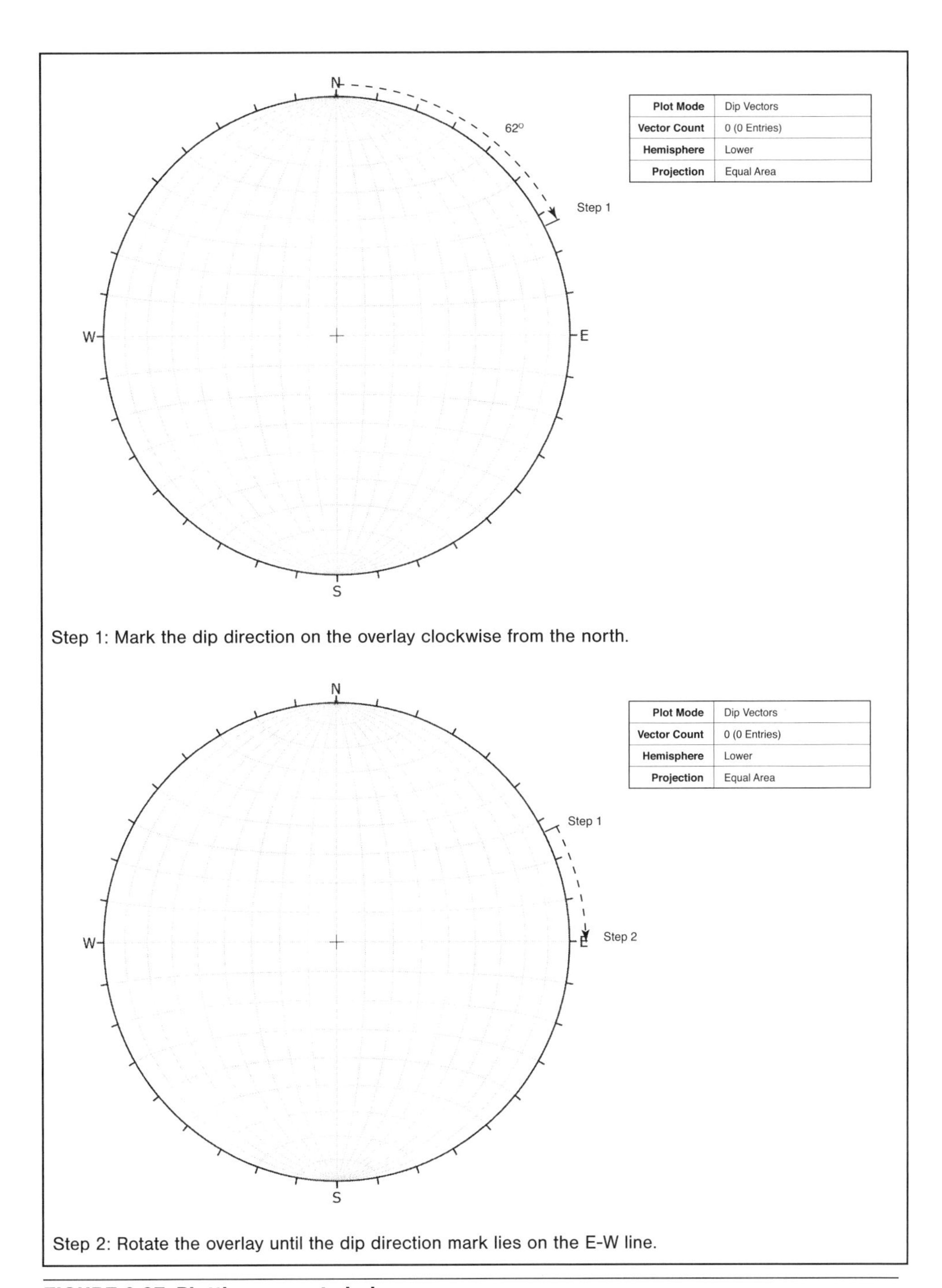

Step 1: Mark the dip direction on the overlay clockwise from the north.

Step 2: Rotate the overlay until the dip direction mark lies on the E-W line.

FIGURE 2.37 Plotting a great circle

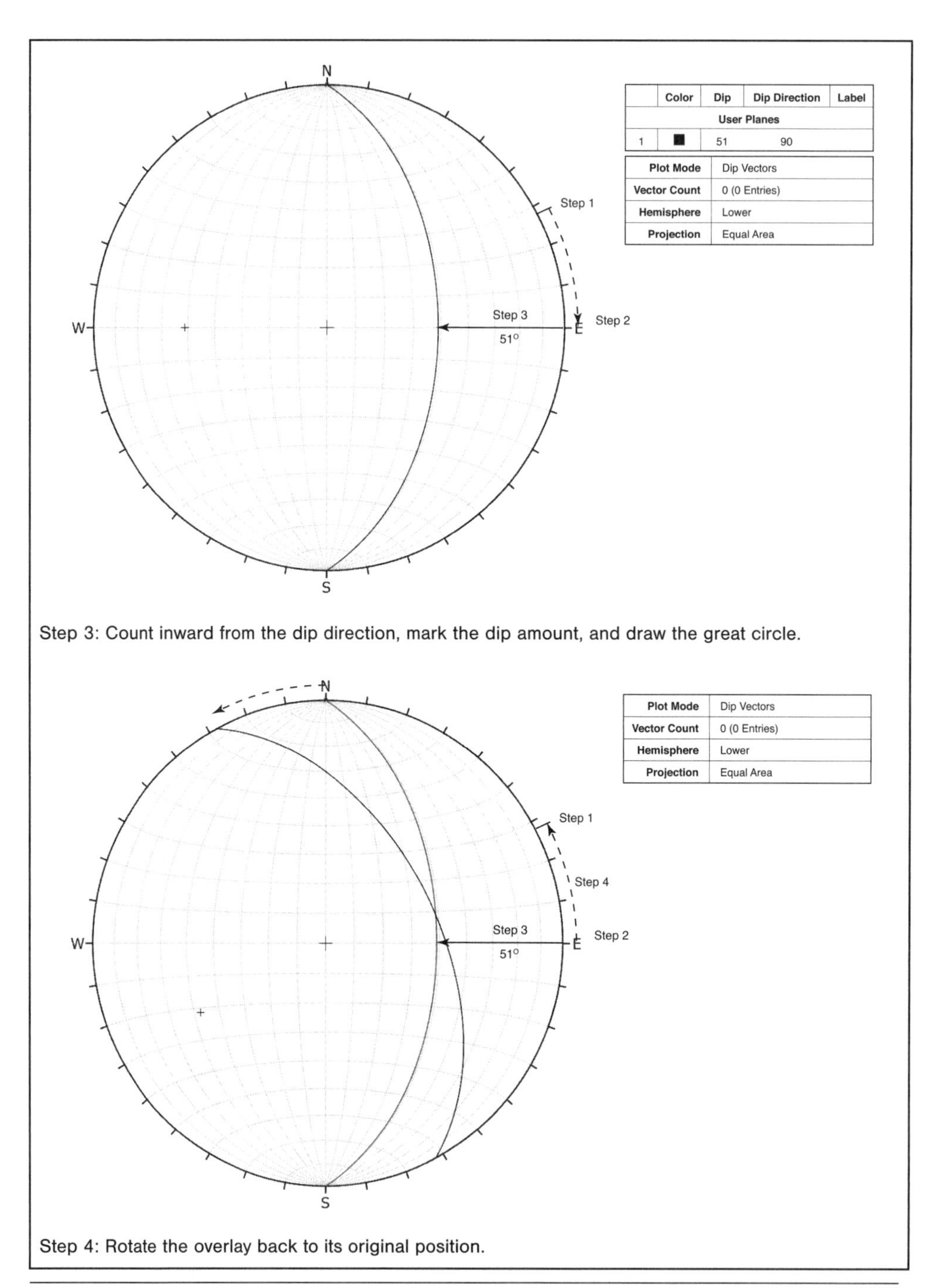

FIGURE 2.37 Plotting a great circle (continued)

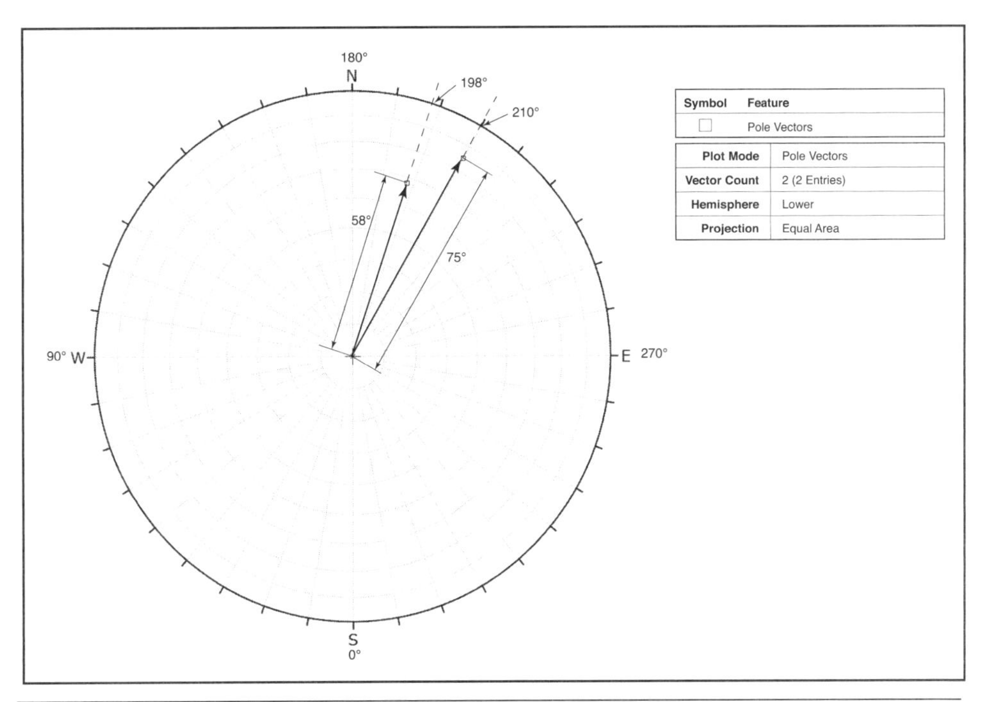

FIGURE 2.38A Polar plot of (198°, 58°) and (210°, 75°)

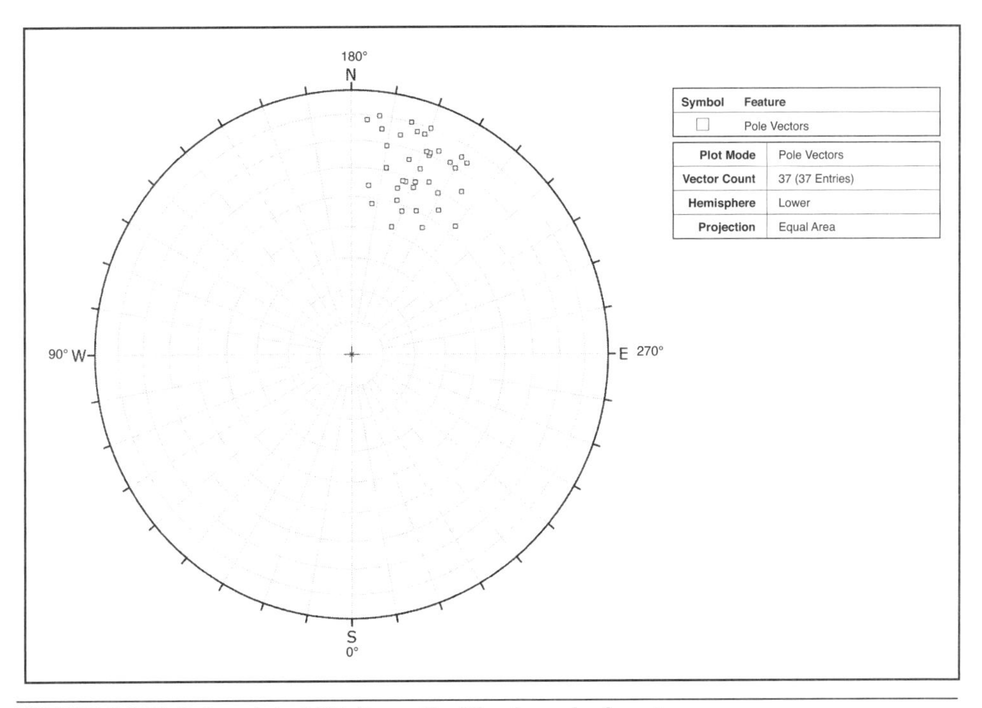

FIGURE 2.38B Polar plot of 37 discontinuities in a single set

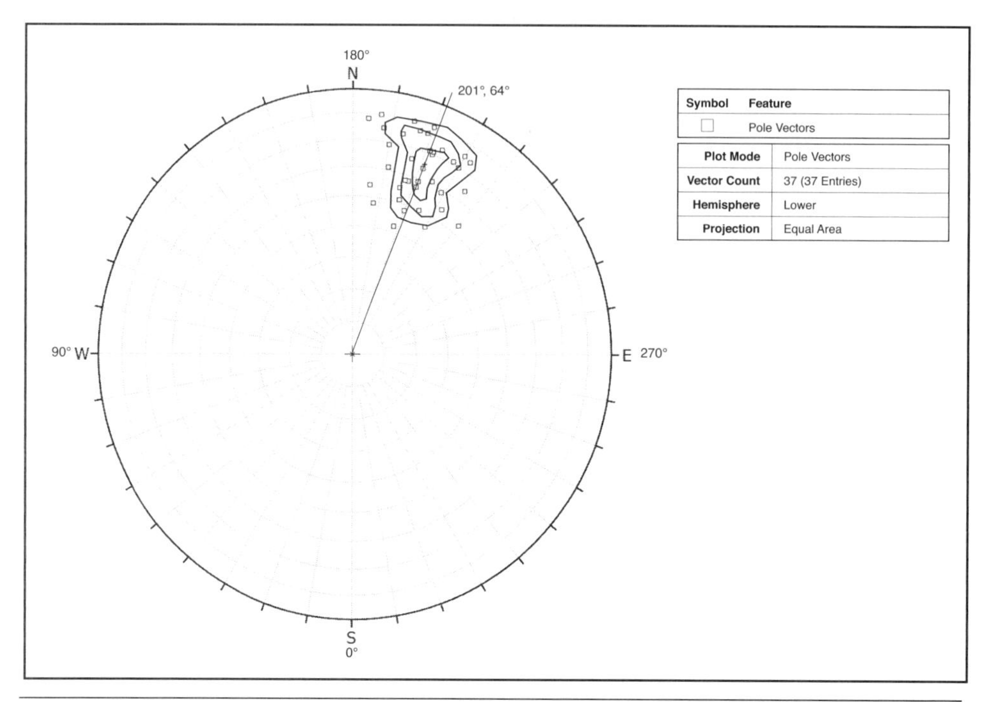

FIGURE 2.38C Contour of pole densities to determine the predominant orientation of the 37 measurements

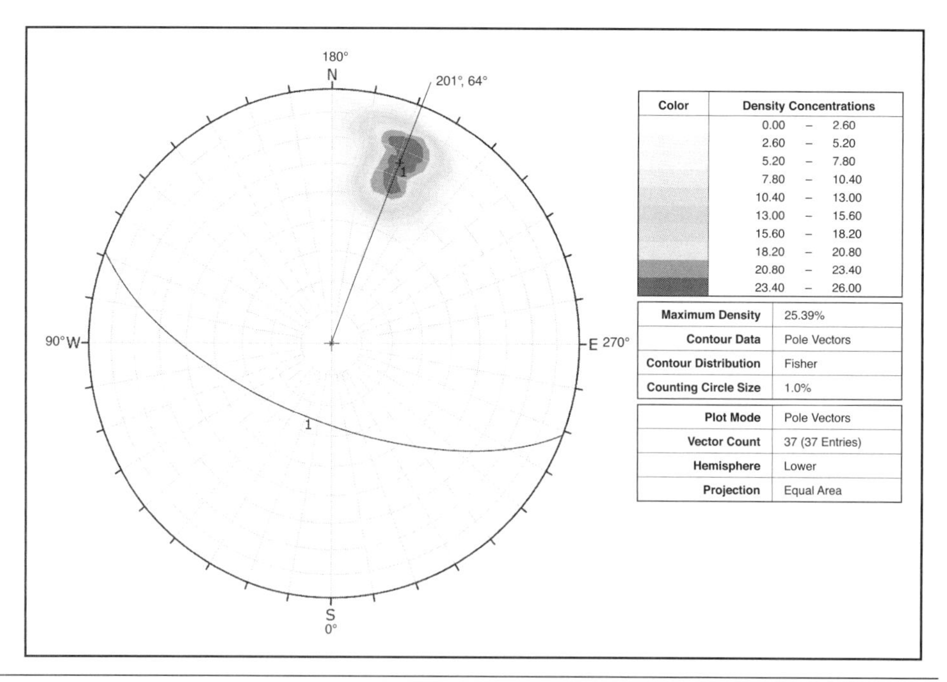

FIGURE 2.38D Great circle representing the predominant orientation of the 37 discontinuities

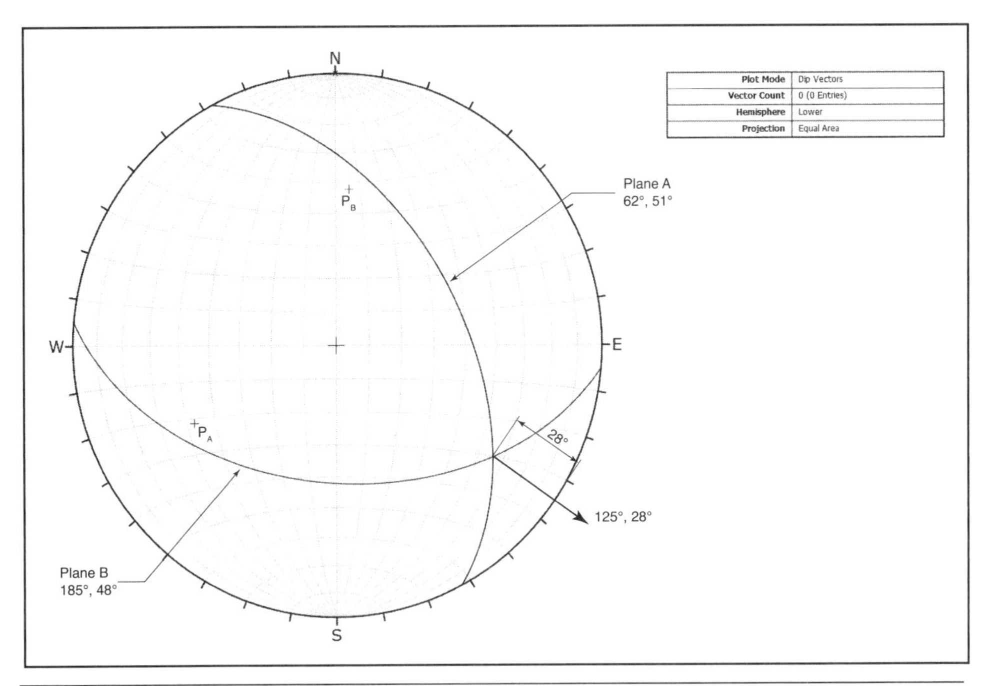

FIGURE 2.39 Plot of the great circle representing A = (62°, 51°) and B = (185°, 48°)

If a polar net is being used, poles can be plotted and contoured to determine the predominant orientation(s) of the structure(s). For plotting poles, the procedure is as follows:

1. Mark the dip direction angle, in degrees, clockwise from north on the overlay. (Remember, N is at the bottom of the overlay.)
2. Count outward from the center of the polar net, toward the dip direction mark, the dip angle amount (in degrees), and place a mark or symbol. This mark or symbol represents the pole of the structure (see Figure 2.38A).
3. Repeat steps 1 and 2 for the remaining orientation data. After all the data are plotted, pole clusters (such as the one shown in Figure 2.38B) can be contoured by using a technique such as a counting net to generate a density contour like the one shown in Figure 2.38C.
4. The predominant orientation(s) of the centroid of the cluster(s) can be determined by noting the azimuth angle from north and the deflection angle from the center of the polar net. The great circle(s) representing the predominant orientation(s) can be plotted (see Figure 2.38D).

Lines. Lines (i.e., borehole orientations of trend and plunge) can be plotted on an equatorial stereonet in the same manner as great circles. The only difference is that a *point* representing the orientation of the line is plotted instead of the great circle representing a plane. The point representing the orientation of the line of intersection of two planes is shown in Figure 2.39. This point occurs where the traces of the two great circles intersect on the stereonet.

The Difference Between Equal-Area and Equal-Angle Stereonets

The equal-area and equal-angle nets are the most common nets used in rock engineering. They can be used for analyzing discontinuous structural rock orientations and interactions, and they are occasionally used for statistical analysis of predominant orientations of discontinuities. The method of plotting data on both projections is identical, but a difference exists in interpreting the results from both types of plots. An example of this difference would be to plot a series of randomly oriented lines, represented by points on the stereonets. Someone might expect that the result would be a random distribution of points on the projections. However, this result is not the case with the equal-angle plot, which would show a clustering effect toward the center of the net, giving a false indication of more near-vertical orientations if one were not aware that the points were plotted on an equal-angle plot. Because of the method of construction of the equal-angle net (see Figures 2.24 and 2.25), the great circles and small circles (analogous to meridians of longitude and latitude on a global projection) get closer together as the dip angle gets steeper (i.e., the lines get closer to the center of the net). This can be seen by inspection of Figures 2.21 and 2.22. On the other hand, the great circles and small circles on the equal-area stereonet do not get closer together as the dip angle gets steeper. The reason for this false indication is that, for example, if a 10°-by-10° section is plotted near the center of an equal-angle stereonet, it would not have the same area as the same section plotted near the margin. In this respect, an equal-area net should be used for statistical contouring of orientation points, although some authors claim the difference between equal-area and equal-angle nets is insignificant in most cases. Aside from this distinction, the only major difference between these two types of nets is that small circles are not represented by circular areas on the equal-area net.

Polar nets are available in both equal-angle and equal-area varieties and are used predominantly for analyzing the orientation of points. The analysis of planar structures is not possible with polar nets, and either a meridional or equatorial net should be used in such cases.

Counting nets are used for statistical analysis of orientations, represented by points, which result in contoured diagrams showing predominant orientations of structural discontinuities. The technique of plotting poles on a polar stereonet, contouring the polar densities, and determining the predominant orientation of the discontinuity set is illustrated in Figures 2.38A through 2.38D.

The 1% Counting Net

The 1% counting net is convenient to use and easy to make for the statistical analysis of pole concentrations. The purpose is to determine by some easy technique the contours of pole density to determine the orientation(s) of the major pole concentration(s).

These pole density contours represent lines of equal pole numbers counted within some window. The window we use in this case is the 1% of the area of the stereonet window, moved around the stereonet area in some predetermined method or pattern. The pattern we will use is a square grid that will have as its spacing the diameter of the 1% window.

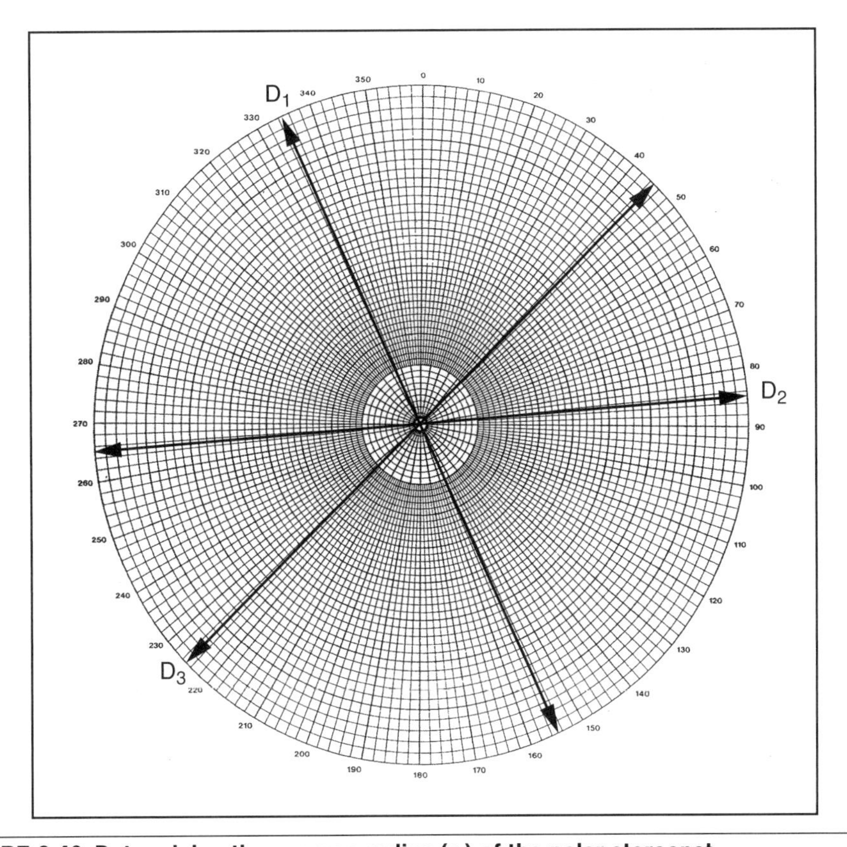

FIGURE 2.40 Determining the average radius (r_s) of the polar stereonet

Follow these steps to develop the 1% window counter and counting grid:

1. Measure the diameter of the polar stereonet you are using (equal area or equal angle). Measure two or three diameters, or more, and take an average (Figure 2.40). This is done because the stereonet, if it is a copier-type copy, will distort a bit. Calculate the radius (r_s) of the stereonet. Calculate the area (A_s) of the stereonet: $A_s = \pi r_s^2$

$$D_{Avg} = (D_1 + D_2 + D_3)/3$$

and

$$r_s = D_{Avg}/2$$

2. Determine 1% of the area of the stereonet: $A_c = 0.01(A_s)$. This is the area of the 1% counting circle.
3. Calculate the diameter of the 1% counting circle: $d_c = 2 \cdot (A_c/\pi)^{0.5}$.
4. Create a counting grid overlay. The spacing of the grid lines should be at r_c. The grid should be laid out N to S and E to W over the stereonet. That is, the N to S gridlines should start exactly on the W mark of the stereonet and end exactly on the E mark,

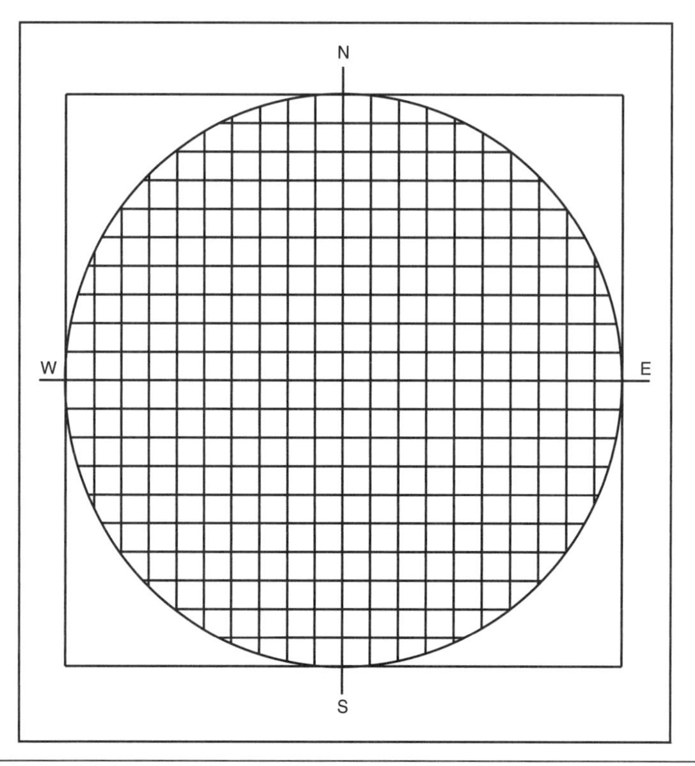

FIGURE 2.41 Counting grid overlay for the 1% counting circle

with distance between gridlines at exactly r_c. Likewise, the E to W gridlines should start exactly on the S mark of the stereonet and end exactly on the N mark, oriented exactly 90° from the E to W gridlines and spaced again at exactly r_c (Figure 2.41).

5. Construct this grid overlay on thin or translucent paper (drafting vellum works best).
6. Construct a small counting circle of radius r_c on clear or transparent paper or Mylar (Figure 2.42A).
7. Construct a paired set of 1% counting circles, each with radius of r_c separated from each other by distance, center-to-center, of d_s (Figure 2.42B). That is, construct one 1% counting circle with radius r_c, then measure from its center a distance of d_s, then construct another 1% counting circle of radius r_c centered at the end of the line of length d_s that originated at the center of the first 1% counting circle. This paired set of 1% counting circles is used to count pole occurrences at the perimeter of the stereonet. Mark the center between the two circles (one-half the distance between the center of the first circle and the center of the second circle) along a line connecting the centers of the two circles.

Follow these steps to use the grid overlay and the counting circles:

1. Plot the poles of the discontinuities you measured in the field on an overlay of the polar stereonet. It is best to use a semitransparent sheet such as drafting vellum

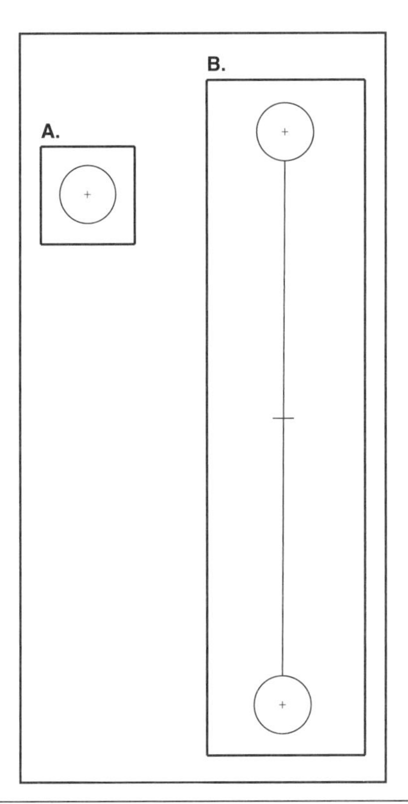

FIGURE 2.42 1% counting circles: (A) single circle for use with interior cells; (B) double circle for use with cells along the edge of the stereonet

(although thin, unlined paper can be used too) for the overlay. Make sure that the circular outline of the polar net (i.e., a circle of radius d_s) is constructed on the overlay. Center the overlay circle on the stereonet and mark the N, S, E, and W points on the overlay (this is to ensure that the overlay does not slip and rotate inadvertently during the plotting).

2. Once all the poles are plotted in their correct locations on the overlay, remove the overlay from the stereonet underlay and place the counting grid you just made atop the pole plot overlay. Make sure to align the N to S and E to W centerlines of the grid with the N, S, E, and W points you marked on the overlay, respectively.
3. Using the small 1% counting circle, place it centered at an intersection of a N to S and E to W gridline and count the number of pole occurrences within the area of the 1% circle (Figure 2.43). Write that number at the intersection point (center of the small 1% counting circle) of the N-S and E-W lines. (NOTE: Instead of cluttering up your gridded overlay with a bunch of numbers, it might be best to place another clear overlay atop the gridded overlay so you can jot the number of occurrences on that overlay.) If the number of occurrences is zero, then either write "0" or leave it blank.

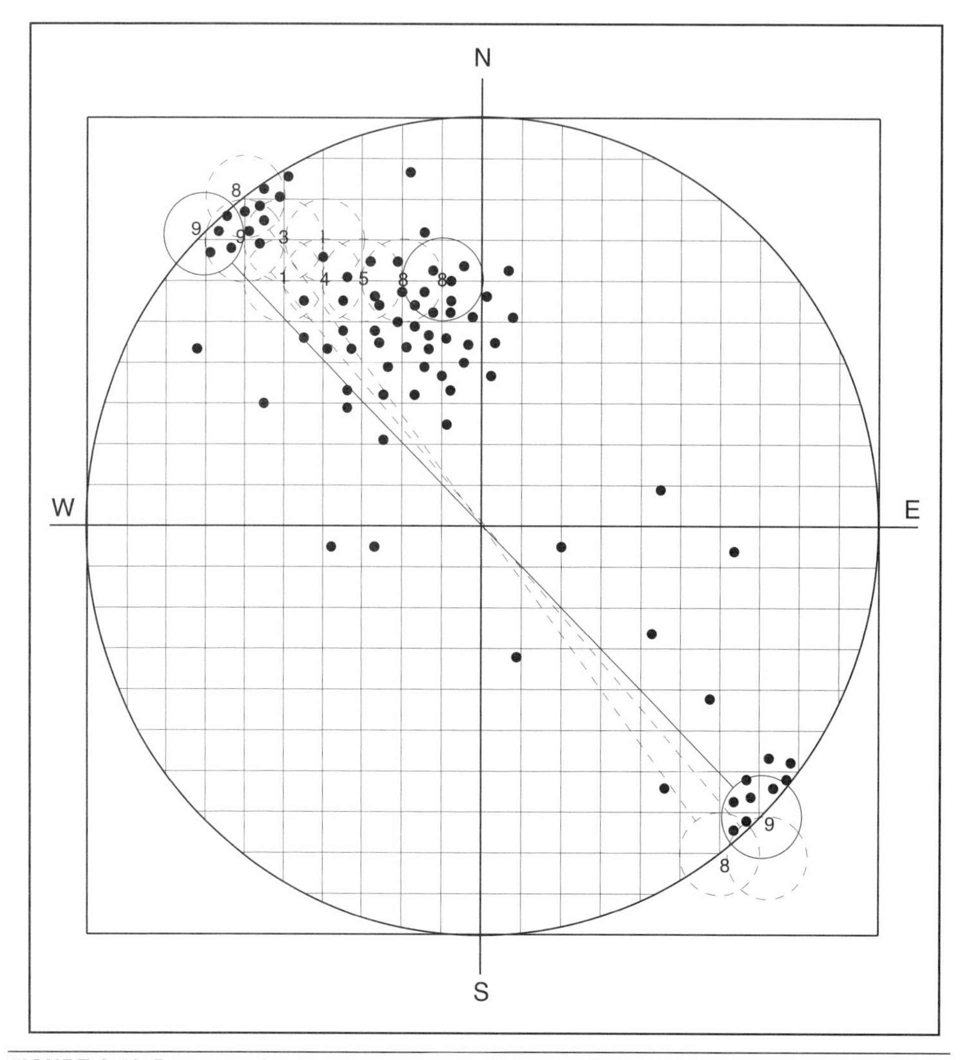

FIGURE 2.43 Process of using the counting circles and the overlay grid to determine the pole point density for contouring

4. Move to the next adjacent intersection and repeat step 3. Continue right to left and top to bottom across the entire stereonet, counting the number of occurrences of poles within the 1% counting circle at each intersection. For intersections along the edge of the great circle (around the edge of the polar net), you will have to use the paired 1% counting circles with them centered at the center of the great circle (i.e., center the mark you made halfway between the two small circles at the center of the great circle).
5. Once all of the grid points are populated with counts (number of occurrences), contour the occurrences the same as if one were contouring elevations for a topographical map, using the same rules for contouring. The most pertinent rules for contouring are as follows:

- Contour lines must close on themselves, either on or off a map. They cannot dead-end.
- Contours are perpendicular to the direction of maximum slope (water flows downhill perpendicular to contours).
- The slope between adjacent contour lines is assumed to be uniform. (Thus it is necessary that breaks [changes] in grade be located in topographic surveys.)
- The distance in between contours dictates the steepness of a slope. Wide separation denotes gentle slopes; close spacing, steep slopes; even and parallel spacing, uniform slopes.
- Irregular contours signify rough, rugged country. Smooth lines imply more uniformly rolling terrain.
- Concentric closed contours that increase in elevation represent hills (or high spots).
- Contours of different elevations never meet.
- A contour cannot branch into two contours of the same elevation.
- Contour lines go in pairs up valleys and along the sides of ridge tops.
- A single contour of a given elevation cannot exist between two equal-height contours of higher or lower elevation.

6. Determine the predominant orientation of every pole cluster. That is, find the point at the top of the “hills” represented by each cluster of poles. Then place the overlay back on the original polar stereonet and determine the dip and dip direction of each peak. (Remember, with 0° at the bottom of the polar net and 180° at the top, count clockwise from 0° to the point where a line from the center of the polar net passing through the predominant orientation point intersects the outside edge of the polar net [this is the dip direction], then count inward from the edge of the polar net to the point representing the predominant orientation [this is the dip amount] along that line from the center of the polar net).
7. One can construct the great circle representing each predominant orientation (peak) by placing the contoured pole plot overlay with the predominant pole denoted on the corresponding hemispherical stereonet (equal angle or equal area) and rotating each pole to the E-W line, then constructing the great circle along the appropriate hemispherical stereonet’s great circle on the opposite side of the center of the stereonet at the determined dip. That is, rotate the pole to the center-to-E line and construct the great circle on the opposite side of the stereonet with its dip vector located on the center-to-W line. Repeat this step for each cluster.

ENGINEERING ROCK MASS CLASSIFICATION SCHEMES

Since Ritter (1879) first attempted to develop a formal approach to determining support requirements for tunnel design more than 120 years ago, numerous engineering rock mass classification schemes have been developed and have found widespread use in design. A

TABLE 2.3 Rock mass classification systems

Name of Classification	Originator	Country of Origin	Applications
1. Rock load	Terzaghi 1946	United States	Tunnels with steel supports
2. Stand-up time	Lauffer 1958	Austria	Tunneling
3. Intact rock strength	Deere and Miller 1966	United States	Communications
4. Rock quality designation (RQD)	Deere et al. 1967	United States	Core logging, tunneling
5. Rock structure rating (RSR)	Wickham et al. 1972	United States	Tunneling
6. Geomechanics classification (rock mass rating, or RMR)	Bieniawski 1973	South Africa and United States	Tunnels, mines, foundations
7. Rock tunneling quality index (Q)	Barton et al. 1974	Norway	Tunneling, large chambers
8. Basic geomechanical classification	ISRM 1981	Canada, international	Tunneling, general
9. Geological strength index (GSI)	Hoek 1994	Canada, international	Support in underground excavations
10. Rock mass index (RMi)	Palmström 1995	Norway	Rock design and engineering, communications

Adapted from Shroff and Shaw 2003; Cosar 2004; Dev and Sharma 2011

note of caution is often put forward to those who intend to use the classification schemes for design: "While the classification schemes are appropriate for their original applications, especially if used within the bounds of the case histories from which they were developed, considerable caution must be exercised in applying rock mass classifications to other rock engineering problems" (Hoek 2006). Table 2.3 summarizes some of the more important classification systems.

Considering that the design of excavations in rock currently uses three main approaches—analytical, observational, and empirical—the most predominant design approach is the empirical design method, which relates the experience encountered at previous projects to the conditions anticipated at a proposed site (Bieniawski 1990). Indeed, on many underground or surface construction and mining projects, rock mass classifications have been providing the only systematic design aid in an otherwise haphazard "trial-and-error" procedure.

Objectives of the Rock Mass Classification Schemes

Rock mass classifications were developed to create some order out of the chaos in site investigation procedures. They were not intended to replace analytical studies, field observations, measurements, or engineering judgment. The objectives of rock mass classifications are to (Bieniawski 1990)

1. Identify the most significant parameters influencing the behavior of a rock mass,
2. Divide a particular rock mass formulation into groups of similar behavior—rock mass classes of varying quality,
3. Provide a basis of understanding the characteristics of each rock mass class,
4. Relate the experience of rock conditions at one site to the conditions and experience encountered at others,
5. Derive quantitative data and guidelines for engineering design, and
6. Provide common basis for communication between engineers and geologists.

These objectives can be fulfilled by ensuring that a classification scheme has the following characteristics (Bieniawski 1990):

1. It is simple, easily remembered, and understandable.
2. Each term is clear and the terminology used widely acceptable by the professional user.
3. Only the most significant properties of rock masses are included.
4. The system is based on measurable parameters that can be determined quickly and cheaply in the field by relevant tests.
5. It is based on a rating system that can weigh the relative importance of the classification parameters.
6. It is functional by providing quantitative data for the design of tunnel or rock support.
7. It is general enough so that the same rock mass will possess the same basic classification regardless whether it is being used for a tunnel, a slope, or a foundation.

Some of the More Useful Rock Mass Classification Schemes

Of the many classification systems, the most commonly known and used ones are those proposed by Terzaghi (1946), Lauffer (1958), Deere et al. (1967), Wickham et al. (1972), Bieniawski (1973), Barton et al. (1974) and Hoek (1994). These classification schemes are discussed in detail here.

1. **Terzaghi's rock load classification method.** The earliest reference to the use of rock mass classification for the design of tunnel support is in a paper by Terzaghi (1946) in which rock loads, carried by steel sets, are estimated on the basis of a descriptive classification. Terzaghi's rock mass descriptions are as follows:
   - *Intact rock* contains neither joints nor hair cracks. Hence, if it breaks, it breaks across sound rock. On account of the injury to the rock due to blasting, spalls may drop off the roof several hours or days after blasting. This is known as a spalling condition. Hard, intact rock may also be encountered in the popping condition involving the spontaneous and violent detachment of rock slabs from the sides or roof.

- *Stratified rock* consists of individual strata with little or no resistance against separation along the boundaries between the strata. The strata may or may not be weakened by transverse joints. In such rock, the spalling condition is quite common.
- *Moderately jointed rock* contains joints and hair cracks, but the blocks between joints are locally grown together or so intimately locked that vertical walls do not require lateral support. In rocks of this type, both spalling and popping conditions may be encountered.
- *Blocky and seamy rock* consists of chemically intact or almost intact rock fragments that are entirely separated from each other and imperfectly interlocked. In such rock, vertical walls may require lateral support.
- *Crushed but chemically intact rock* has the character of crusher run. If most or all of the fragments are small as fine sand grains and no re-cementation has taken place, crushed rock below the water table exhibits the properties of water-bearing sand.
- *Squeezing rock* slowly advances into the tunnel without perceptible volume change. A prerequisite for squeeze is a high percentage of microscopic and sub-microscopic particles of micaceous minerals or clay minerals with low swelling capacity.
- *Swelling rock* advances into the tunnel chiefly on account of expansion. The capacity to swell seems to be limited to those rocks that contain clay minerals such as montmorillonite, with high swelling capacity.

2. **Lauffer's stand-up time classification method.** Lauffer's stand-up time classification method was based on the earlier work of Josef Stini (1950), considered the father of the "Austrian School" of tunneling and rock mechanics, who emphasized the importance of structural defects in rock masses. Lauffer put forth the idea that the stand-up time for any unsupported rock span was related to the various rock mass classes, A through G, as shown in Figure 2.44. *Stand-up time* is defined as the length of time that an underground opening will stand unsupported after excavation and scaling, whereas the *active unsupported span* is the width of the tunnel or the distance from the face to the nearest support, if this is greater than the tunnel span. Lauffer's original classification has since been modified by a number of others and now forms a part of the general tunneling approach known as the new Austrian tunneling method, or NATM. This concept of stand-up time has also been used by others, including Bieniawski and Barton, to translate classification ratings into meaningful engineering criteria.
3. **Rock quality designation (RQD).** The RQD was introduced by Deere et al. in 1967 to provide a quantitative estimate of rock mass quality from drill-core logs. It measures the percentage of "good" rock with a borehole (Deere and Deere 1988). It is now used as a standard parameter in drill-core logging and forms a basic input in several rock mass classification systems.

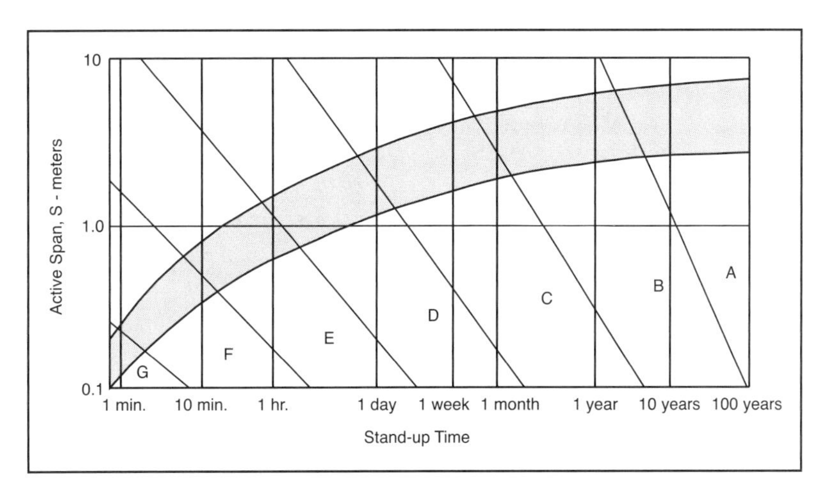

Source: Lauffer 1958

FIGURE 2.44 Lauffer's relationship between active span and stand-up time for different classes of rock mass: A = very good rock; G = very poor rock

The RQD is a modified core recovery percentage in which all the pieces of sound core more than 100 mm (4 in.) in length are summed and divided by the length of the core run. The correct procedure for determining the RQD is illustrated in Figure 2.45.

Deere and Deere (1988) state that drill-core sizes between BQ and PQ with core diameters of 36.5 mm (1.44 in.) and 85 mm (3.35 in.), respectively, are applicable for measuring RQD, as long as proper drilling techniques are used that do not cause excess core breakage and/or poor recovery. However, the RQD was originally developed for NX-sized core (54.7 mm [2.16 in.]), and this is still the recommended optimal core size for measuring RQD.

Different people may measure the different core fragments another way—along the centerline, from tip to tip, or along the fully circular barrel section. The recommended procedure (Deere and Deere 1988) is to measure the core length along the centerline (Figure 2.45). Core breaks caused by the drilling process should be fitted together and counted as one piece. As it is often difficult to determine if a break is natural or caused by the drilling process for schistose and laminated rocks, the break should be considered as natural, to be more conservative in the calculation of RQD.

The RQD index is sensitive to the length of the core run. Therefore, Deere and Deere (1988) recommend that, in general, the calculation of RQD be based on the actual drilling-run length used in the field, preferably no greater than 1.5 m (5 ft).

An important consideration is that RQD is a directionally dependent parameter, and its value may change significantly, depending on the borehole orientation.

Table 2.4 shows the estimated rock mass classification based on RQD.

Currently, the most important use for RQD is as a vital component of the rock mass rating (RMR) and rock tunneling quality (Q) systems.

Palmström (1982) suggested that, when no core is available but discontinuity traces are visible in surface exposures or exploration adits, the RQD may be estimated from the number of discontinuities per unit volume. The suggested relationship for clay-free rock masses is

$$\text{RQD} = 115 - 3.3 \cdot J_v$$

where J_v is the sum of the number of joints per unit length for all joint (discontinuity) sets, known as "the volumetric joint count."

4. **Rock structure rating (RSR).** The RSR system, a ground-support prediction model, was developed in the United States by Wickham et al. (1972). The scheme presents a quantitative method, rather than a qualitative method, for describing the quality of a rock and for selecting the appropriate support densities for tunnel drives. It was based on a study of the performance of 33 tunnels in various strata (Wickham et al. 1972).

   Besides being a quantitative rock classification method, it was also a step forward beyond Terzaghi's (1946) rock load classification method or Deere et al.'s (1967)

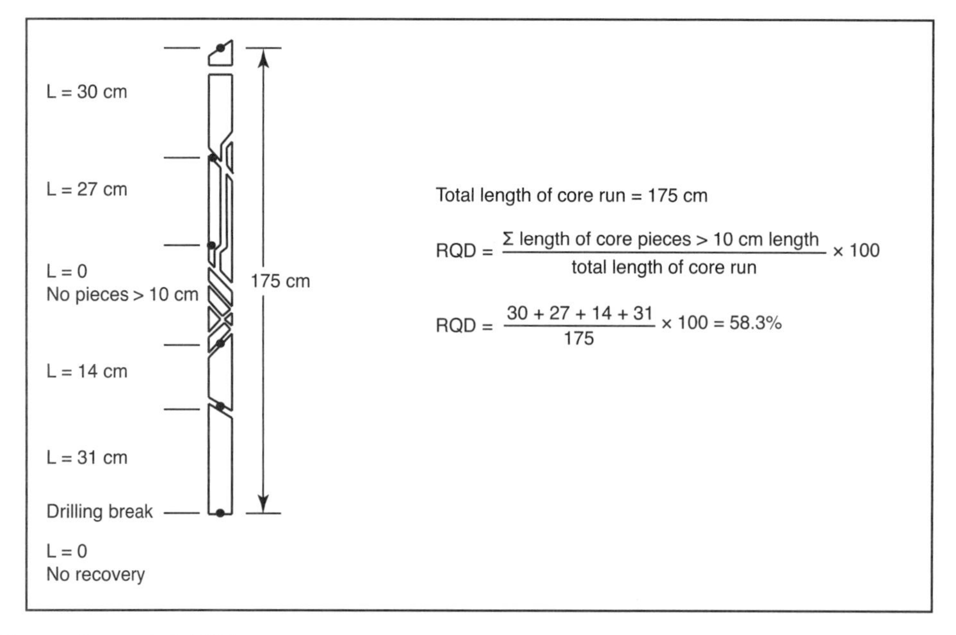

Source: Deere et al. 1967

FIGURE 2.45 Procedure for measurement and calculation of RQD

TABLE 2.4 Rock mass classification based on RQD

RQD, %	Rock Quality Classification
<25	Very poor
25–50	Poor
50–75	Fair
75–90	Good
90–100	Excellent

Source: Deere et al. 1967

RQD method in that it incorporated multiple parameters and generated an output value from multiple inputs.

The main contribution of the RSR concept, however, was that it introduced a rating system for rock masses. This was the sum of the weighted values of the three individual parameters considered in this classification scheme. The three parameters—A, B, and C, defining the general geology, the joint pattern related to the drive direction, and the groundwater and joint condition—are selected and allocated the values in Tables 2.5, 2.6, and 2.7. The rating is then obtained as follows:

$$\mathrm{RSR} = \mathrm{A} + \mathrm{B} + \mathrm{C}$$

NOTE: Maximum RSR value is 100.

The most remarkable omission from this rating system is rock strength, the basic assumption being that, at the depths encountered, deformation is unlikely to result from intact rock fracture.

The inputs—three parameters for the RSR—are the general categories of geology, geometry, and groundwater inflow and joint condition. The values of A, B, and C are obtained from the respective tables:

(1) **Parameter A, Geology** (Table 2.5): General appraisal of geological structure on the basis of
   a. Rock type origin (igneous, metamorphic, sedimentary),
   b. Rock hardness (hard, medium, soft, decomposed), and
   c. Geologic structure (massive, slightly faulted/folded, moderately faulted/folded, intensely faulted/folded).

(2) **Parameter B, Geometry** (Table 2.6): Effect of discontinuity pattern with respect to the direction of the tunnel drive on the basis of
   a. Joint spacing,
   b. Joint orientation (strike and dip), and
   c. Direction of tunnel drive.

TABLE 2.5 Rock structure rating parameter A (general area geology)

	Basic Rock Type				Geologic Structure			
	Hard	Medium	Soft	Decomposed	Massive	Slightly Folded or Faulted	Moderately Folded or Faulted	Intensively Folded or Fautled
Igneous	1	2	3	4				
Metamorphic	1	(2)	3	4				
Sedimentary	2	3	4	4				
Type 1					30	22	15	9
Type 2					27	20	(13)	8
Type 3					24	18	12	7
Type 4					18	15	10	6

Source: Wickham et al. 1972

TABLE 2.6 Rock structure rating parameter B (joint pattern, direction of drive)

	Strike Perpendicular to Axis					Strike Parallel to Axis		
	Direction of Drive					Direction of Drive		
	Both	With Dip		Against Dip		Either Direction		
	Dip of Predominant Joints*					Dip of Predominant Joints		
Average Joint Spacing	Flat	Dipping	Vertical	Dipping	Vertical	Flat	Dipping	Vertical
1. Very closely jointed, <50 mm	9	11	13	10	12	9	9	7
2. Closely jointed, 50–150 mm	13	16	19	15	17	14	14	11
3. Moderately jointed, 150–300 mm	23	24	28	19	22	23	23	19
4. Moderate to blocky, 0.3–0.7 m	30	32	36	25	28	(30)	28	24
5. Blocky to massive, 0.7–1.3 m	36	38	40	33	35	36	34	28
6. Massive, >1.3 m	40	43	45	37	40	40	38	34

Source: Wickham et al. 1972

*-Dip: flat = 0°–20°; dipping = +20°–50°; vertical = +50°–90°

(3) **Parameter C** (Table 2.7): Effect of groundwater inflow and joint condition on the basis of

a. Overall rock mass quality on the basis of A and B combined,

b. Joint condition (good, fair, poor), and

c. Amount of water inflow (in gallons per minute per 1,000 ft of tunnel).

As an example, let us determine the appropriate RSR for the following: Medium, metamorphic rock, moderately folded or faulted (Table 2.5); joints spaced about 45.7 cm (1.5 ft) apart on the average and the discontinuities are relatively flat and strike approximately parallel to the drive (Table 2.6); and moderate water inflow (Table 2.7) is assumed.

Therefore, the sum of A + B = 43, so with "moderate" water inflow and with assumed "fair" joint conditions, we get A + B + C = 65.

TABLE 2.7 Rock structure rating parameter C (groundwater, joint condition)

	Sum of Parameters A + B					
	13–44			45–75		
	Joint Condition*					
Anticipated Water Inflow gpm/1,000 ft of Tunnel	**Good**	**Fair**	**Poor**	**Good**	**Fair**	**Poor**
None	22	18	12	25	22	18
Slight, <200 gpm	19	15	9	23	19	14
Moderate, 200–1,000 gpm	15	(22)	7	21	16	12
Heavy, >1,000 gpm	10	8	6	18	14	10

Source: Wickham et al. 1972

*Joint condition: good = tight or cemented; fair = slight weathering or altered; poor = severely weathered, altered, or open

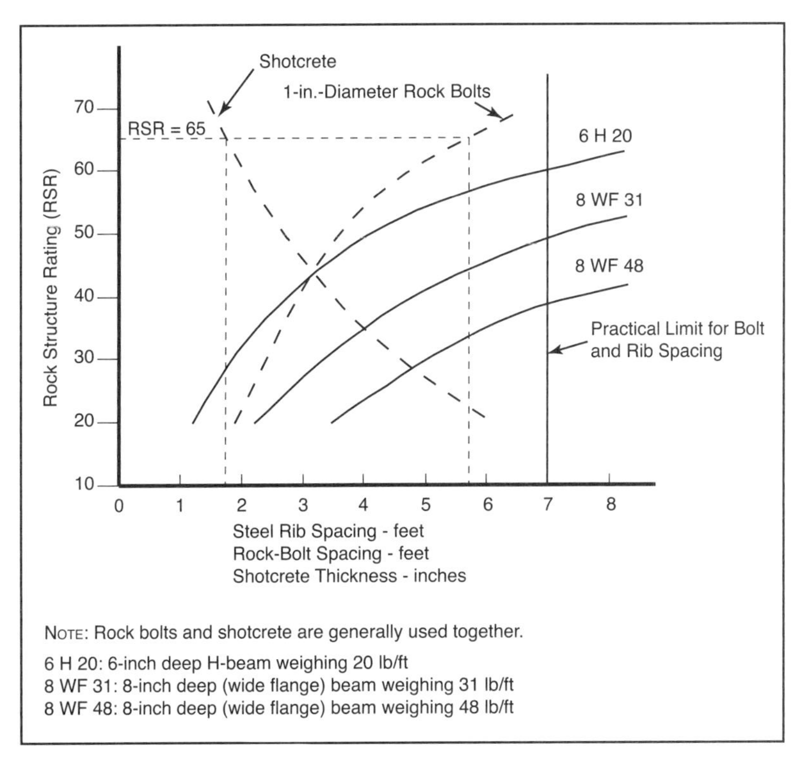

Source: Wickham et al. 1972

FIGURE 2.46 RSR support estimates for a 6.1-m (20-ft)-diameter circular tunnel

Thus, from Figure 2.46, the support requirements for a 7.3-m (24-ft)-diameter tunnel with RSR = 65 will be 4.3 cm (1.7 in.) of shotcrete over a 5.8 by 5.8 spacing of 2.5-cm (1-in.)-diameter rock bolts.

Bieniawski (1990) stated that his experience with the RSR concept led him to believe that it was a very useful method for selecting steel rib support for rock

TABLE 2.8 Rock mass rating for the geomechanics classification system

Class	Description	Range of RMR
I	Very good rock	81–100
II	Good rock	61–80
III	Fair rock	41–60
IV	Poor rock	21–40
V	Very poor rock	0–20

Source: Bieniawski 1989

tunnels but that he did not recommend it for selection of rock-bolt and shotcrete support. Wickham's work with RSR, however, played a significant role in the development of the later, more widely used and accepted, rock mass classification schemes (Wickham et al. 1972).

5. **Rock mass rating (RMR).** The geomechanics classification, also known as the RMR system, was developed by Z.T. Bieniawski in 1973. Bieniawski subsequently has made many changes to the RMR system, so the following discussion is based on the 1989 system (Bieniawski 1989). The output from the classification system is a single number, the RMR value, which represents the quality of the rock mass ranging from very good rock on the high end to very poor rock on the low end (Table 2.8).

   This engineering classification of rock masses uses the following six parameters (Bieniawski 1990), all of which can be measured or otherwise objectively determined:

   1. Uniaxial compressive strength of intact rock material
   2. RQD
   3. Spacing of discontinuities
   4. Orientation of discontinuities
   5. Condition of discontinuities
   6. Groundwater conditions

   The geomechanics classification (or RMR system) is presented in Table 2.9. In part A of the classification, five very important parameters—the strength of the intact rock material, the drill-core quality (RQD), the discontinuity spacing, the condition of the discontinuities, and the groundwater—are grouped into five ranges of values and given a rating for each. A higher rating indicates a better rock mass, and a lower rating is a less good rock situation. The 1973 ratings were determined from 49 case histories investigated by Bieniawski (1973), while the initial ratings were based on the studies by Wickham et al. (1972).

   The sixth item on the list of parameters—orientation of discontinuities—was subsequently removed from Bieniawski's 1973 list of basic parameters; and, instead, in part B of the classification, a "rating adjustment" number is determined for tunnels and mines, foundations or slopes, depending on the discontinuity strike and dip orientations (Bieniawski 1989).

TABLE 2.9 Bieniawski's geomechanics classification (RMR system)

A. CLASSIFICATION PARAMETERS AND THEIR RATINGS

	Parameter		**Range of Values**				
1	Strength of intact rock material	Point-load strength index	>10 MPa	4–10 MPa	2–4 MPa	1–2 MPa	For this low range, uniaxial comp. test is preferred
		Uniaxial comp. strength	>250 MPa	100–250 MPa	50–100 MPa	25–50 MPa	5–25 MPa 1–5 MPa <1 MPa
	Rating		15	12	7	4	2 1 0
2	Drill-core quality (RQD)		90–100%	75–90%	50–75%	25–50%	<25%
	Rating		20	17	13	8	3
3	Spacing of discontinuities		>2 m	0.6–2 m	200–600 mm	60–200 mm	<60 mm
	Rating		20	15	10	8	5
4	Condition of discontinuities (See E)		Very rough surfaces, not continuous, no separation, unweathered wall rock	Slightly rough surfaces, separation <1 mm, slightly weathered walls	Slightly rough surfaces, separation <1 mm, slightly weathered walls	Slickensided surface or gouge <5 mm thick or separation 1–5 mm, continuous	Soft gouge >5 mm thick or separation >5 mm, continuous
	Rating		30	25	20	10	0
5	Groundwater	Inflow per 10 m tunnel length (l/m)	None	<10	10–25	25–125	>125
		(Joint water press.)/(Major principal σ)	0	<0.1	0.1–0.2	0.2–0.5	>0.5
		General conditions	Completely dry	Damp	Wet	Dripping	Flowing
	Rating		15	10	7	4	0

B. RATING ADJUSTMENT FOR DISCONTINUITY ORIENTATIONS (See F)

Strike and Dip Orientations		**Very Favorable**	**Favorable**	**Fair**	**Unfavorable**	**Very Unfavorable**
Ratings	Tunnels and mines	0	−2	−5	−10	−12
	Foundations	0	−2	−7	−15	−25
	Slopes	0	−5	−25	−50	

C. ROCK MASS CLASSES DETERMINED FROM TOTAL RATINGS

Rating	100 ← 81	80 ← 61	60 ← 41	40 ← 21	<21
Class number	I	II	III	IV	V
Description	Very good rock	Good rock	Fair rock	Poor rock	Very poor rock

D. MEANING OF ROCK CLASSES

Class number	I	II	III	IV	V
Average stand-up time	20 yr for 15-m span	1 yr for 10-m span	1 week for 5-m span	10 hr for 2.5-m span	30 min for 1-m span
Cohesion of rockk mass (kPa)	>400	300–400	200–300	100–200	<100
Friction angle of rock mass (degrees)	>45	35–45	25–35	15–25	<15

E. GUIDELINES FOR CLASSIFICATION OF DISCONTINUITY conditions*

Discontinuity length (persistence)	<1 m	1–3 m	3–10 m	10–20 m	>20 m
Rating	6	4	2	1	0
Separation (aperture)	None	<0.1 mm	0.1–10 mm	1–5 mm	>5 mm
Rating	6	5	4	1	0
Roughness	Very rough	Rough	Slightly rough	Smooth	Slickensided
Rating	6	5	3	1	0
Infilling (gouge)	None	Hard filling <5 mm	Hard filling >5 mm	Soft filling <5 mm	Soft filling <5 mm
Rating	6	4	2	2	0
Weathering	Unweathered	Slightly weathered	Moderately weathered	Highly weathered	Decomposed
Rating	6	5	3	1	0

Table continues next page

TABLE 2.9 Bieniawski's geomechanics classification (RMR system) (continued)

F. EFFECT OF DISCONTINUITY STRIKE AND DIP ORIENTATION IN TUNNELLING†			
Strike perpendicular to tunnel axis		Strike parallel to tunnel axis	
Drive with dip–Dip 45–90°	Drive with dip–Dip 20–45°	Dip 45–90°	Dip 20–45°
Very unfavorable	Favorable	Very unfavorable	Fair
Drive against dip–Dip 45–90°	Drive against dip–Dip 20–45°	Dip 0–20°, irrespective of strike	
Fair	Unfavorable	Fair	

Source: From *Support of Underground Excavations in Hard Rock* by E. Hoek, P.K. Kaiser, and W.F. Bawden, Copyright © 1995, A.A. Balkema, reproduced by permission of Taylor & Francis Books UK

*Some conditions are mutually exclusive. For example, if infilling is present, the roughness of the surface will be overshadowed by the influence of the gouge. In such cases, use A.4 directly.

†Modified after Wickham et al. 1972.

FIGURE 2.47 Chart for determining RMR value, R_1

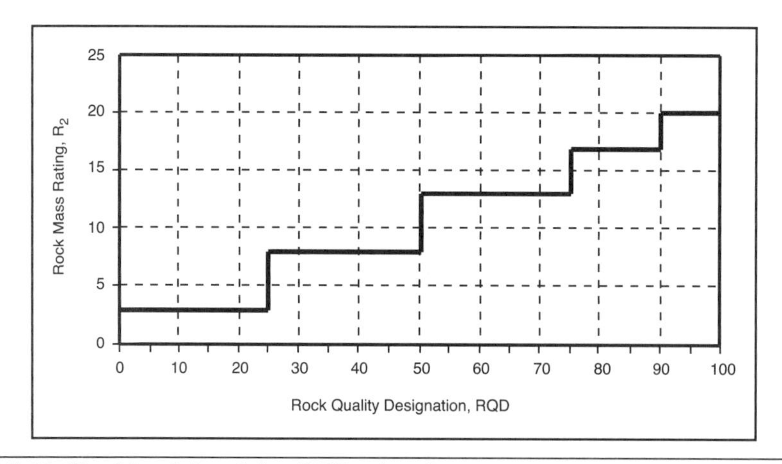

FIGURE 2.48 Chart for determining RMR value, R_2

In applying the RMR system, the rock mass is divided into geotechnical sectors and each region is classified separately. The boundaries of the sectors usually coincide with major structural features, such as a fault or with a change in rock type. In some cases, significant changes in one or more of the major parameters, such as

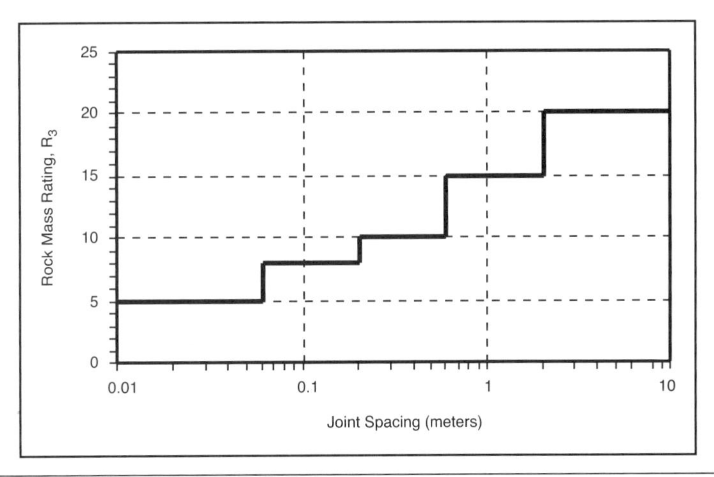

FIGURE 2.49 Chart for determining RMR value, R_3

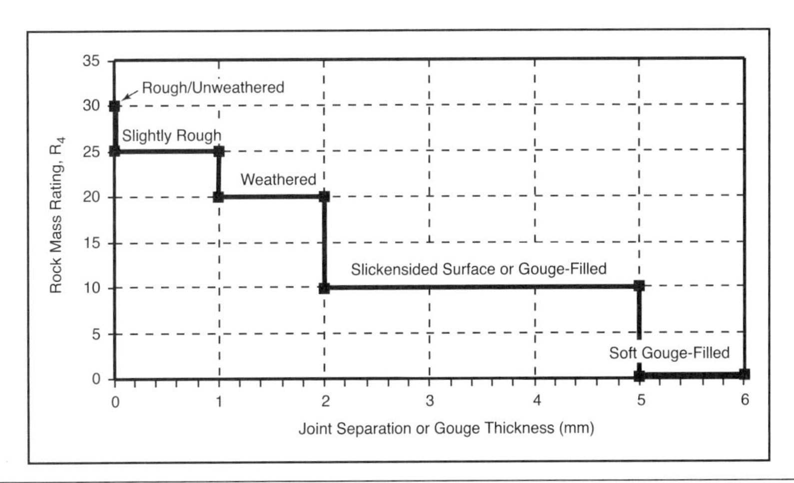

FIGURE 2.50 Chart for determining RMR value, R_4

discontinuity spacing or groundwater conditions, may necessitate the division of the rock mass into a number of sectors or subsectors.

The RMR = $R_1 + R_2 + R_3 + R_4 + R_5 \pm$ adjustments for joint orientation, where R_1 through R_5 can be determined off Table 2.9 or from Figures 2.47–2.51 and the adjustment for joint orientation is determined in part B of Table 2.9. The rock mass then can be classified according to the ranges in Table 2.8.

In the case of tunnels and chambers, the output from the geomechanics classification can be used to estimate the stand-up time and the maximum stable rock span for a given rock mass rating. Bieniawski (1989) has related his rock mass rating (or total rating score for the rock mass) to the stand-up time of an active unsupported

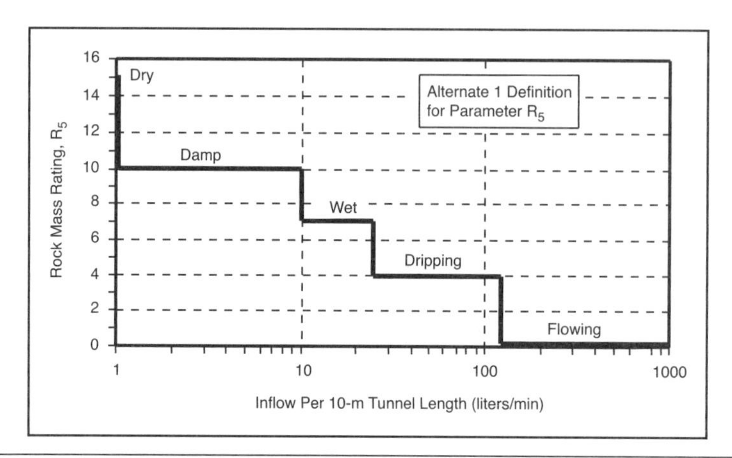

FIGURE 2.51A Chart for determining RMR value, R_5

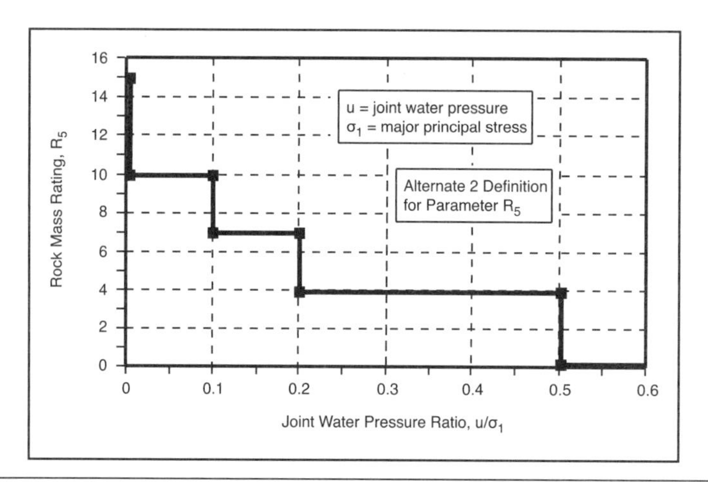

FIGURE 2.51B Alternate chart for determining RMR value, R_5

span as originally proposed by Lauffer. A chart for determining the stand-up time versus unsupported span is shown in Figure 2.52.

Support load can be determined as follows:

$$P = \frac{(100 - \text{RMR})}{100} \cdot \gamma B$$

where P is the support load (kg), RMR is the rock mass rating, γ is the unit weight of the rock (kg/m^3), and B is the tunnel width (m).

The output from the RMR system tends to be rather conservative, resulting in overdesign of the support system (Bieniawski 1989). A cost is always associated with overconservatism in design.

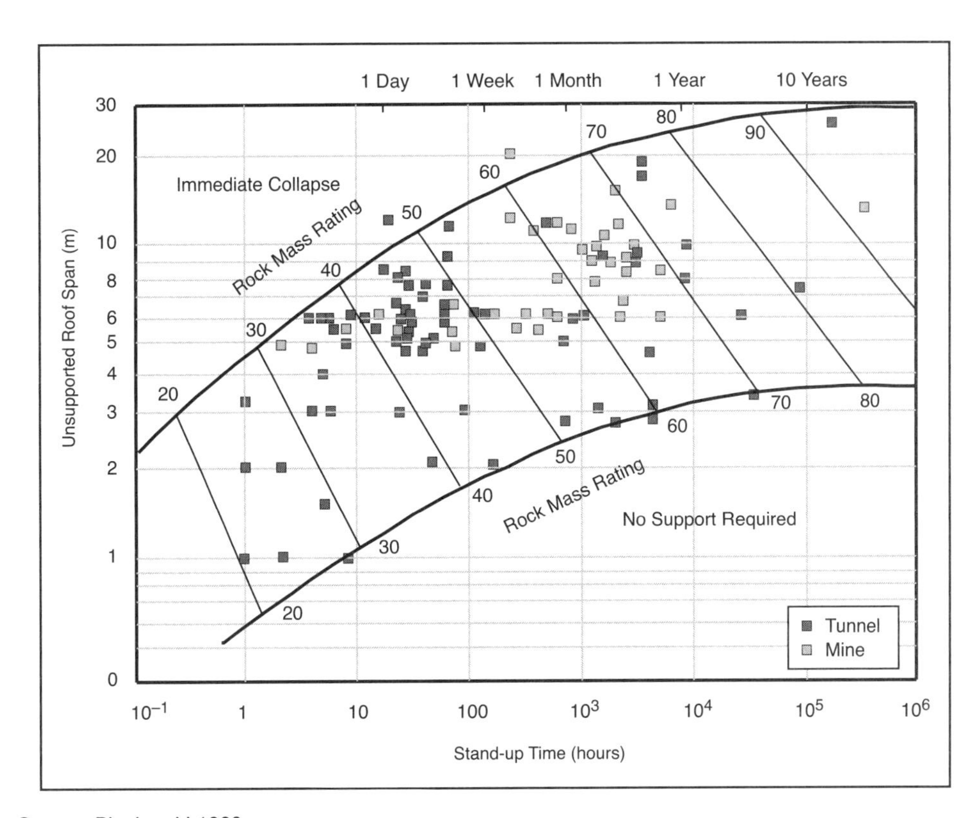

Source: Bieniawski 1989

FIGURE 2.52 Stand-up time versus unsupported active span as a function of RMR

6. **Rock tunneling quality index (Q system).** The rock mass quality system (Q system) is more complex than the other systems discussed previously, and the reasoning behind it is more cogent. It was introduced by Barton et al. (1974) and was based on the analysis of more than 200 Scandinavian tunnel support case histories collected by Cecil (1970). The Q system is based on a numerical assessment of the rock mass quality using six different parameters (Bieniawski 1990): (a) RQD; (b) number of joint sets, J_n; (c) roughness of the most unfavorable joint or discontinuity, J_r; (d) degree of alteration or filling along the weakest joint, J_a; (e) water inflow, J_w; and (f) stress reduction factor, SRF.

   The ratings for each of these factors can be determined from Table 2.10, and the rock mass quality, Q, is defined as follows:

$$Q = \left(\frac{\text{RQD}}{J_n}\right)\left(\frac{J_r}{J_a}\right)\left(\frac{J_w}{\text{SRF}}\right)$$

TABLE 2.10 Classification of individual parameters used in the tunneling quality index Q

Description	Value	Notes
1. Rock Quality Designation	***RQD***	
A. Very poor	0–25	1. Where RQD is reported or measured as ≤10 (including 0), a nominal value of 10 is used to evaluate *Q*.
B. Poor	25–50	2. RQD intervals of 5 (i.e., 100, 95, 90, etc.) are sufficiently accurate.
C. Fair	50–75	
D. Good	75–90	
E. Excellent	90–100	
2. Joint Set Number	J_n	
A. Massive, no or few joints	0.5–1.0	1. For intersections, use (3.0 × J_n).
B. One joint set	2	2. For portals, use (2.0 × J_n).
C. One joint set plus random	3	
D. Two joint sets	4	
E. Two joint sets plus random	6	
F. Three joint sets	9	
G. Three joint sets plus random	12	
H. Four or more joint sets, random, heavily jointed, "sugar cube," etc.	15	
I. Crushed rock, earthlike	20	
3. Joint Roughness Number	J_r	
a. Rock wall contact		1. Add 1.0 if the mean spacing of the relevant joint set is greater than 3 m.
b. Rock wall contact before 10-cm shear		2. J_r = 0.5 can be used for planar, slickensided joints having lineations, provided that the lineations are oriented for minimum length.
A. Discontinuous joints	4	
B. Rough and irregular, undulating	3	
C. Smooth undulating	2	
D. Slickensided undulating	1.5	
E. Rough or irregular, planar	1.5	
F. Smooth, planar	1	
G. Slickensided, planar	0.5	
c. No rock wall contact when sheared		
H. Zones containing clay minerals thick enough to prevent rock wall contact	1.0 (nominal)	
I. Sandy, gravely, or crushed zone thick enough to prevent rock wall contact	1.0 (nominal)	

Table continues next page

TABLE 2.10 Classification of individual parameters used in the tunneling quality index Q (continued)

Description	Value		Notes
4. Joint Alteration Number	J_a	ϕ_r degrees (approx.)	
a. Rock wall contact			1. Values of ϕ_r, the residual friction angle, are intended as an approximate guide to the mineralogical properties of the alteration products, if present.
A. Tightly healed, hard, nonsoftening, impermeable filling	0.75		
B. Unaltered joint walls, surface staining only	1.0	25–35	
C. Slightly altered joint walls, nonsoftening mineral coatings, sandy particles, clay-free disintegrated rock, etc.	2.0	25–30	
D. Silty- or sandy-clay coatings, small clay fraction (nonsoftening)	3.0	20–25	
E. Softening or low-friction clay mineral coatings (i.e., kaolinite, mica). Also chlorite, talc, gypsum, graphite, etc., and small quantities of swelling clays (discontinuous coatings, 1–2 mm or less).	4.0	8–16	
b. Rock wall contact before 10-cm shear			
F. Sandy particles, clay-free, disintegrating rock, etc.	4.0	25–30	
G. Strongly over-consolidated, nonsoftening clay mineral fillings (continuous <5 mm thick)	6.0	16–24	
H. Medium or low over-consolidation, softening clay mineral fillings (continuous <5 mm thick)	8.0	12–16	
I. Swelling clay fillings (i.e., montmorillonite; continuous <5 mm thick). Values of J_a depend on percentage of swelling clay-size particles and access to water.	8.0–12.0	6–12	
c. No rock wall contact when sheared			
J. Strongly over-consolidated, nonsoftening clay mineral fillings (continuous <5 mm thick)	6.0		
K. Medium or low over-consolidation, softening clay mineral fillings (continuous <5 mm thick)	8.0		
L. Swelling clay fillings (i.e., montmorillonite; continuous <5 mm thick). Values of J_a depend on percentage of swelling clay-size particles and access to water.	8.0–12.0	6–24	
M. Zones or bands of silty- or sandy-clay, small clay fraction, nonsoftening	5.0		
N. Thick continuous zones or bands of clay	10.0–13.0		
O. and P. (see G, H, and I for clay conditions)	6.0–24.0		

Table continues next page

TABLE 2.10 Classification of individual parameters used in the tunneling quality index Q (continued)

Description	Value		Notes
5. Joint Water Reduction	J_w	**Approx. Water Pressure (kgf/cm²)**	
A. Dry excavation or minor inflow (i.e., <5 L/m locally)	1.0	<1.0	1. Factors C to F are crude estimates; increase J_W if drainage installed. 2. Special problems caused by ice formation are not considered.
B. Medium inflow or pressure, occasional outwash of joint fillings	0.66	1.0–2..5	
C. Large inflow or high pressure in competent rock with unfilled joints	0.5	2.5–10.0	
D. Large inflow or high pressure	0.33	2.5–10.0	
E. Exceptionally high inflow or pressure at blasting, decaying with time	0.2–0.1	>10	
F. Exceptionally high inflow or pressure	0.1–0.05	>10	
6. Stress Reduction Factor	***SRF***		
a. Weakness zones intersecting excavation, which may cause loosening of rock mass when tunnel is excavated			1. Reduce these values of SRF by 25%–50%, but only if the relevant shear zone influences do not intersect the excavation. 2. For strongly anisotropic virgin stress field (if measured): when 5 $\leq \sigma_1/\sigma_3 \leq$ 10, reduce σ_c to 0.8 σ_c and σ_t to 0.8 σ_t. When $\sigma_1/\sigma_3 >$ 10, reduce σ_c and σ_t to 0.6 σ_c and 0.6 σ_t, where σ_c = unconfined compressive strength, and σ_t = tensile strength (point load), and σ_1 and σ_3 are the major and minor principal stresses. 3. Few case records available where depth of crown below surface is less than span width. Suggest SRF increase from 2.5 to 5 for such cases (see H).
A. Multiple occurrences of weakness zones containing clay or chemically disintegrated rock, very loose surrounding rock (any depth)	10.0		
B. Single weakness zones containing clay, or chemically disintegrated rock (excavation depth <50 m)	5.0		
C. Single weakness zones containing clay, or chemically disintegrated rock (excavation depth >50 m)	2.5		
D. Multiple shear zones in competent rock (clay free), loose surrounding rock (any depth)	7.5		
E. Single shear zone in competent rock (clay free) (depth of excavation <50 m)	5.0		
F. Single shear zone in competent rock (clay free) (depth of excavation >50 m)	2.5		
G. Loose open joints, heavily jointed or "sugar cube" (any depth)	5.0		

Table continues next page

TABLE 2.10 Classification of individual parameters used in the tunneling quality index Q (continued)

Description	Value			Notes
6. Stress Reduction Factor (continued)				
b. Competent rock, rock stress problems	σ_c/σ_1	σ_t/σ_c	*SRF*	
H. Low stress, near surface	>200	>13	2.5	
I. Medium stress	200–10	13–0.66	1.0	
J. High stress, very tight structure (usually favorable to stability; may be unfavorable to wall stability)	10–5	0.66–0.33	0.5–2	
K. Mild rockburst (massive rock)	5–2.5	0.33–0.16	5–10	
L. Heavy rockburst (massive rock)	<2.5	<0.16	10–20	
c. Squeezing rock, plastic flow of incompetent rock under influence of high rock pressure				
M. Mild squeezing rock pressure			5–10	
N. Heavy squeezing rock pressure			10–20	
d. Swelling rock, chemical swelling activity depending on presence of water				
O. Mild swelling rock pressure			5–10	
P. Heavy swelling rock pressure			10–15	

Source: From *Support of Underground Excavations in Hard Rock* by E. Hoek, P.K. Kaiser, and W.F. Bawden, Copyright © 1995, A.A. Balkema, reproduced by permission of Taylor & Francis Books UK

Notes: When making estimates of the rock mass quality (*Q*), the following guidelines should be followed in addition to the notes listed in the table:

1. When borehole core is unavailable, RQD can be estimated from the number of joints per unit volume, in which the number of joints per meter for each joint set are added. A simple relationship can be used to convert this number to RQD for the case of clay-free rock masses: RQD = 115 − 3.3J_v (approx.), where J_v = total number of joints per cubic meter (0 < RQD < 100 for 35 > J_v > 4.5).
2. The parameter J_n, representing the number of joints sets, will often be affected by foliation, schistosity, slaty cleavage or bedding, etc. If strongly developed, these parallel "joints" should obviously be counted as a complete joint set. However, if there are few "joints" visible, or if only occasional breaks in the core are due to these features, then it will be more appropriate to count them as "random" joints when evaluating J_n.
3. The parameters J_r and J_a (representing shear strength) should be relevant to the weakest significant joint set or clay-filled discontinuity in the given zone. However, if the joint set or discontinuity with the minimum value of J_r/J_a is favorably oriented for stability, then a second, less favorably oriented joint set or discontinuity may sometimes be more significant, and its higher value of J_r/J_a should be used when evaluating *Q*. The value of J_r/J_a should in fact relate to the surface most likely to allow failure to initiate.
4. When a rock mass contains clay, the factor SRF appropriate to loosening loads should be evaluated. In such cases, the strength of the intact rock is of little interest. However, when jointing is minimal and clay is completely absent, the strength of the intact rock may become the weakest link, and the stability will then depend on the ratio rock-stress/rock-strength. A strongly anisotropic stress field is unfavorable for stability and is roughly accounted for as in the second note in the table for stress reduction factor evaluation.
5. The compressive and tensile strengths (σ_c and σ_t) of the intact rock should be evaluated in the saturated condition if this is appropriate to the present and future in situ conditions. A very conservative estimate of the strength should be made for those rocks that deteriorate when exposed to moist or saturated conditions.

TABLE 2.11 Norwegian Geotechnical Institute Q-system rating for rock masses

Q Value	Quality of Rock Mass
<0.01	Exceptionally poor
0.01–0.10	Extremely poor
0.1–1	Very poor
1–4	Poor
4–10	Fair
10–40	Good
40–100	Very good
100–400	Extremely good
>400	Exceptionally good

Adapted from Barton et al. 1974

where the numerators and denominators are defined previously. The numerical value of the index Q varies on a logarithmic scale from <0.01 to >400, as shown in Table 2.11.

The reasoning behind the equation is that each of the quotients represents an important quantity of the rock mass. Thus, (RQD/J_n) represents the *structure* of the mass and is a crude measure of block or particle size. Barton et al. (1974) suggest that the range of values from 200 to 0.5 represents a slightly truncated size range for blocks, if expressed in centimeters.

The quotient (J_r/J_a) represents the *interblock shear resistance* or friction coefficient. The range of values is from 5 to about 0.1 and represents the range from strongly dilatant rock to the residual frictional coefficient of clay infill.

The final quotient (J_w/SRF) can be said to represent the *active stresses*, but it is complex, because SRF represents different stress parameters, depending on the continuity of the rock. Thus, it is the *loosening load* when the excavation is through shear zones or clay-bearing rock; the *squeezing load* in incompetent, plastic rocks; and the *rock stress* in competent rocks.

7. **Empirical rock slope design.** Because of the complexity of rock masses, several researchers have attempted to correlate rock slope design with rock mass parameters. Many of these methods have been subsequently modified by others and are now currently being used in practice for preliminary and sometimes final design. Some of the empirical RMR techniques that can be used in the design of slopes include the following:
   - RMR, rock mass rating (Bieniawski 1976, 1989)
   - MRMR, mining rock mass rating (Laubscher 1975, 1990)
   - RMS, rock mass strength (Selby 1980)
   - SMR, slope mass rating (Romana 1985)

- SRMR, slope rock mass rating (Robertson 1988)
- GSI, geological strength index (Hoek et al. 1995)

The majority of the methods require the determination of a basic RMR. The rating is usually calculated as the summation of a number of rating values that account for intact rock strength, block size, defect condition, and possibly groundwater. Several methods adjust this value based on such factors as defect orientation, excavation method, weathering, induced stresses, and the presence of major planes of weakness.

Bieniawski's RMR is probably the most commonly used system for estimating rock mass strength (discussed previously). Initially created to assess the stability and support requirements of tunnels, it has been found to be useful in assessing the strength of rock masses for slope stability.

Hoek et al. (1995) modified the RMR system to make it more applicable to assessing the strength of rock masses. The result of this was the GSI.

Geological strength index (GSI). The GSI is a system of rock mass characterization that has been developed in engineering rock mechanics to meet the need for reliable input data, particularly those related to rock mass properties required as inputs into numerical analysis or closed-form solutions for designing tunnels, slopes, or foundations in rocks (Marinos et al. 2005). The GSI, introduced by Hoek (1994) and Hoek et al. (1995), provides a system for estimating the reduction in rock mass strength for different geological conditions (Hoek and Karzulovic 2001). It is an extension of the generalized Hoek–Brown criterion as noted previously (Equations 2.4–2.12). This system is presented in Figure 2.53 for blocky rock masses and in Figure 2.54 for schistose metamorphic rocks.

To use the Hoek–Brown criterion for estimating the strength and deformability of jointed rock masses, the "properties" of the rock mass have to be estimated. They are (1) the uniaxial compressive strength, σ_{ci}, of the intact rock elements; (2) the value of the Hoek–Brown constant, m_i, for these intact rock elements; and (3) the value of the GSI for the rock mass.

Once the GSI has been estimated, the parameters that describe the RMS characteristics are calculated as follows:

$$m_b = m_i \cdot \exp\left(\frac{\text{GSI}-100}{28}\right) \qquad \text{(EQ 2.11)}$$

For GSI > 25, that is, rock masses of good to reasonable quality:

$$s = \exp\left(\frac{\text{GSI}-100}{9}\right) \qquad \text{(EQ 2.12)}$$

and

$$a = 0.5$$

GEOLOGICAL STRENGTH INDEX FOR BLOCKY JOINTED ROCKS

From a description of the structure and surface conditions of the rock mass, pick an appropriate box in this chart. Estimate the average value of GSI from the contours. Do not attempt to be too precise. Quoting a range from 36 to 42 is more realistic than stating that GSI = 38. It is also important to recognize that the Hoek-Brown criterion should only be applied to rock masses where the size of individual blocks or pieces is small compared with the size of the excavation under consideration. When the individual block size is more than about one quarter of the excavation size, the failure will be structurally controlled and the Hoek-Brown criterion should not be used.

SURFACE CONDITIONS — DECREASING SURFACE QUALITY ⟹

STRUCTURE — DECREASING INTERLOCKING OF ROCK PIECES ⟸

STRUCTURE	VERY GOOD Very rough, fresh unweathered surfaces	GOOD Rough, slightly weathered, iron stained surfaces	FAIR Smooth, moderately weathered and altered surfaces	POOR Slickensided, highly weathered surfaces with compact coatings or fillings or angular fragments	VERY POOR Slickensided, highly weathered surfaces with soft clay coatings or fillings
INTACT OR MASSIVE - intact rock specimens or massive in situ rock with few widely spaced discontinuities			N/A	N/A	N/A
BLOCKY - well interlocked un-disturbed rock mass consisting of cubical blocks formed by three intersecting discontinuity sets					
VERY BLOCKY- interlocked, partially disturbed mass with multi-faceted angular blocks formed by 4 or more joint sets					
BLOCKY/DISTURBED - folded and/or faulted with angular blocks formed by many intersecting discontinuity sets					
DISINTEGRATED - poorly inter-locked, heavily broken rock mass with mixture of angular and rounded rock pieces					
FOLIATED/LAMINATED - folded and tectonically sheared. Lack of blockiness due to schistosity prevailing over other discontinuities	N/A	N/A			

GSI contours: 90, 80, 70, 60, 50, 40, 30, 20, 10

Source: Hoek, Marinos, and Benissi 1998, as cited in Hoek and Karzulovic 2001

FIGURE 2.53 Characterization of blocky rock masses based on the particle interlocking and discontinuity condition

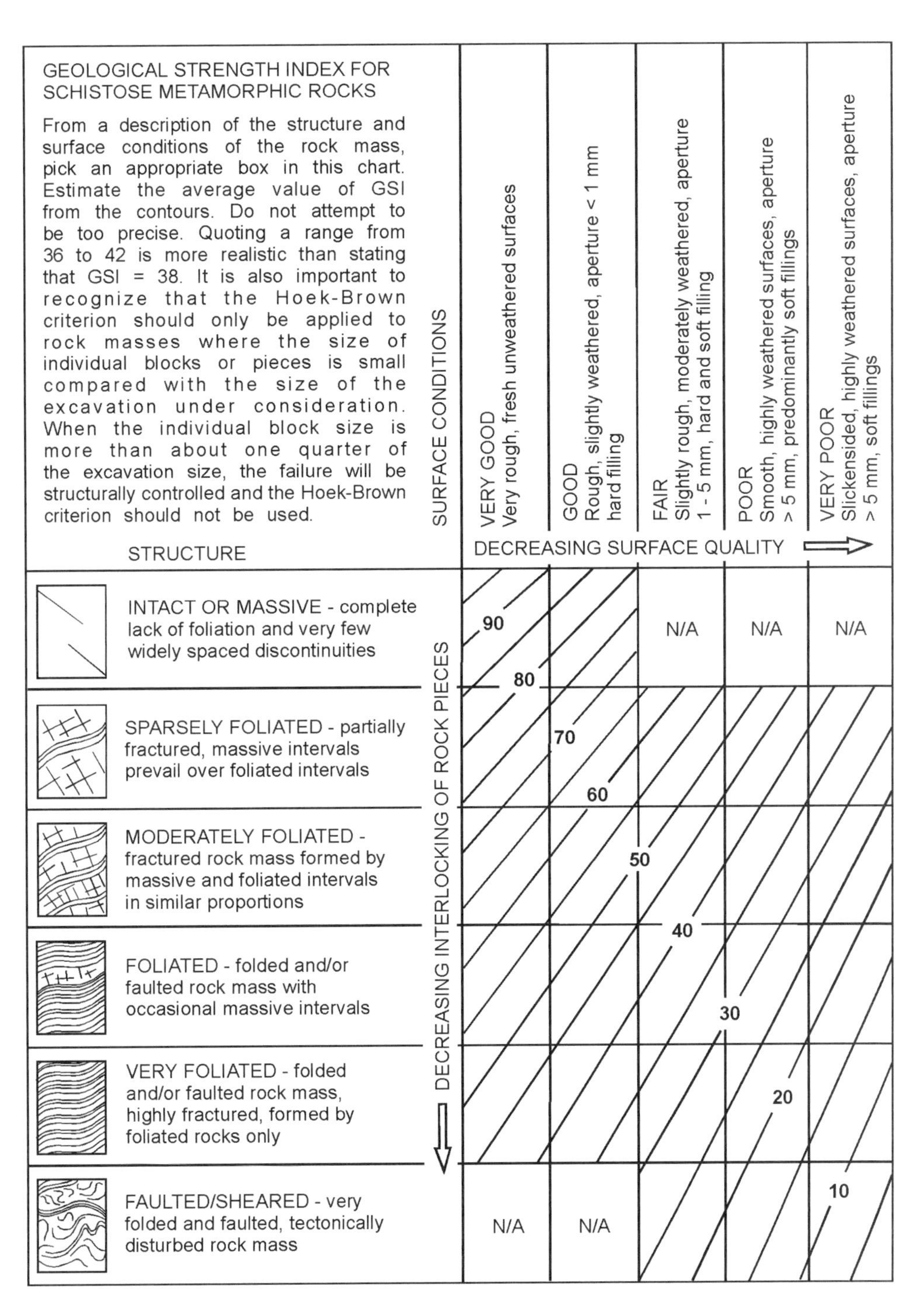

Source: Truzman 1999, as cited in Hoek and Karzulovic 2001

FIGURE 2.54 Characterization of schistose metamorphic rock masses based on foliation and discontinuity condition

For GSI < 25, that is, rock masses of very poor quality:

$$s = 0$$

and

$$a = 0.65 - (\text{GSI}/200)$$

For better-quality rock masses (GSI > 25), the GSI value can be estimated directly from the 1976 version of Bieniawski's RMR, with the groundwater rating set to 10 (dry) and the adjustment for joint orientation set to 0 (very favorable) (Bieniawski 1976). For very-poor-quality rock masses, the RMR value is very difficult to estimate, and the balance between the ratings no longer gives a reliable basis for estimating RMS. Consequently, Bieniawski's RMR classification should not be used for estimating GSI values for poor-quality rock masses (RMR < 25) and the GSI charts should be used directly (Hoek and Karzulovic 2001).

If the 1989 version of Bieniawski's RMR classification is used, then GSI = RMR_{89} – 5, because RMR_{89} has the groundwater rating set to 15 and the adjustment for joint orientation is set to zero (Hoek and Karzulovic 2001).

A reduced value of the Hoek–Brown material constant m_i, which takes into account the influence of blast damage on the rock mass, was presented in the 2002 version of the Hoek–Brown criterion (Hoek et al. 2002) as follows:

$$m_b = m_i \cdot \exp\left(\frac{\text{GSI} - 100}{28 - 14D}\right) \quad \text{(EQ 2.13)}$$

$$s = \exp\left(\frac{\text{GSI} - 100}{9 - 3D}\right) \quad \text{(EQ 2.14)}$$

and

$$a = \frac{1}{2} + \frac{1}{6}\left(e^{-\text{GSI}/15} - e^{-20/3}\right) \quad \text{(EQ 2.15)}$$

where

D = a factor that depends on the degree of disturbance to which the rock mass has been subjected by blast damage and stress relaxation. It varies from 0 for undisturbed in situ rock masses to 1 for very disturbed rock masses. The factor D applies only to the blast-damaged zone and it should not be applied to the entire rock mass. For example, in tunnels, the blast damage is generally limited to a 1-to-2-m (3.3-to-6.6-ft)-thick zone around the tunnel and this should be incorporated into numerical models as a different and weaker material than the surrounding rock mass. Applying the blast damage factor D to the entire rock mass is inappropriate and can result in misleading and unnecessarily pessimistic results.

For small-scale blasting in civil engineering slopes resulting in modest rock mass damage, particularly if controlled blasting is used, Hoek (2006) recommends a D value of 0.7 for good blasting and 1.0 for poor blasting. For very large open-pit mine slopes that may suffer significant disturbances due to heavy production blasting and also from stress relief, Hoek (2006) recommends a D value of 1.0; whereas for the case of softer rocks excavated by ripping and/or dozing, the degree of damage would be less, and Hoek recommends a D value of 0.7.

GSI = geological strength index, values of which are related to both the degree of fracturing and the condition of fracture surfaces. The GSI ranges from >90 for very good rock with very rough, fresh, unweathered surfaces to <10 for very poor rock with slickensided, highly weathered surfaces with soft clay coatings or fillings. For GSI > 25, the rock mass is of good to reasonable quality; for GSI < 25, the rock mass is of very poor quality.

REFERENCES

3GSM GmbH. 2015. ShapeMetriX^{3D} Measurement and assessment of rock and terrain surfaces by metric #D images. White Paper. Graz, Austria: 3GSM GmbH.

Afana, A., G. Hunter, J. Davis, N. Rosser, R. Hardy, and J. Williams. 2013. Integration of full waveform terrestrial laser scanners into a slope monitoring system. In *Slope Stability 2013*. Edited by P.M. Dight. Perth: Australian Centre for Geomechanics.

Barton, N.R. 1973. Review of a new shear strength criterion for rock joints. *Engineering Geology*, 7(4):287–332.

Barton, N.R. 1982. Shear strength investigations for surface mining. In *Stability in Surface Mining*, Vol. 3. Edited by C.O. Brawner. New York: SME-AIME.

Barton, N.R., and S. Bandis. 1980. Some effects of scale on the shear strengths of joints. *International Journal of Rock Mechanics and Mining Sciences and Geomechanics Abstracts*, 17(1):69–76.

Barton, N.R., and V. Choubey. 1977. The shear strength of rock joints in theory and practice. *Rock Mechanics*, 10(1):1–54.

Barton, N.R., R. Lien, and J. Lunde. 1974. Engineering classification of rock masses for the design of tunnel support. *Rock Mechanics*, 6(4):189–239.

Bieniawski, Z.T. 1973. Engineering classification of jointed rock masses. *Transactions of the South African Institution of Civil Engineers*, 15:335–344.

Bieniawski, Z.T. 1976. Rock mass classification in rock engineering. In *Exploration for Rock Engineering, Proceedings of the Symposium*, Vol. 1 Edited by Z.T. Bieniawski. Cape Town: Balkema. pp. 97–106.

Bieniawski, Z.T. 1989. *Engineering Rock Mass Classifications*. New York: Wiley.

Bieniawski, Z.T. 1990. *Tunnel Design by Rock Mass Classifications*. Update of Technical Report GL-79-19. Vicksburg, MS: U.S. Army Corps of Engineers, U.S. Army Engineer Waterways Experiment Station.

Booth, P.W., and G.E. Meyer. 2013. Quarry wall stability and design optimisation using photogrammetric mapping and analysis techniques. In *Slope Stability 2013*. Perth: Australian Centre for Geomechanics.

Brown, E.T., ed. 1981. *Rock Characterization Testing and Monitoring: ISRM Suggested Methods.* Commission on Testing Methods, International Society of Rock Mechanics. New York: Pergamon Press.

Cecil, O.S. 1970. Correlation of rockbolt–shotcrete support and rock quality parameters in Scandinavian tunnels. PhD thesis, University of Illinois, Urbana.

Cosar, S. 2004. Application of rock mass classification systems for future support design of the Dim Tunnel near Alanya. Thesis, Middle East Technical University, Ankara, Turkey.

Deere, D.U., and D.W. Deere. 1988. The rock quality designation (RQD) index in practice. In *Rock Classification Systems for Engineering Purposes, ASTM STP 984.* Edited by L. Kirkaldie. Philadelphia: American Society for Testing and Materials. pp. 91–101.

Deere, D.U., and R.P. Miller. 1966. *Engineering Classification and Index Properties for Intact Rock.* Technical Report No. AFWL-TR-65-116. Kirkland Air Force Base, New Mexico: Air Force Weapons Laboratory, Research and Technology Division, Air Force Systems Command.

Deere, D.U., A.J. Hendron Jr., F.D. Patton, and E.J. Cording. 1967. Design of surface and near-surface construction in rock. In *Failure and Breakage of Rock.* Edited by C. Fairhurst. New York: SME-AIME. pp. 237–302.

Dev, H., and S.K. Sharma. 2011. *Monograph on Rock Mass Classification Systems and Applications.* New Delhi: Central Soil and Materials Research Station.

Eberhardt, E. 2012. The Hoek–Brown failure criterion. *Rock Mechanics and Rock Engineering,* 45(6):981–988. doi 10.1007/s00603-012-0276-4

Farmer, I.W. 1983. *Engineering Behaviour of Rocks.* New York: Chapman and Hall.

F.W. Breithaupt and Sohn. 2015 *Cocla Stratum Compass* (brochure). Kassel, Germany: F.W. Breithaupt and Sohn. www.breithaupt.de/ (accessed Aug. 9, 2015).

Goodman, R.E. 1976. *Methods of Geological Engineering in Discontinuous Rocks.* St. Paul, MN: West Publishing.

Goodman, R.E. 1980. *Introduction to Rock Mechanics.* New York: John Wiley and Sons.

GPS.gov. 2017. GPS accuracy. https://www.gps.gov/systems/gps/performance/accuracy/. Accessed November 2017.

Griffith, A.A. 1920. The phenomena of rupture and flow in solids. *Philosophical Transactions of the Royal Society of London, Series A: Mathematical, Physical and Engineering Sciences,* 221(587):163–198.

Griffith, A.A. 1924. The theory of rupture. In *Proceedings of the First International Congress for Applied Mechanics.* Edited by C.B. Biezeno and J.M. Burgers. Delft: J. Waltman Jr. pp. 55–63.

Herget, G. 1977. Structural geology. In *Pit Slope Manual.* Report 77–41. Edited by D.F. Coates. Ottawa, ON: Canada Centre for Mineral and Energy Technology.

Hoek, E. 1968. Brittle failure of rock. In *Rock Mechanics in Engineering Practice.* Edited by K.G. Stagg and O.C. Zienkiewic. New York: Wiley. pp. 99–124.

Hoek, E. 1994. Strength of rock and rock masses. *International Society for Rock Mechanics News Journal,* 2(2):4–16.

Hoek, E. 2006. *Practical Rock Engineering.* Toronto: Rocscience. https://www.rocscience.com/hoek/corner/Practical_Rock_Engineering.pdf

Hoek, E., and J.W. Bray. 1977. *Rock Slope Engineering.* London: Institution of Mining and Metallurgy.

Hoek, E., and E.T. Brown. 1980a. Empirical strength criteria for rock masses. In *Journal of the Geotechnical Engineering Division, ASCE,* 106(GT9):1013–1035.

Hoek, E., and E.T. Brown. 1980b. *Underground Excavations in Rock.* London: Institution of Mining and Metallurgy.

Hoek, E., and A. Karzulovic. 2001. Rock mass properties for surface mines. In *Slope Stability in Surface Mining*. Edited by W.A. Hustrulid, M.K. McCarter, and D.J.A. van Zyl. Littleton, CO: SME. pp. 59–70.

Hoek, E., P.K. Kaiser, and W.F. Bawden. 1995. *Support of Underground Excavations in Hard Rock*. Rotterdam: A.A. Balkema.

Hoek, E., C.T. Carranza-Torres, and B. Corkum. 2002. Hoek–Brown failure criterion—2002 edition. In *Proceedings of the 5th North American Rock Mechanics Symposium*. Toronto: University of Toronto Press. pp. 267–273.

Hunt, R.E. 1984. *Geotechnical Engineering Investigation Manual*. New York: McGraw-Hill.

ISRM (International Society for Rock Mechanics). 1981. *Rock Characterization Testing and Monitoring*. New York: Oxford.

John, K.W. 1965. Civil engineering approach to evaluate strength and deformability of regularly jointed rock. *Rock Mechanics*, 15(2):69–80.

Krahn, J., and N.R. Morgenstern. 1979. The ultimate frictional resistance of rock discontinuities. *International Journal of Rock Mechanics and Mining Sciences and Geomechanics Abstracts*, 16:127–133.

Laubscher, D.H. 1975. Class distinction in rock masses. *Coal, Gold, and Base Minerals of Southern Africa*, Vol. 23.

Laubscher, D.H. 1990. A geomechanics classification system for the rating of rock mass in mine design. *Journal of the South African Institute of Mining and Metallurgy*, 90(86):257–273.

Lauffer, H. 1958. Gebirgsklassifizierung für den Stollenbau. *Geol. Bauwesen*, 24(1):46–51.

Maptek, KRJA Systems Inc. Letter to author. Aug. 25, 2015.

Marinos, V., P. Marinos, and E. Hoek. 2005. The geological strength index: Applications and limitations. *Bulletin of Engineering Geology and the Environment*, 64(1):55–65. doi 10.1007/s10064-004-0270-5

Palmström, A. 1982. The volumetric joint count—A useful and simple measure of the degree of rock jointing. In *Proceedings of the 4th Congress of the International Association of Engineering Geology*. New Delhi: Oxford and IBH Publishing. pp. 221–228.

Palmström, A. 1995. RMi—A system for characterizing rock mass strength for use in rock engineering. *Journal of Rock Mechanics and Tunnelling Technology*, 1(2):69–108.

Patton, F.D. 1966. Multiple modes of shear failure in rock and related materials. Ph.D. thesis, University of Illinois, Urbana-Champaign.

Piteau, D.R. 1970. Geological factors significant to the stability of slopes cut in rock. In *Planning Open Pit Mines, Proceedings of the Symposium on the Theoretical Background to the Planning of Open Pit Mines with Special Reference to Slope Stability*. Amsterdam: A.A. Balkema.

Priest, S.P. 1985. *Hemispherical Projection Methods in Rock Mechanics*. London: George Allen and Unwin.

Ragan, D.M. 1973. *Structural Geology: An Introduction to Geometrical Techniques*. New York: John Wiley and Sons.

Rees, K.M, and J. Graaf. 2013. Structural geology modelling: A summary on data integrity and modelling methods. In *Slope Stability 2013*. Perth: Australian Centre for Geomechanics.

Ritter, W. 1879. *Die Statik der Tunnelgewölbe*. Berlin: Springer.

Roberds, W.J., and H.H. Einstein. 1978. Comprehensive model for rock discontinuities. *Journal of the Geotechnical Engineering Division, ASCE*, 104(5):553–569.

Robertson, A.M. 1988. Estimating weak rock strength. In *Proceedings of the AIME Annual General Meeting*. New York: AIME.

Romana, M. 1985. New adjustment rating for application of the Bieniawski classification to slopes. In *Proceedings of the International Symposium on Rock Mechanics Mining Civil Works*. Zacatecas, Mexico: ISRM. pp. 59–63.

Selby, M.J. 1980. A rock mass strength classification for geomorphic purposes: With tests from Antarctica and New Zealand. *Zeitschrift für Geomorphologie*, 23:31–51.

Shroff, A.V., and D.L. Shah. 2003. *Soil Mechanics and Geotechnical Engineering*. India: A.A. Balkema.

Stini, J. 1950. *Tunnelbaugeologie*. Vienna: Springer.

Terzaghi, K. 1946. Rock defects and loads on tunnel supports. In *Rock Tunneling with Steel Supports*, Vol. 1. Edited by R.V. Proctor and T.L. White. Youngstown, OH: Commercial Shearing and Stamping Company. pp. 17–99.

Tse, R., and D.M. Cruden. 1979. Estimating joint roughness coefficient. *International Journal of Rock Mechanics and Mining Sciences and Geomechanics Abstracts*, 16:303–307.

Wickham, G.E., H.R. Tiedemann, and E.H. Skinner. 1972. Support determination based on geologic predictions. In *Proceedings of the North American Rapid Excavation and Tunneling Conference*. Edited by K.S. Lane and L.A. Garfield. New York: SME-AIME. pp. 43–64.

CHAPTER 3

Groundwater

Groundwater can be defined as water below the water table (i.e., in the zone of saturation). Replenishment of groundwater is termed *recharge*, its fundamental source being precipitation. Loss of groundwater (e.g., that which escapes through springs and evaporation) is termed *discharge*. Groundwater makes up one element of the hydrologic cycle (Figure 3.1), which also includes precipitation, evaporation, runoff, transpiration, and channel flow.

Sources of recharge, which greatly influence conditions in open pits or other excavations, include the following:

- Infiltration of rainfall and melting snow
- Surface water bodies (e.g., lakes, rivers, tailings dams, and reservoirs)
- Water in storage (i.e., groundwater contained in rocks and soils, which will move toward an excavation)

GROUNDWATER FLOW WITHIN ROCK MASSES

In rock masses, groundwater flow is often controlled by the secondary permeability (i.e., the permeability of the structure—faults, folds, discontinuities, etc.) instead of the primary permeability (i.e., that of intact material), as is the case for soils. It is known that only relatively few cracks of remarkably small aperture are needed to increase the effective permeability of a rock mass by several orders of magnitude (Morgenstern 1971).

Permeability Conditions

Soils and rocks are classified with respect to groundwater flow on the basis of their permeability. Permeability—more correctly called hydraulic conductivity—is the rate at which water will flow through a material under a given pressure differential. For homogeneous materials, it is expressed via Darcy's law (Figure 3.2):

$$Q = KAi \qquad \text{(EQ 3.1)}$$

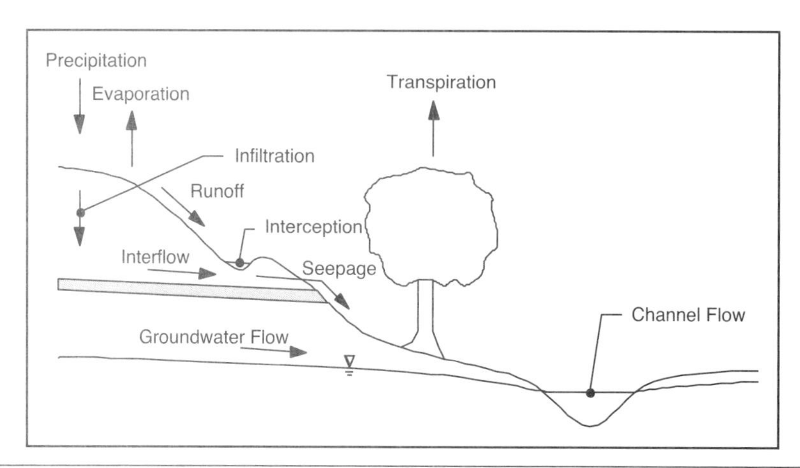

FIGURE 3.1 Features of the hydrologic cycle

where

Q = flow rate

K = coefficient of permeability (generally considered a material constant)

A = cross-sectional area through which flow takes place

i = hydraulic gradient

Note that K is designated as the permeability in the preceding expression for Darcy's law. Many references and texts still use *permeability* instead of the more accepted term, *hydraulic conductivity*. The permeability of a rock or soil defines the latter's ability to transmit a fluid. It is a property of the medium only and is independent of fluid properties. The hydraulic conductivity, K, is a material constant that serves as a measure of the permeability of the porous medium. It is related to the permeability, k, but is generally not the same. A medium has a unit hydraulic conductivity if it will transmit in unit time a unit volume of groundwater at the prevailing kinematic viscosity through a cross section of unit area, measured at right angles to the direction of flow, under a unit hydraulic gradient.

The hydraulic gradient, i, is the change in hydraulic head, or potential, with distance; it is measured in the direction of flow. The hydraulic head, h, is defined for a given point as the sum of (1) the elevation or height above datum and (2) the pressure component of head, or pressure expressed as head of water for the point concerned. Consider an experimental apparatus like the one shown in Figure 3.2. A container of cross section A is filled with sand and stopped at each end with a screen so that the sand will not drain out. Attached to each end are water columns, situated such that the water is free to flow through the container of sand. If the column on the left is filled with water to a level d_1 and the column on the right is filled to a level d_2, where $d_1 > d_2$, then the water will flow from 1 to 2 because of the hydraulic head, $d_1 - d_2$, over the hydraulic gradient, i.

Experiments carried out by H. Darcy showed that the specific discharge, v, through the container of area A is directly proportional to $h_1 - h_2$ when the distance, ΔL, is constant

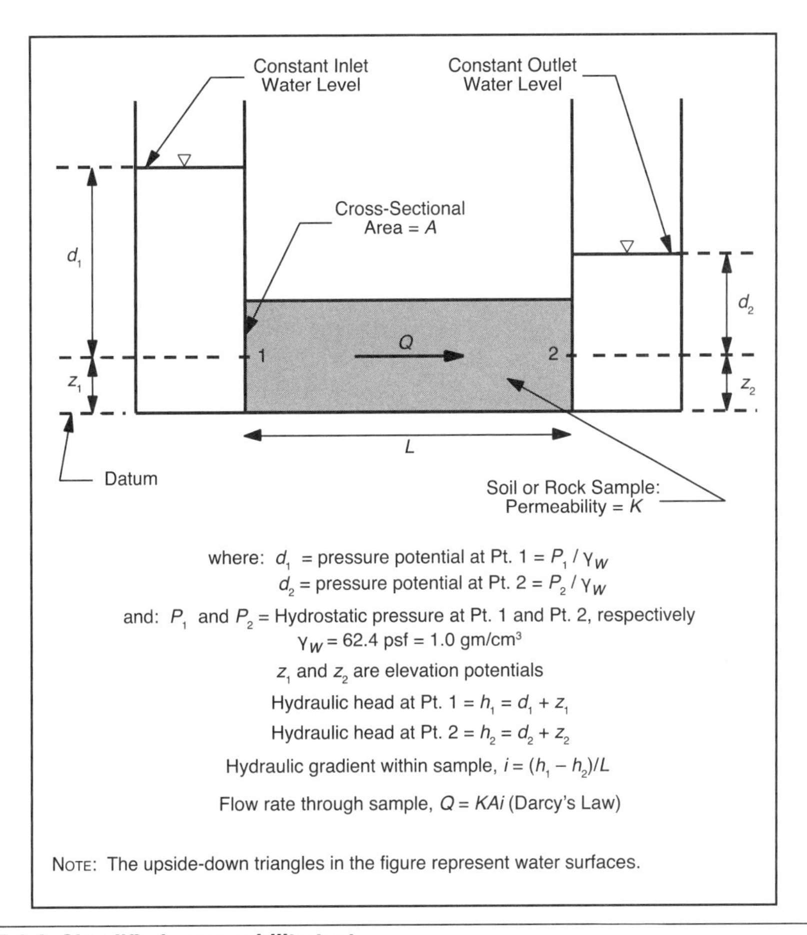

FIGURE 3.2 Simplified permeability test

and inversely proportional to ΔL when $h_1 - h_2$ is constant. If Δh is defined as $h_2 - h_1$, then $v \propto -\Delta h$ and $v \propto 1/\Delta L$, where $\Delta h = \phi_1 - \phi_2$ in Figure 3.2.

Darcy's law can now be written as

$$v = K(\Delta h/\Delta L)$$

or, in differential form,

$$v = -K(dh/dL) \qquad \text{(EQ 3.2)}$$

where

h = hydraulic head

dh/dL = hydraulic gradient

K = a constant of proportionality that is a property of the material in the cylinder

TABLE 3.1 Typical permeability values

Type of Material	Hydraulic Conductivity (Permeability) (K), cm/sec
Soils (intact, primary permeability)	
Gravel	$1–10^{2}$
Clean sands	$10^{-3}–1$
Clayey sands	$10^{-6}–10^{-3}$
Clays	$10^{-9}–10^{-6}$
Rocks (intact, primary permeability)	
Limestone, 25% porosity	8.5×10^{-8}
Limestone, 16% porosity	10^{-4}
Silty sandstone	2×10^{-6}
Sandstone, 29% porosity	2×10^{-3}
Granite	5×10^{-9}
Slate	10^{-10}
Rock mass (mass permeability)	
With clay-filled joints	10^{-5}
Moderately jointed	$10^{-4}–10^{-2}$
Well jointed	$10^{-2}–10$
Heavily fractured	$10–10^{2}$

Source: Coates 1977

The parameter K is known as permeability (also known as hydraulic conductivity); it has high values for sand and gravel and lower values for clays and rock.

If we define the specific discharge through the container of area A as $v = Q/A$, then Equation 3.2 can be rewritten as

$$Q = -K\left(dh/dL\right)A$$

or

$$Q = KiA, \text{ with the negative sign disregarded}$$

In soils, the permeability is influenced by both the particle size and the distribution of particle sizes or degree of homogeneity. Typical soil, rock, and rock mass hydraulic conductivities (permeabilities) are shown in Table 3.1.

The permeability of intact rock is usually low. In most rock masses, groundwater flow occurs largely through the discontinuities. The nature and orientation of these discontinuities thus determine the permeability of the rock mass (Hunt 1984). Because the discontinuities generally fall into "sets" having a given mean orientation, the permeability of the rock mass varies with direction (i.e., the rock mass exhibits anisotropic permeability). The permeability of the rock mass can be defined by its joint conductivity, K_j, which is the flow characteristics of a single specific fissure or joint. Figure 3.3 illustrates a joint, or fissure, in

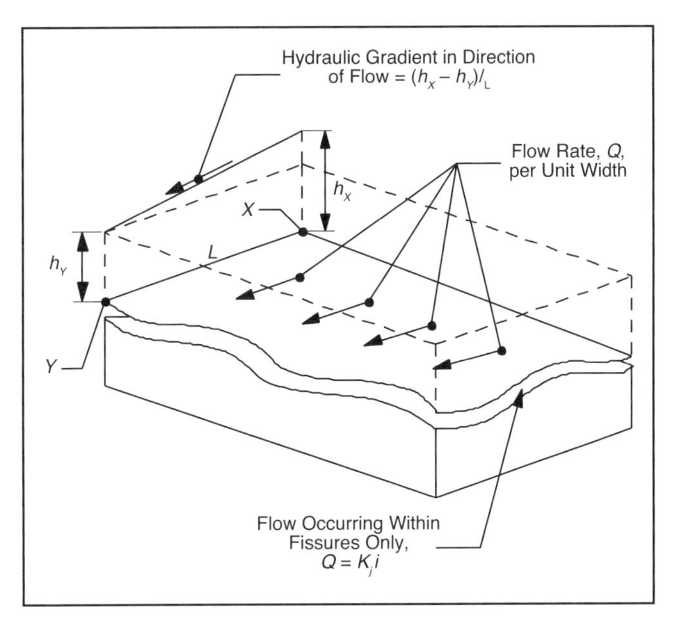

Source: Coates 1977

FIGURE 3.3 Definition of joint conductivity, K_j

a rock mass inclined at hydraulic gradient $(h_X - h_Y)/L$, where h_X and h_Y are the heads at points X and Y along the joint and L is the length of the joint. The flow through the joint, according to Darcy's law, can be determined as

$$Q = K_j iA$$

For a unit width of the joint or fissure, the flow can be determined as

$$Q = K_j i$$

where

Q = flow per unit width of the fissure
K_j = conductivity of the fissure
i = hydraulic gradient in the direction of flow

Owing to the complexity of most rock types and the practical difficulties of performing detailed measurements in rock mass at depth, the groundwater flow characteristics of the rock must usually be described by mass or bulk permeability.

Influence of Geology on Water Flow Through Slopes

More permeable regions of soil or rock masses (e.g., in heavily jointed or porous rock) are commonly termed *aquifers*, whereas impermeable zones (e.g., intact rock and gouge-filled

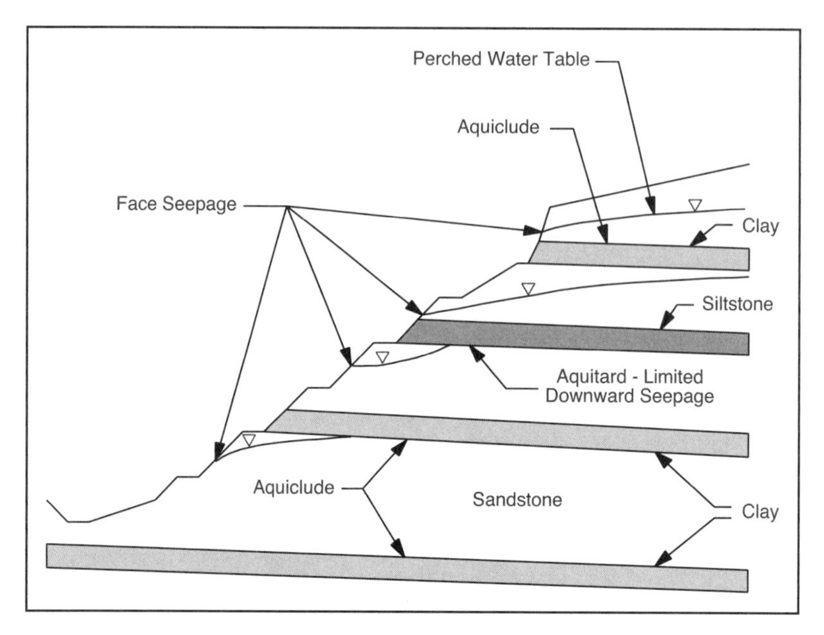

Source: Coates 1977

FIGURE 3.4 Groundwater patterns in a stratified slope

faults) are termed *aquicludes*, or confining layers (Figure 3.4). Formations that do yield some water, but usually not enough to meet even modest demands, are termed *aquitards*. In reality, most formations yield some water and are either aquifers or aquitards.

Water can exist in aquifers under two completely different physical conditions: confined or unconfined. Confined groundwater is isolated from the atmosphere at the point of discharge by impermeable geologic formations, and a confined aquifer is generally subject to pressures greater than atmospheric (Driscoll 1986). An unconfined aquifer occurs when the water table is exposed to the atmosphere through openings in the overlying geologic strata. This is not to say that the water table is open to the atmosphere; it is, however, at atmospheric pressure at its surface.

The occurrence of faults in rock slopes can have a dramatic influence on both water flow and pressure. Low-permeability faulting often occurs in association with ore-body formations and obstructs flow, causing a buildup in groundwater pressure. Alternatively, shatter or breccia zones associated with faults can act as preferential flow paths.

Groundwater Pressure in Slopes

The excavation of an open pit causes groundwater to flow into the pit, setting up hydraulic gradients. The pattern of vertical and lateral variations in hydraulic head is called the groundwater pressure distribution. The flow patterns—and thus the groundwater pressure distribution that will be generated within a slope—will depend on the following factors:

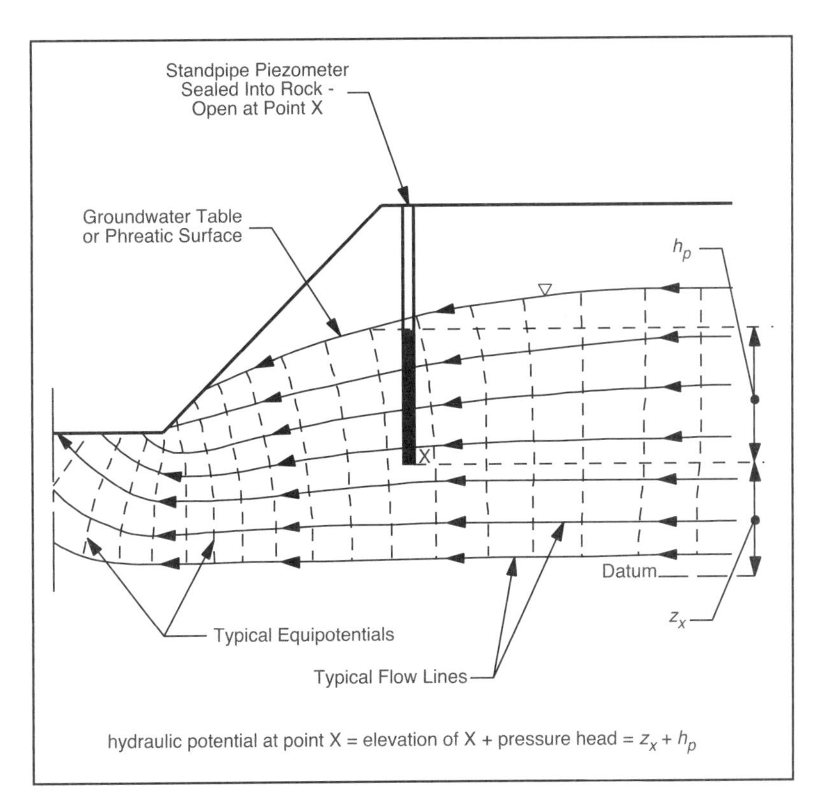

Source: Coates 1977

FIGURE 3.5 Flow net for seepage through a slope

- Geometry of the slope
- Permeability characteristics of the slope material
- Recharge from the surrounding rock mass
- Water storage within the slope
- Local precipitation, runoff, and infiltration characteristics

The overall flow pattern and pressure distribution in a slope can best be represented by a flow net (Figure 3.5). Flow nets are plots of flow lines (the paths that molecules of water will follow when flow is taking place) and of equipotentials (contours of equal hydraulic potential). They represent a solution to the Laplace equation of flow for steady-state flow through a homogeneous, isotropic medium. In rectangular coordinates, the equation takes the following form:

$$\frac{\partial^2 h}{\partial x^2} + \frac{\partial^2 h}{\partial y^2} + \frac{\partial^2 h}{\partial z^2} = 0 \qquad \text{(EQ 3.3)}$$

The solution of the equation is a function $h(x, y, z)$ that describes the value of hydraulic head, h, at any point in a three-dimensional flow field. A solution to Equation 3.3 allows us to produce a contoured equipotential map of h and, with the addition of flow lines, a flow net (Freeze and Cherry 1979).

To construct a flow net, the groundwater pressure at a number of representative points must be either measured directly or computed. From the flow net, the flow through a given zone, as well as the groundwater pressure at any given point, can be determined. Only in the case of nearly horizontal flow in isotropic materials can direct measurement of the water table, as determined from water level measurements in open boreholes, provide sufficient data to determine the groundwater flow and pressure distribution.

Transmissivity and Storage Coefficient

The principal factors of aquifer performance are the transmissivity, T, and the storage coefficient, S. These terms will appear frequently in the equations of this chapter, so a brief introduction is warranted here. The transmissivity is related to hydraulic conductivity and saturated thickness (i.e., $T = Kb$). It is defined as the rate at which water of the prevailing kinematic viscosity is transmitted through a vertical strip of the aquifer of unit width and extending through the full saturated thickness, b, under a unit hydraulic gradient. Transmissivity has dimensions of volume per time per length. The *storage coefficient*, also known as the coefficient of storage or the storativity, is defined as the volume of water that an aquifer releases from or takes into storage per unit surface area of aquifer per unit change in the component of head normal to that surface. It has units of volume per area per length (i.e., it is dimensionless and normally expressed as a decimal number or with scientific notation).

The concepts of transmissivity and storage coefficient were developed primarily for the analysis of well hydraulics in confined aquifers (Freeze and Cherry 1979), but they can also be used for the unconfined case. For transmissivity in the unconfined case, b is the saturated thickness of the aquifer or the height of the water table above the top of the underlying aquitard that bounds the aquifer. The storage term for unconfined aquifers is known as the specific yield, S_y. It is defined as the ratio of (1) the volume of water that the rock or soil, after being saturated, will yield by gravity to (2) the volume of the rock or soil. It has the same dimensions (i.e., it is dimensionless and normally expressed as a decimal number) as storage coefficient.

INFLUENCE OF GROUNDWATER ON SLOPE STABILITY

The primary effect of groundwater pressure in reducing the stability of rock slopes is the resulting decrease in effective shear strength of discontinuities. This phenomenon is described by the effective stress principle, which is fundamental to understanding the influence of groundwater on rock slope stability.

The stresses at a given point in a saturated rock mass depend on the influence of gravity, on tectonic and other stresses, and on water pressure. Consider a discontinuity, as shown in Figure 3.6A, that is critically oriented with respect to stability. The stresses acting on a portion of the joint can be resolved into components normal and parallel to the joint plane (remember the concept of the block sitting on an inclined plane at limiting equilibrium; see Figure 1.16). In a simple, normal-stress versus shear-strength relationship, ignoring water pressure effects, the magnitude of normal stresses governs the shear strength that can be mobilized. This relationship is given by

$$S = \sigma_n \tan\phi$$

where

S = shear strength
σ_n = normal stress acting across the joint
ϕ = angle of frictional resistance

In this case, the normal stress, σ_n, is wholly transmitted through the asperities or contact points of the rock, which in turn accounts for the frictional resistance of the joint.

Consider now full saturation of the rock, including the joint, where no drainage of water is allowed. If we assume that water is incompressible and that no flow of water into or out of the joint is allowed, then we see that the volume of the test specimen, including the joint, must remain constant. For this to happen, the water must sustain stresses sufficient to prevent volume change of the specimen. The total applied stress across the joint will be transmitted by the rock asperities and by the water. If the water carries some of the normal stress, then the rock asperities carry less normal load and therefore have less shear strength than they would if drained. The normal stress transmitted by the water is equal to the joint water pressure. The stress transmitted through the rock asperities is therefore equal to the total applied stress minus the joint water pressure. The joint shear strength will now be reduced proportionally. The reduced normal stress acting through the rock contacts is termed the *effective normal stress* and is given by (see Figure 3.6B)

$$\sigma'_n = \sigma_n - u$$

where

σ'_n = effective normal stress
σ_n = normal stress
u = water pressure

The corresponding shear strength that can be mobilized is

$$\begin{aligned} S &= \sigma'_n \tan\phi \\ &= (\sigma_n - u)\tan\phi \end{aligned}$$

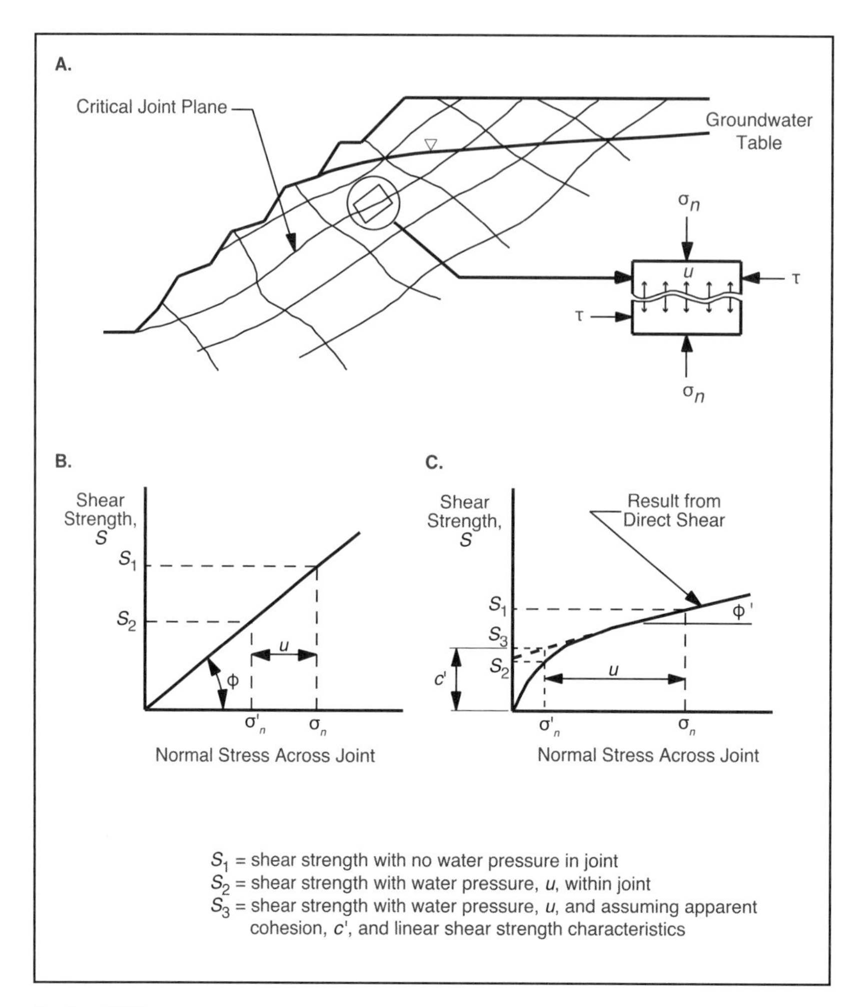

Source: Coates 1977

FIGURE 3.6 Influence of water pressure on the strength of rock discontinuities

Obviously, when the water pressure is equal to the applied normal stress, the effective shear strength of the joint reduces to zero.

Since the stability analysis techniques currently available cannot readily determine shear strength from the nonlinear relationship between shear strength and normal stress, an alternative approach is often used. In this approach, the curved strength envelope is approximated by a straight line that is defined by two well-known Coulomb parameters: effective cohesion, c', and effective angle of shearing resistance, ϕ'. The shear strength is then given by

$$S = c' + (\sigma_n - u)\tan\phi$$

Although this approximation of the curved strength envelope by a straight-line envelope is commonly used, it can lead to errors unless the straight-line approximation corresponds closely to the curved envelope in the range of water pressures that will occur; see Figure 3.6C. For the given water pressure, u, the straight-line approximation overestimates the shear strength, S, by the amount $(S_3 - S_2)$. This can cause serious problems with estimates of the strength of a slope for stability purposes.

EVALUATION OF GROUNDWATER CONDITIONS IN SLOPES

The most important groundwater parameter for stability purposes is the groundwater pressure distribution within slopes. This distribution can be obtained in two ways: (1) by direct measurement of pressure via piezometers or (2) by determining pressures from an analysis of the hydraulic properties of the rock mass (e.g., geology and permeability characteristics). The most satisfactory approach is usually to measure groundwater pressures with piezometers at representative locations; these data are then correlated with analytical studies based on a thorough understanding of the geology, as well as on selected permeability or conductivity measurements of representative soil and rock strata.

Measurement of Groundwater Pressures by Piezometers

Measurement of groundwater pressure is required not only for stability analysis but also to monitor changes in groundwater conditions resulting from seasonal variations, mining activity, and drainage. Piezometers are common means of making such measurements. Piezometers are transducers that convert pressure to some more readily measurable form of output (e.g., elevation head, electrical voltage, or electrical current). They are generally installed in boreholes and should be sealed into the holes so that pressures are measured at the sampling points only. Since one of the primary purposes of piezometer readings is to determine the distribution of groundwater pressures with depth below the phreatic surface, poor sealing would result in improper groundwater pressure readings and thus would cause an error in the determination of groundwater flow characteristics. If the output information is to be of value, maintaining a completely effective seal of the piezometer is essential. A general classification of piezometers, listing the most important characteristics of each class, is shown in Table 3.2.

Piezometers are used to conduct permeability tests. These tests can be divided into four general types (Bureau of Reclamation 1995):

1. Pressure tests
2. Constant-head gravity tests
3. Falling-head gravity tests
4. Slug or bail tests

In pressure tests and falling-head gravity tests, one or two packers are used to segregate the test section in the hole. Inflatable packers are generally capable of expanding to a larger

TABLE 3.2 Piezometer types—general classification

Broad Classification	Relative Volume Demand*	Readout Equipment	Major Limitations	Major Advantages
Open tube (standpipe)	High	Water level finder	Longer time lag in most rock types. Tube must be straight for whole length. Difficulties likely in small-diameter tubes if water levels significantly below 30 m (100 ft) or dip less than 45°.	Cheap; simple to read
Closed tube (hydraulic)	Medium to low	Usually Bourdon gauge or mercury manometer	Requires readout location not significantly above lowest water level. Therefore, not suitable for general borehole use.	Relatively cheap
Mechanical diaphragm (pneumatic or hydraulic)	Low to very low	Pressure transducer	Hydraulic types require periodic "de-airing" of monitoring system.	Excellent long-term stability; can be made very small; simple to install
Electrical	Negligible	Electronic readout	Relatively expensive, especially if cable lengths are large. Some zero drift possible. Certain types may be susceptible to blast damage.	Ideal for remote monitoring; simple installation

Source: Coates 1977

*The volume of water required to operate this device. High-volume-demand piezometers are not well suited for measuring rapid changes in groundwater conditions in low-permeability materials. Low-volume-demand types can accurately reflect even small changes in groundwater pressures.

diameter than the hole diameter. The packer is placed in the proper location above or below the intake screen and inflated by injecting gas, water, or a solidifying liquid into the balloon-like apparatus. Inflatable packers are often used for only a short time and then retrieved for use elsewhere, but some can be installed permanently. In pressure tests, water is usually forced into the test section through a combination of applied pressure and gravity head, although the tests can be performed under conditions of gravity head only. In falling-head tests, only gravity head is used. In constant-head gravity tests, no packers are used and a constant water level is maintained. Slug or bail tests use only small changes in water level, generally over a short time period (although changes can sometimes be relatively large and can return to normal quickly, especially if in permeable material). When water is removed, the test is called a bail test; when water is added, the test is called a slug test.

Several methods exist to conduct and analyze permeability tests (see, for instance, Bureau of Reclamation 1995). For the purposes of this chapter, the Hvorslev piezometer test is described below (Bureau of Reclamation 1995; Freeze and Cherry 1979). The Hvorslev

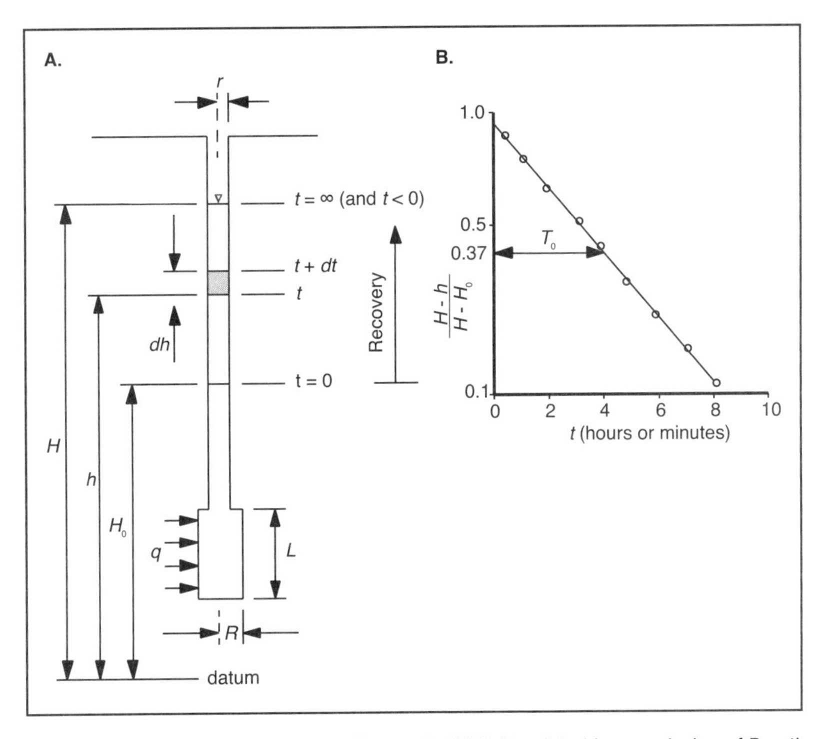

Source: *Groundwater* by R.A. Freeze and J.A. Cherry. © 1979. Reprinted by permission of Prentice-Hall, Inc., Upper Saddle River, New Jersey.

FIGURE 3.7 Hvorslev piezometer test (bail test shown): (A) geometry; (B) method of analysis

analysis assumes a homogeneous, isotropic, infinite medium in which both soil and water are incompressible (Figure 3.7). It neglects wellbore storage. The method is for point piezometers that are open only over a short interval at their base. It can be used for either bail or slug tests. The geometry illustrated in Figure 3.7A is for a bail test where water is removed from the initial height, H, to the height at the start of the test, H_0. Recovery over time starts at $t = 0$ where $h = H_0$ and commences for a period of time, t, at which the water level recovers to height h. Hvorslev reasoned that the rate of inflow, q, at the piezometer tip at any time t is proportional to the hydraulic conductivity, K, of the soil and to the unrecovered head difference, $H - h$, so that

$$q(t) = \pi r^2 \frac{dh}{dt} = FK(H - h)$$

where

r = radius of a circular piezometer

h = height of water at some time t

F = a factor that depends on the shape and dimensions of the piezometer intake

K = hydraulic conductivity

H = initial height of the water at $t < 0$

$(H - h)$ = unrecovered head difference, which is the difference between the initial water level, H, and the water level at time t (it is a measure of the water level recovery)

If $q = q_0$ at $t = {}_0$, then $q(t)$ will decrease asymptotically toward zero as time goes on. A term known as the basic time lag, T_0, may be defined as

$$T_0 = \frac{\pi r^2}{FK} \qquad \text{(EQ 3.4)}$$

When we substitute this term into the earlier equation for $q(t)$ and solve at the initial condition, $h = H_0$ at $t = 0$, we get

$$\frac{H - h}{H - H_0} = e^{-t/T_0} \qquad \text{(EQ 3.5)}$$

Plotting well-head changes—i.e., $(H - h)/(H - H_0)$—versus time on a semilog graph should approximate a straight line (see Figure 3.7B). At the point where $(H - h)/(H - H_0) = 0.37$, we find that $\ln[(H - h)/(H - H_0)] = -1$. (NOTE: $\ln(0.37) = -1$.) Utilizing Equation 3.5, we find that $T_0 = t$. This, then, defines the basic time lag, T_0 (see Figure 3.7B).

To interpret a set of field recovery data from the Hvorslev piezometer test, the data are plotted in log-linear form as in Figure 3.7B. The value of T_0 is measured graphically, and K is determined from Equation 3.4. For a piezometer intake of length L and radius R, with $L/R > 8$, Hvorslev determined that hydraulic conductivity, K, can be given by

$$K = \frac{r^2 \ln(L/R)}{2LT_0}$$

This equation, or others developed by Hvorslev for different ratios of L/R, can be used to determine in situ hydraulic conductivity values by means of simple tests carried out in a single piezometer.

Determination of Rock Mass Hydraulic Characteristics from Well Tests

Field tests (i.e., well tests) provide quantitative data on hydraulic characteristics of aquifers, including transmissivities, storage coefficients, and boundary conditions. This information is essential to understanding and solving many aquifer problems. Well tests involve the removal of water from a well (discharge) or the addition of water to a well (recharge) and the subsequent observation of the reaction of the aquifer to the change. Well tests may be conducted to determine (1) the performance characteristics of a well and (2) the hydraulic parameters of the aquifer.

For a well performance test, well yield and drawdown are recorded so that the specific capacity can be calculated. These data, taken under controlled conditions, give a measure of the productive capacity of the completed well and also provide information needed for the selection of pumping equipment. The terms *well yield*, *drawdown*, and *specific capacity* can be defined as follows (Driscoll 1986):

- Well yield is the volume of water per unit of time discharged from a well, either by pumping or free flow. It is commonly measured in units of cubic meters per day (m^3/d) or gallons per minute (gpm).
- Drawdown is the difference, measured in meters or feet, between the original pre-pumping water table or potentiometric surface and the pumping water level. This difference represents the head of water (force) that causes water to flow through an aquifer toward a well at the rate that water is being withdrawn from the well. As a result of pumping, the lowered water surface develops a continually steeper slope toward the well. The form of this surface resembles a cone and is called the cone of depression or drawdown cone. Also, the outer limit of the cone of depression defines the area of influence of the well.
- A well's specific capacity is its yield per unit of drawdown, usually expressed as cubic meters per day per meter ($m^3/d/m$) or gallons of water per minute per foot (gpm/ft) of drawdown, after a given time has elapsed, usually 24 hours. Dividing the yield of a well by the drawdown, when each is measured at the same time, gives the specific capacity.

The second purpose of pumping tests is to provide data from which the principal factors of aquifer performance—transmissivity and storage coefficient—can be calculated. This type of test is called an aquifer test because it is primarily the aquifer characteristics that are being determined, even though the specific capacity of the well can also be calculated.

Aquifer tests will predict the

- Effect of new withdrawals on existing wells,
- Drawdowns in a well at future times and different discharges, and
- Radius of the cone of depression for individual or multiple wells.

An aquifer test consists of pumping a well at a certain rate and recording the drawdown in the pumping well and in nearby observation wells at specific times. The two primary types of aquifer tests are constant-rate tests and step-drawdown tests. In the constant-rate test, the well is pumped for a significant length of time at one rate, whereas in a step-drawdown test, the well is pumped at successively greater discharges for relatively shorter periods.

Measurements required for both well tests and aquifer tests include the static water levels just before the test is started, time since the pump started, pumping rate, pumping levels or dynamic water levels at various intervals during the pumping period, time of any change in discharge rate, and time the pump stopped. Measurements of water levels after the pump is stopped (recovery) are extremely valuable in verifying the aquifer coefficients calculated during the pumping phase of the test.

Steady unidirectional flow in an aquifer. Steady flow implies that no change occurs with time. Flow conditions differ for the confined and the unconfined cases. For the confined case, head decreases linearly with flow in the x-direction as

$$h = -\left(vx/K\right)$$

where

h = head

v = velocity of flow in the aquifer

x = direction of flow

K = hydraulic conductivity

For steady, unidirectional flow in an unconfined aquifer, direct analytic solution of the Laplace equation (Equation 3.3) is not possible; hence, the flow net cannot be directly determined. The difficulty arises because the water table in the two-dimensional (2-D) case represents a flow line. The shape of the water table determines the flow distribution, but at the same time the flow distribution governs the water table shape. To obtain a solution, J. Dupuit made the following assumptions (Todd 1980):

1. Flow is homogeneous-isotropic.
2. Flow through the system is constant.
3. The drawdown cone is stable.
4. The hydraulic gradient can be described (i.e., closely approximated, in most cases) by the tangent instead of the sine. In general, Darcy's law, in rectangular coordinates, can be written as $v = -K(\partial h/\partial L)$, where h is the head, L is the distance along the average direction of flow, and $\partial h/\partial L$ is the hydraulic gradient. In this equation, $\partial h/\partial L$ describes the sine of the angle of drawdown (in differential form), as does $\Delta h/\Delta L$ in Figure 3.8. For small angles θ, $\sin\theta \cong \tan\theta$. Therefore, it is necessary only to measure the drawdown (or thickness, h, of the aquifer) at a distance r to describe the hydraulic gradient.
5. Flow is everywhere horizontal and uniform in vertical section.
6. The flow pattern is perfectly radial.

In an aquifer, the flow toward a sink (or well) is a function of the aquifer constants (K, T, and S) and the hydraulic gradient, dh/dr. In the case of a well pumping from an aquifer, the cone of depression will expand until the volume of flow into the sink (or well) just equals the flow out of the well, Q. The equations to solve for the well constants, therefore, have been developed in terms of Q, the constant pumping rate. For a given pumping rate, therefore, we must rearrange the appropriate equation and solve for the desired well attribute. This is the methodology used in the next few sections.

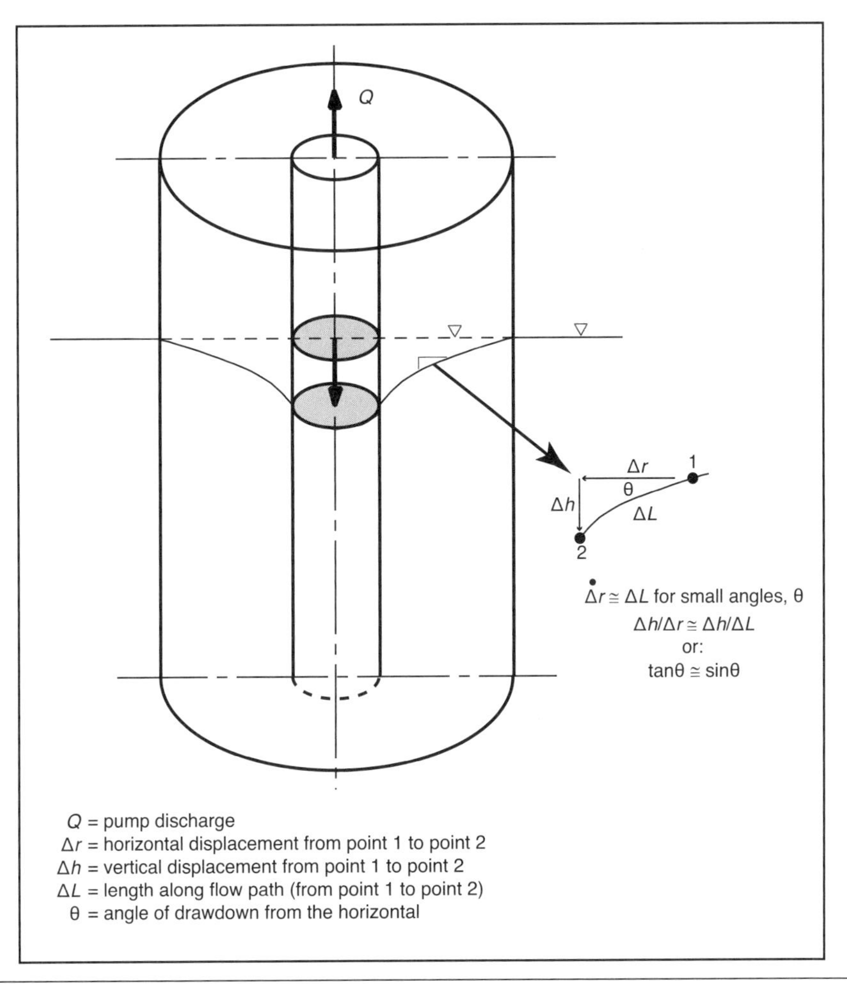

FIGURE 3.8 Representation of the Dupuit assumption of hydraulic gradient for a confined aquifer

Steady radial flow to a well in a confined aquifer. The flow is assumed to be 2-D to a well centered on a circular island and penetrating a homogeneous and isotropic aquifer (Figure 3.8). Because flow in a confined aquifer is everywhere horizontal, the Dupuit assumptions apply without error. If plane polar coordinates are used, with the well as the origin (Figure 3.8), the applicable 2-D form of the differential equation is

$$\frac{\partial^2 h}{\partial r^2}+\frac{1}{r}\frac{\partial h}{\partial r}=\frac{S}{T}\frac{\partial h}{\partial t} \qquad \text{(EQ 3.6)}$$

where

h = head, or piezometric pressure (dimensions of length)

r = radial distance from the pumped well

S = storage coefficient

T = transmissivity

t = time since beginning of pumping

(For the derivation of Equation 3.6 from Equation 3.3, the reader is referred to Freeze and Cherry [1979] and Todd [1980].) The well discharge, Q (units of volume per time), at any distance r is as follows (Todd 1980):

$$Q = AV = 2\pi rbK\frac{dh}{dr}$$

where

A = radial area of influence around the well through which flow occurs

V = Specific discharge assuming that flow occurs through the entire cross section of the material without regard to solids and pores (Todd 1980)

b = saturated thickness of the aquifer (units of length)

K = hydraulic conductivity (units of length per time)

Rearranging and integrating for the boundary conditions at the well, $h = h_w$ and $r = r_w$, and at the edge of the boundary (at circumference of the area of influence), $h = h_0$ and $r = r_0$, yields

$$h_0 - h_w = \frac{Q}{2\pi Kb}\ln\frac{r_0}{r_w}$$

Rewriting and solving for Q (and neglecting the minus sign), we get

$$Q = 2\pi Kb\frac{h_0 - h_w}{\ln(r_0/r_w)} \quad \text{(EQ 3.7)}$$

For all real numbers x greater than zero, $\log(x)$ is equal to $0.4343 \times \ln(x)$. Therefore, we can convert the above equation from natural log to base-10 log to yield the following:

$$Q = \frac{2.73Kb(h_0 - h_w)}{\log(r_0/r_w)}\ \mathrm{m^3/d}, \text{ in metric units} \quad \text{(EQ 3.8)}$$

$$Q = \frac{Kb(h_0 - h_w)}{528\log(r_0/r_w)}\ \text{gpm, in U.S. customary units} \quad \text{(EQ 3.9)}$$

Steady radial flow to a well in an unconfined aquifer. For an unconfined aquifer, assuming the well completely penetrates the aquifer to the horizontal base and that a concentric boundary of constant head surrounds the well, the well discharge is (Todd 1980)

$$Q = -2\pi Kh\frac{dh}{dr} \qquad \text{(EQ 3.10)}$$

This equation, when integrated between the limits $h = h_w$ at $r = r_w$ and $h = h_0$ at $r = r_0$, rearranged, and converted to base-10 logarithm, yields

$$Q = \frac{1.366K\left(h_0^2 - h_w^2\right)}{\log\left(r_0/r_w\right)}\ \text{m}^3/\text{d, in metric units} \qquad \text{(EQ 3.11)}$$

$$Q = \frac{K\left(h_0^2 - h_w^2\right)}{1{,}055\log\left(r_0/r_w\right)}\ \text{gpm, in U.S. customary units} \qquad \text{(EQ 3.12)}$$

where the terms in the equations are the same as for a confined aquifer.

Applying the Thiem equations. Equations 3.8, 3.9, 3.11, and 3.12—also known as the "equilibrium" equations—yield values of hydraulic conductivity for an unconfined aquifer and hydraulic conductivity or transmissivity for a confined aquifer. For years they were the only equations used for the analysis of discharging well tests. The general test procedure associated with these equations involves (1) pumping from a test well at a constant known rate and (2) periodically measuring drawdown in two or more nearby observation wells (at different distances from the test well). For each observation well, the time since the start of pumping is plotted (on a log scale) versus drawdown (on an arithmetic scale); the test normally continues until each plot falls on a straight line.

In the general case of a well penetrating an extensive confined aquifer, there is no external limit for r (Figure 3.9). In this case, Q becomes (from Equation 3.7)

$$Q = 2\pi Kb\frac{h - h_w}{\ln\left(r/r_w\right)} \qquad \text{(EQ 3.13)}$$

which shows that h increases indefinitely with increasing r. However, the maximum h is the initial uniform head h_0. From a practical standpoint, h approaches h_0 with distance from the well.

Because any two points define the logarithmic drawdown curve, the method consists of measuring drawdown in at least two observation wells at different distances from a well pumped at a constant rate (Todd 1980). We can then use Equation 3.13 to obtain the transmissivity, T.

$$T = \frac{Q}{2\pi\left(h_2 - h_1\right)}\ln\frac{r_2}{r_1}$$

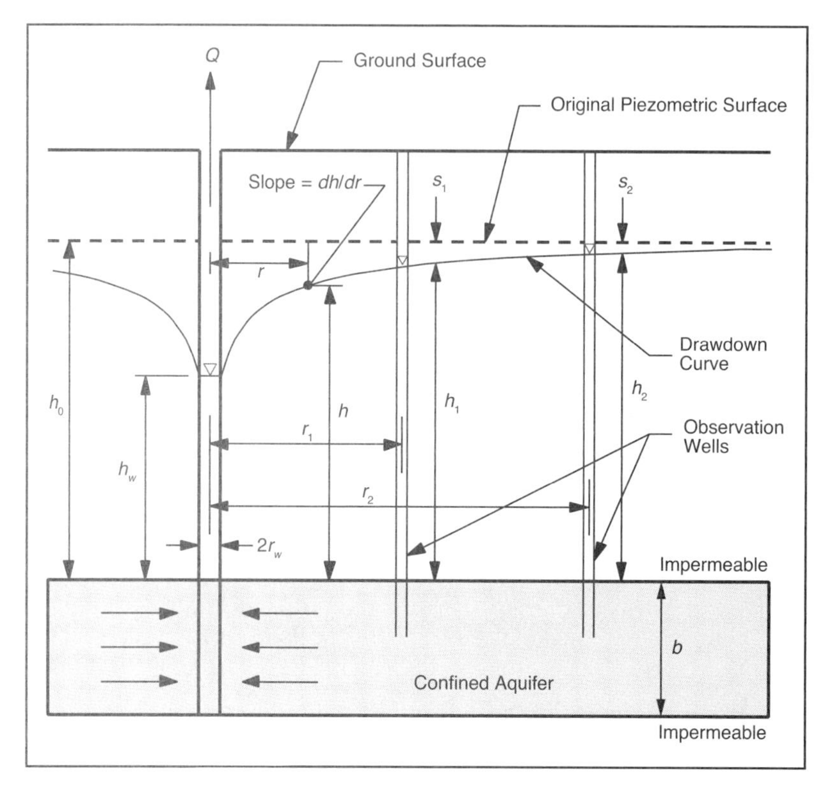

Source: *Groundwater Hydrology* by D.K. Todd. © 1980. Reprinted by permission of John Wiley and Sons, Inc.

FIGURE 3.9 Radial flow to a well penetrating an extensive confined aquifer

where

h_1, h_2 = heads of the observation wells

r_1, r_2 = distances to the observation wells

By measuring drawdown at the observation wells instead of head, the Thiem equation becomes

$$T = \frac{Q}{2\pi(s_1 - s_2)} \ln\frac{r_2}{r_1} \qquad \text{(EQ 3.14)}$$

where s_1 and s_2 are the drawdowns at the observation wells.

Equation 3.8 or 3.9 can be rewritten in terms of h_1, h_2, r_1, and r_2 for a pumping well and two observation wells to solve for the hydraulic conductivity for a confined aquifer:

$$K = \frac{Q \log(r_2/r_1)}{2.73b(h_2 - h_1)} \text{ m/d, in metric units} \qquad \text{(EQ 3.15)}$$

$$K = \frac{528Q\log(r_2/r_1)}{b(h_2 - h_1)} \text{ gpd/ft}^2\text{, in U.S. customary units} \quad \text{(EQ 3.16)}$$

An equation for the hydraulic conductivity in an unconfined aquifer can also be developed in terms of h_1, h_2, r_1, and r_2 for a pumping well and two observation wells by rewriting Equation 3.11 or 3.12:

$$K = \frac{Q\log(r_2/r_1)}{1.366(h_2^2 - h_1^2)} \text{ m/d, in metric units} \quad \text{(EQ 3.17)}$$

$$K = \frac{1{,}055Q\log(r_2/r_1)}{(h_2^2 - h_1^2)} \text{ gpd/ft}^2\text{, in U.S. customary units} \quad \text{(EQ 3.18)}$$

NOTE: If drawdown at the observation wells is measured, instead of saturated thickness, then $(s_1 - s_2)$ can be substituted for $(h_2 - h_1)$ in Equation 3.15 or 3.16. In the case of Equations 3.17 and 3.18, where drawdowns are appreciable, the heads h_1 and h_2 can be replaced by $(h_0 - s_1)$ and $(h_0 - s_2)$, respectively, where h_0 is the original saturated thickness of the aquifer.

The storage coefficient, S, cannot be determined by using the Thiem equation. Instead, the Theis nonequilibrium equation or the Cooper–Jacob approximation to the Theis equation—both of which are discussed later in this chapter—can be used.

Example. As an example of using the Thiem equilibrium equation, suppose an aquifer test has been completed on the Tall Butte Aquifer. This aquifer is composed of laminated, clean fine sand and is overlain and underlain by clayey units of relatively low permeability (confined aquifer). Drilling indicates that the aquifer averages 50 m (160 ft) thick in the vicinity of the pumped well, but preliminary geophysical work shows that the aquifer thins slightly to the west of this well. During the test, the hydraulic head of the aquifer remained above the base of the underlying clayey unit.

A fully penetrating well was pumped at a rate of 1,750 m³/d (320 gpm), and drawdowns were measured in two fully penetrating observation wells located 25 m (82 ft) and 75 m (250 ft) to the west of the pumping well. The drawdowns for the two observation wells are given in Table 3.3.

If the Thiem equilibrium equation where drawdown, s, was measured (Equation 3.14) is used to calculate the transmissivity of the aquifer, then

$$T = \frac{Q}{2\pi(s_1 - s_2)}\ln\frac{r_2}{r_1} = \frac{1{,}750}{2\pi(2.69 - 1.59)}\ln\frac{75}{25} = 278.2 \text{ m}^2\text{/d}$$
$$\cong 22{,}400 \text{ gpd/ft}$$

TABLE 3.3 Time–drawdown data for observation wells 1 and 2 (OW 1 and OW 2)

Time (days) Since Pump Started	Drawdown (meters) OW 1 (r = 25 m)	OW 2 (r = 75 m)
2.18×10^{-3}	0.12	–
3.12×10^{-3}	0.20	–
4.38×10^{-3}	0.29	–
5.63×10^{-3}	0.375	–
7.19×10^{-3}	0.48	–
9.38×10^{-3}	0.57	0.02
1.31×10^{-2}	0.70	0.05
2.06×10^{-2}	0.90	0.10
3.13×10^{-2}	1.08	0.16
5.00×10^{-2}	1.30	0.20
8.13×10^{-2}	1.50	0.41
0.125	1.75	0.65
0.194	2.01	0.90
0.313	2.25	1.15
0.500	2.50	1.42
0.625	2.69	1.59

If so desired, the hydraulic conductivity, K, can be calculated from the preceding data by using Equation 3.15.

Theis nonequilibrium equation. C.V. Theis obtained a solution for the Laplace equation based on the analogy between groundwater flow and heat conduction. This approach assumes that the well is replaced by a mathematical sink of constant strength; it imposes the following boundary conditions: for time $t = 0$, head $h = h_0$; for $t \geq 0$, $h \rightarrow h_0$ as radius $r \rightarrow \infty$. The solution is expressed as follows:

$$s = \frac{Q}{4\pi T} \int_u^\infty \frac{e^{-u}\,du}{u} \qquad \text{(EQ 3.19)}$$

where

s = drawdown

Q = constant well discharge

The term u is given by the following equation:

$$u = \frac{r^2 S}{4tT} \text{ or } \frac{r^2}{t} = \left(\frac{4T}{S}\right)u \qquad \text{(EQ 3.20)}$$

where

S = storage coefficient

T = transmissivity

The drawdown integral can be approximated by a convergent series:

$$\text{drawdown} = (s) = \frac{Q}{4\pi T}\left[-0.5772 - \ln(u) + u - \frac{u^2}{2 \cdot 2!} + \frac{u^3}{3 \cdot 3!} - \frac{u^4}{4 \cdot 4!} + \cdots\right] \quad \text{(EQ 3.21)}$$

This convergent series (Equation 3.21) is preferred over the equilibrium equation (Equation 3.13) because, for the series equation,

- A value of S can be determined,
- Only one observation well is required,
- The necessary period of pumping is generally shorter, and
- No assumption of steady-state flow is required.

The convergent series equation may be simplified to

$$s = \left(\frac{Q}{4\pi T}\right) W(u) \quad \text{(EQ 3.22)}$$

where $W(u)$, the well function, is a convenient symbolic form of the exponential integral.

To solve for s, a type curve, or plot on log-log paper, of $W(u)$ versus $1/u$ is prepared (Figure 3.10) based on tabulated values as in Table 3.4. Also prepared on log-log paper is a plot of t/r^2 versus s (Figure 3.11). The observed time–drawdown data are superimposed on the type curve—with the coordinates of the two graphs kept parallel—and are adjusted until a position is found by trial whereby most of the plotted points of the observed data (Table 3.4) fall on a segment of the type curve. Any convenient point on the overlapping graphs—generally $W(u) = 0.1$, 1, and so on, and $1/u = 1$, 10, and so on—is then selected as a match point, and the corresponding values of s and t/r^2 are determined. With values of $W(u)$, u, s, and t/r^2 thus determined, T can be obtained from Equation 3.22, and then S can be determined from Equation 3.20.*

Example. As an example, consider again the data in Table 3.3. If t/r^2 versus s for observation well 1 is plotted on five-cycle horizontally by three-cycle vertically log-log paper, a plot similar to Figure 3.12 would be developed. By the method of graphical superposition, matching the curve of Figure 3.12 with the type curve of Figure 3.10 will yield a solution to the Theis nonequilibrium equation. At the convenient match point of $W(u) = 1$ and $1/u = 1$, the corresponding values on Figure 3.12 are $t/r^2 = 3.3 \times 10^{-6}$ and $s = 0.5$. Applying Equations 3.20 and 3.22 gives

* Various computer packages are available that can greatly simplify and speed up the curve-fitting process.

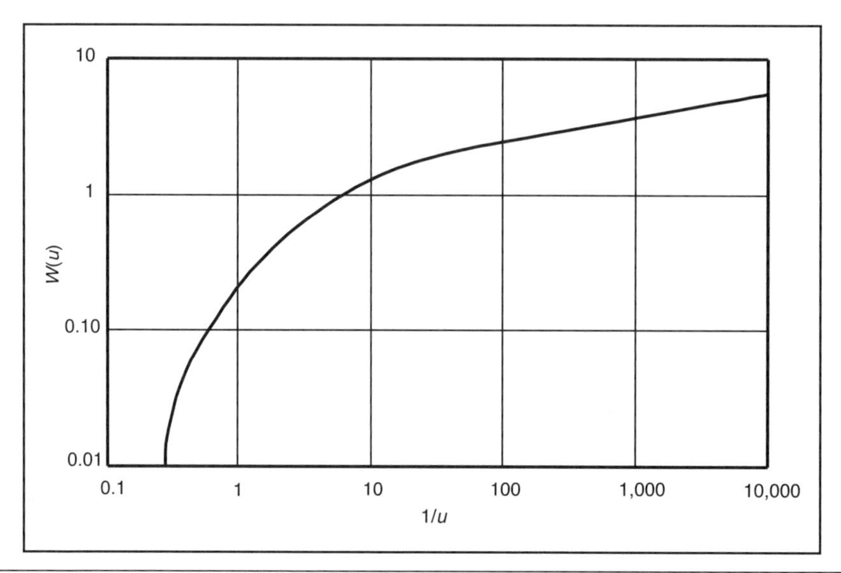

FIGURE 3.10 Type curve of the relationship between *W*(*u*) and 1/*u*

TABLE 3.4 Values of *W*(*u*) for values of *u**

Power of *u* Term	Prefix of *u* Term								
	1.0	2.0	3.0	4.0	5.0	6.0	7.0	8.0	9.0
1	0.219	0.049	0.013	0.0038	0.0011	0.00036	0.00012	0.000038	0.000012
10^{-1}	1.82	1.22	0.91	0.70	0.56	0.45	0.37	0.31	0.26
10^{-2}	4.04	3.35	2.96	2.68	2.47	2.30	2.15	2.03	1.92
10^{-3}	6.33	5.64	5.23	4.95	4.73	4.54	4.39	4.26	4.14
10^{-4}	8.63	7.94	7.53	7.25	7.02	6.84	6.69	6.55	6.44
10^{-5}	10.94	10.24†	9.84	9.55	9.33	9.14	8.99	8.86	8.74
10^{-6}	13.24	12.55	12.14	11.85	11.63	11.45	11.29	11.16	11.04
10^{-7}	15.54	14.85	14.44	14.15	13.93	13.75	13.60	13.46	13.34
10^{-8}	17.84	17.15	16.74	16.46	16.23	16.05	15.90	15.76	15.65
10^{-9}	20.15	19.45	19.05	18.76	18.54	18.35	18.20	18.07	17.95
10^{-10}	22.45	21.76	21.35	21.06	20.84	20.66	20.50	20.37	20.25
10^{-11}	24.75	24.06	23.65	23.36	23.14	22.96	22.81	22.67	22.55
10^{-12}	27.05	26.36	25.96	25.67	25.44	25.26	25.11	24.97	24.86
10^{-13}	29.36	28.66	28.26	27.97	27.75	27.56	27.41	27.28	27.16
10^{-14}	31.66	30.97	30.56	30.27	30.05	29.87	29.71	29.58	29.46
10^{-15}	33.96	33.27	32.86	32.58	32.35	32.17	32.02	31.88	31.76

Source: *Groundwater Hydrology* by D.K. Todd. © 1980. Reprinted by permission of John Wiley and Sons, Inc.

*Each *u* term is a combination of a prefix (see column heads) times a power (see first column); the corresponding *W*(*u*) term is given in the main table body. For example, for $u = 2.0 \times 10^{-5}$, $W(u) = 10.24$.

†This is read as $W(u) = 10.24$ for $u = 2.0 \times 10^{-5}$.

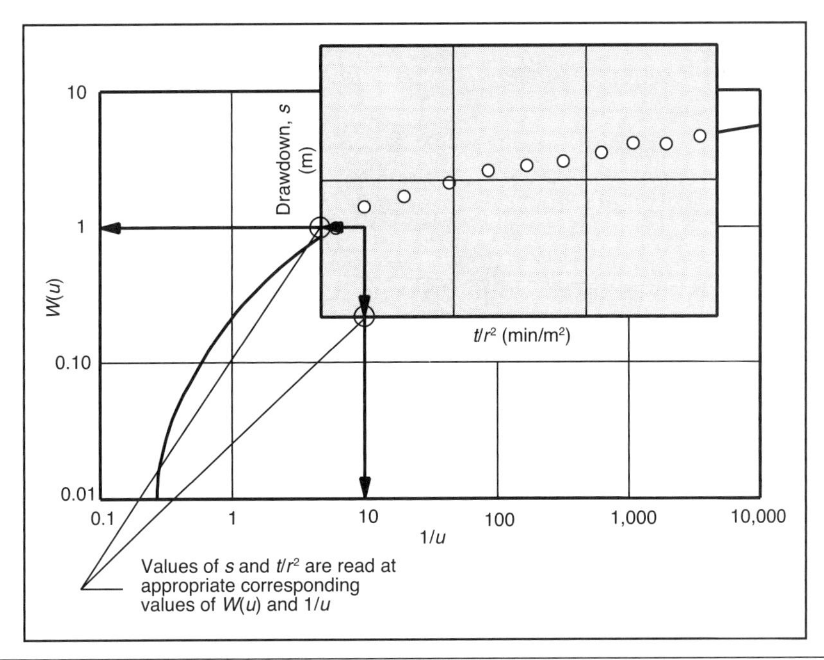

FIGURE 3.11 Method of superposition for solving the Theis nonequilibrium equation

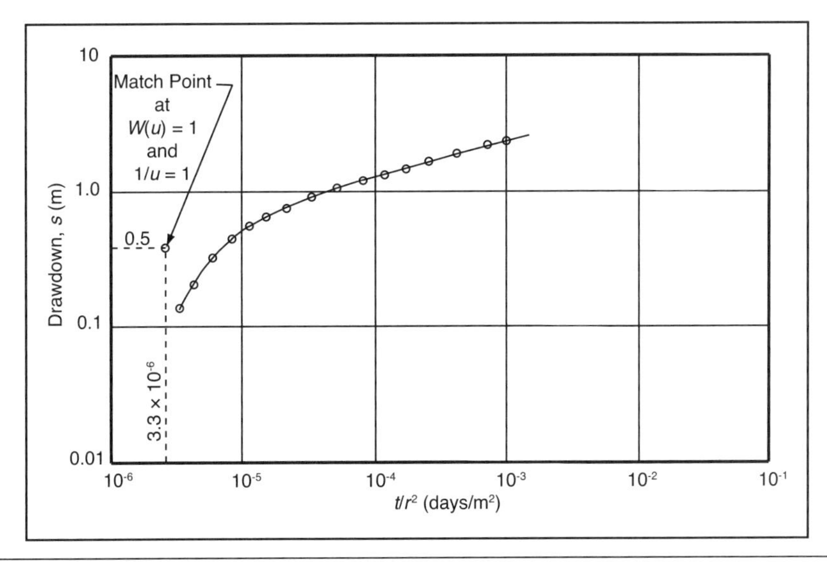

FIGURE 3.12 Drawdown, *s*, versus (*time*/r^2) for observation well 1

from Equation 3.22: $$T = \frac{QW(u)}{4\pi s} = 278.52 \text{ m}^2/\text{d}$$
$$\cong 22{,}400 \text{ gpd/ft}$$

from Equation 3.20: $$S = 4Tu\frac{t}{r^2} = 0.0037$$

Cooper–Jacob approximation for the nonequilibrium equation. H.H. Cooper and C.E. Jacob (1946) noted that for small values of r and large values of t, u is small, so that the convergent series terms become negligible after the first two terms. As a result, the drawdown can be expressed by the asymptote

$$S = \frac{Q}{4\pi T}\left(-0.5772 - \ln\frac{r^2 S}{4\pi T}\right) \quad \text{(EQ 3.23)}$$

Rewriting and changing to base-10 log, we find that this expression reduces to

$$S = \frac{2.30Q}{4\pi T}\log\frac{2.25Tt}{r^2 S} \quad \text{(EQ 3.24)}$$

Therefore, a plot of drawdown, s, versus log-scale t forms a straight line (Figure 3.13). Projecting this line to $s = 0$, where $t = t_0$, gives

$$0 = \frac{2.30Q}{4\pi T}\log\frac{2.25Tt_0}{r^2 S}$$

It follows that

$$\frac{2.25Tt_0}{r^2 S} = 1$$

resulting in

$$S = \frac{2.25Tt_0}{r^2} \quad \text{(EQ 3.25)}$$

A value for T can be obtained by noting that if $t/t_0 = 10$, then $\log(t/t_0) = 1$; therefore, if s is replaced by Δs, where Δs is the difference in drawdown per log cycle of t (i.e., per factor of 10 along the t axis), Equation 3.24 becomes

$$T = \frac{2.30Q}{4\pi \Delta s} \quad \text{(EQ 3.26)}$$

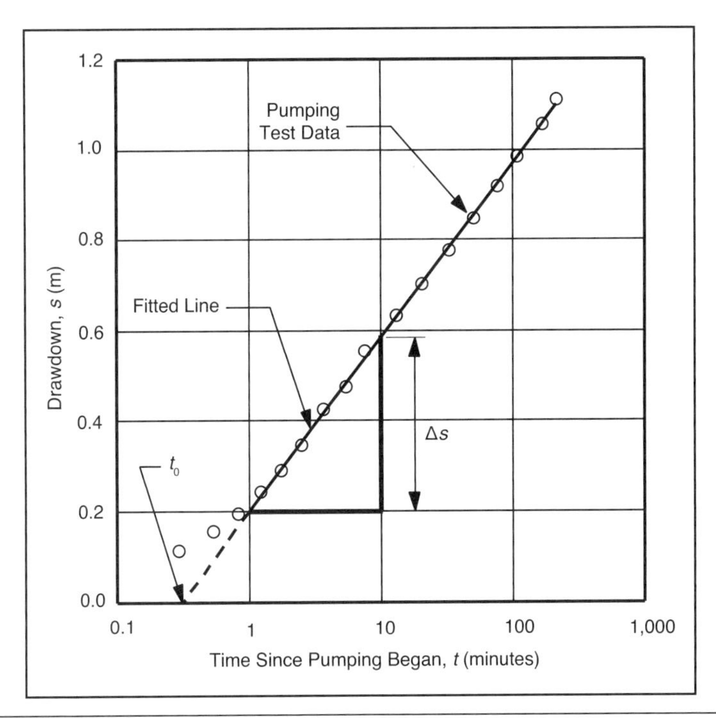

FIGURE 3.13 Cooper–Jacob method for solving the Theis nonequilibrium equation

Thus, the procedure is to solve for T first with Equation 3.26, and then solve for S using Equation 3.25. The straight-line approximation for this method should be restricted to small values of u ($u < 0.01$) to avoid large errors.

SLOPE DEWATERING

Increased stability of an excavation in soil or rock can be achieved by a number of methods, as discussed later in Chapter 9. One of the most effective methods is slope dewatering, where appropriate. Dewatering can be achieved by simply rerouting surface water away from the slope, by installing horizontal drains in the slope face, or by lowering the water table.

The removal of water from behind a rock slope in a mine or other excavation will decrease the driving forces on potentially unstable rock masses and increase the resisting forces by eliminating or decreasing the pore water pressure on the normal component across a potential failure zone. These forces, as well as the role water plays with respect to the forces, are discussed in subsequent chapters.

CONE OF DEPRESSION

Darcy's law (Equation 3.1) indicates that the velocity of flow through porous media varies directly with the hydraulic gradient. As the hydraulic gradient increases, velocity increases

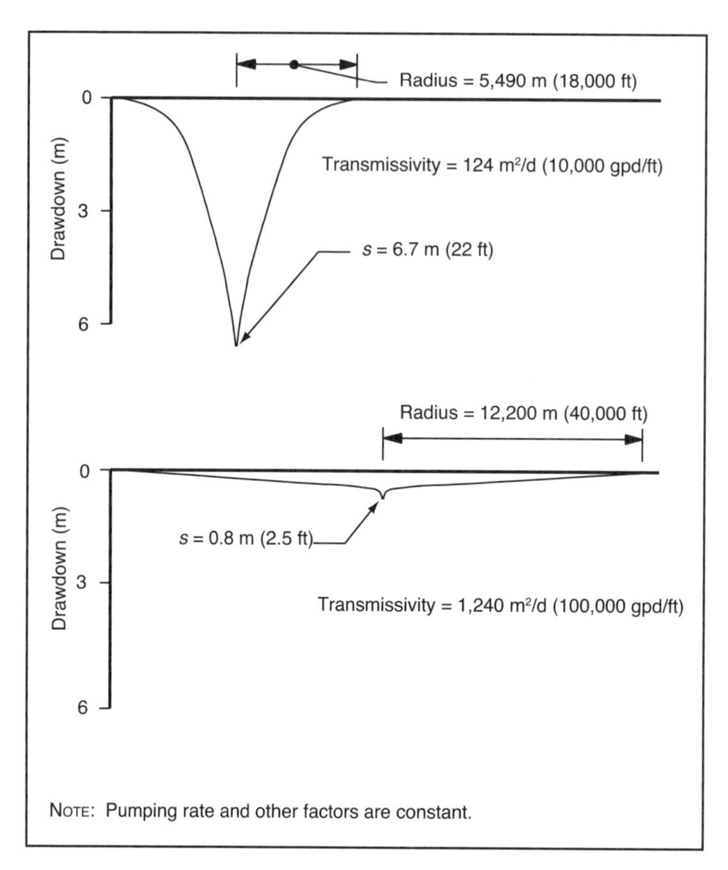

Source: Driscoll 1986

FIGURE 3.14 Effect of different coefficients of transmissibility on the shape, depth, and extent of the cone of depression

as flow converges toward a well. As a result, the lowered water surface develops a continually steeper slope toward the well. The form of this surface resembles a cone and is called the cone of depression. When pumped, all wells are surrounded by a cone of depression. Each cone differs in size and shape depending on the pumping rate, pumping duration, aquifer characteristics, slope of the water table, and recharge within the cone of depression of the well.

Figure 3.14 shows two cones of depression around pumped wells, illustrating how transmissivity of an aquifer affects the shape of the cone. In a formation with low transmissivity, the cone is deep with steep sides and has a small radius. In a formation with high transmissivity, the cone is shallow with flat sides and has a large radius. The explanation for the different cone shapes is that greater hydraulic head (i.e., more meters or feet of head) is required to move water through a less permeable formation than through a more permeable formation.

When water is pumped from a well, the initial discharge is derived from casing storage and aquifer storage immediately surrounding the well. As pumping continues, more water must be derived from aquifer storage at greater distances from the well bore. This means that the cone of depression must expand. The radius of influence of the well increases as the cone expands. Drawdown at any point also increases as the cone deepens to provide the additional head required to move the water from greater distances. The cone expands and deepens more slowly with time, however, because an increasing volume of stored water is available with each additional unit length of horizontal expansion.

The cone of depression will continue to enlarge until one or more of the following conditions is met (Driscoll 1986):

1. It intercepts enough of the flow in the aquifer to equal the pumping rate.
2. It intercepts a body of surface water from which enough additional water will enter the aquifer to equal the pumping rate when combined with all the flow toward the well.
3. Enough vertical recharge from precipitation occurs within the radius of influence to equal the pumping rate.
4. Sufficient leakage occurs through overlying or underlying formations to equal the pumping rate.

When the cone has stopped expanding because one or more of the above conditions have been met, equilibrium exists; there is no further drawdown with continued pumping. In some wells, equilibrium occurs within a few hours after pumping begins. In others, it never occurs, even though the pumping period may be extended for years.

Dewatering Equations

Under equilibrium conditions, the volume of water that a dewatering system will have to pump from an aquifer to produce a certain drawdown is given by the Thiem equations specified earlier. For an unconfined aquifer, Equations 3.11 and 3.12 may be rewritten as follows (Driscoll 1986):

$$Q = \frac{K\left(H^2 - h^2\right)}{0.733\log\left(r_c/r_w\right)} \text{ m}^3\text{/d, in metric units} \qquad \text{(EQ 3.27)}$$

$$Q = \frac{K\left(H^2 - h^2\right)}{1{,}055\log\left(r_c/r_w\right)} \text{ gpm, in U.S. customary units} \qquad \text{(EQ 3.28)}$$

where

Q = discharge, in m^3/d (gpm)
K = hydraulic conductivity, in m/d (gpd/ft^2)
H = saturated thickness of the aquifer before pumping, in m (ft)
h = depth of the water in the well while pumping, in m (ft)

r_c = radius of the cone of depression, in m (ft)
r_w = radius of the well, in m (ft)

The depth of the water table, h, at any distance R from the well (R must be at least $1.5H$)* is then given by

$$h=\sqrt{H^2-\frac{0.733Q\log\left(r_c/r_w\right)}{K}}\ \text{m}=\sqrt{H^2-\frac{1{,}055Q\log\left(r_c/r_w\right)}{K}}\ \text{ft} \qquad \text{(EQ 3.29)}$$

For a confined aquifer, Equations 3.8 and 3.9 may be rewritten as follows (Driscoll 1986):

$$Q=\frac{Kb(H-h)}{0.366\log\left(r_c/r_w\right)}\ \text{m}^3/\text{d, in metric units}$$

$$Q=\frac{Kb(H-h)}{528\log\left(r_c/r_w\right)}\ \text{gpm, in U.S. customary units}$$

where

b = thickness of the aquifer, in m (ft)
H = distance from the static water level to the bottom of the aquifer, in m (ft)

(The other variables are the same as for the unconfined aquifer equations.) The drawdown, s, at any point within the cone is then given by

$$s=\frac{0.366Q\log\left(r_c/r_w\right)}{Kb}\ \text{m}=\frac{528Q\log\left(r_c/r_w\right)}{Kb}\ \text{ft}$$

Time–Drawdown and Distance–Drawdown Graphs

Figure 3.15 shows a pumping well and three observation wells (OWs). Figure 3.16 is a plot of drawdown versus time since the pump started at OW 1; such a plot is called a time–drawdown curve. In addition to its use for calculating the aquifer constants, a time–drawdown graph provides a graphical means of predicting future drawdowns at a constant pumping rate. The straight line on the plot of time (plotted as the abscissa on a log scale) versus drawdown (plotted as the ordinate on a linear scale) for an observation well at some distance r from the pumping well can be extended to estimate the drawdown that would

* Basic assumptions of the equilibrium equations provide that the test well is fully penetrating, that the well has a 100% open hole (or screen), and that flow is horizontal. In many tests these conditions are not met; however, the distribution of flow to a well approaches that of the assumed horizontal condition at a distance from the well equal to approximately 1.5 times the thickness of the aquifer. Therefore, to minimize the effects of convergent flow on test results, the nearest observation well should be located at least 1.5 times the aquifer thickness from the pumping well unless a large aquifer thickness provides for distances that are unreasonably large (Bureau of Reclamation 1995).

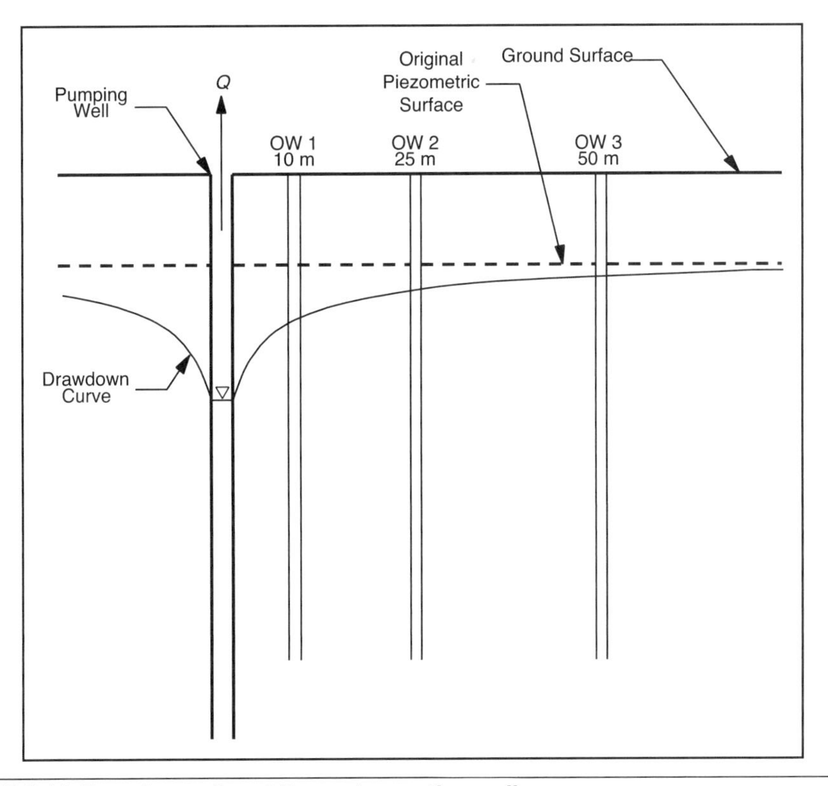

FIGURE 3.15 Pumping well and three observation wells

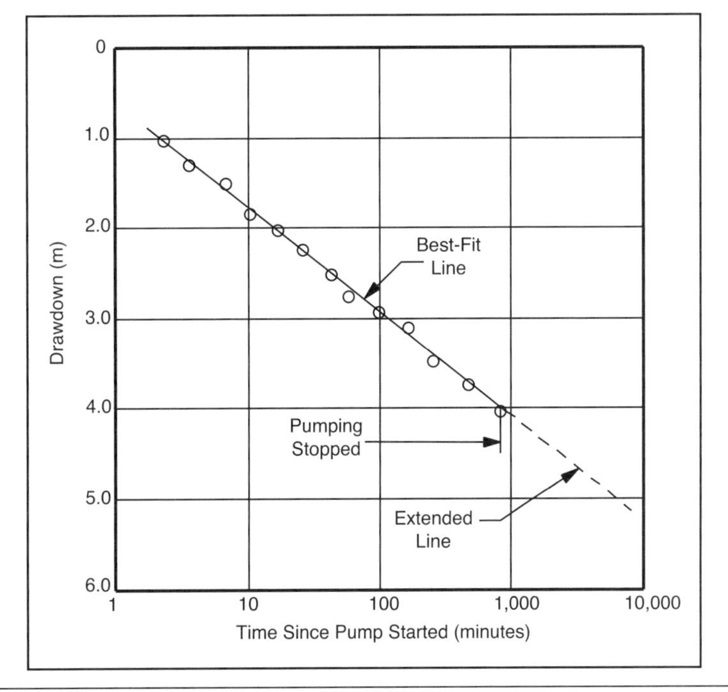

FIGURE 3.16 Time–drawdown curve for OW 1

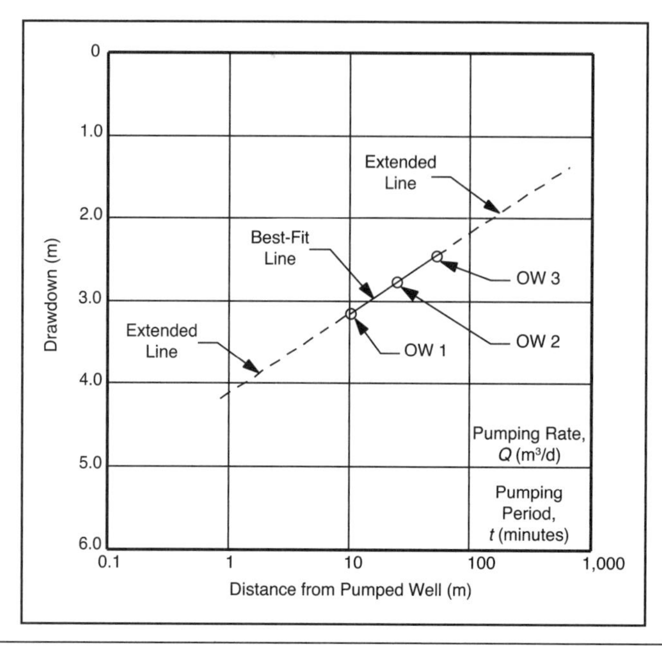

FIGURE 3.17 Drawdown versus distance from the pumped well for the three observation wells

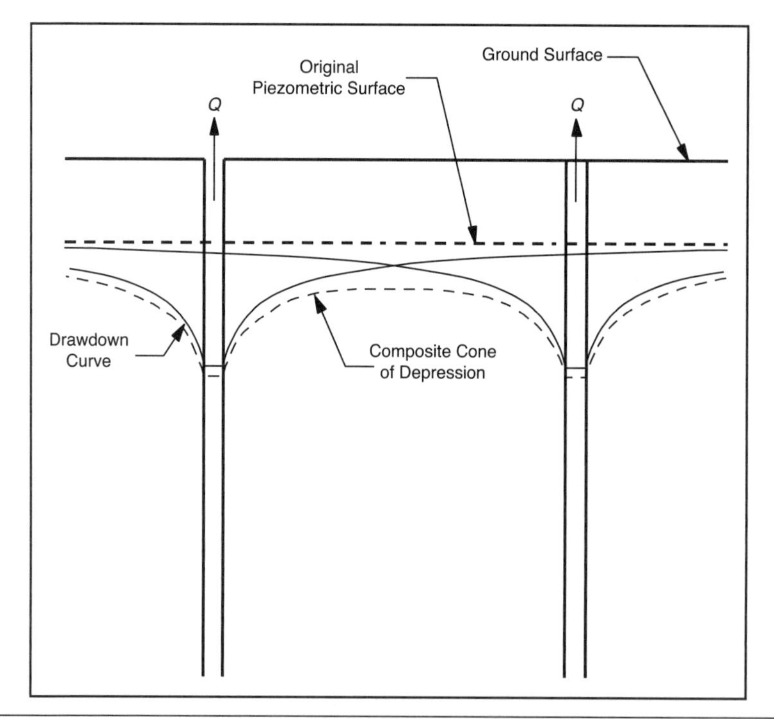

FIGURE 3.18 Composite cone of depression from two pumping wells based on the principle of superposition

occur in the observation well after any period of continuous pumping at a constant rate. The transmissivity and storage coefficient can be calculated by using the Cooper–Jacob approximation for the Theis nonequilibrium equation.

A plot of distance from the pumping well versus drawdown on semilog paper for the three observation wells gives a distance–drawdown curve similar to that shown in Figure 3.17. The distance–drawdown curve can be used to estimate the cumulative drawdown from a well field by the principle of superposition at any point when the wells are all pumping at the same rate. The *principle of superposition* states that "If the transmissivity and storativity [i.e., storage coefficient] of an ideal aquifer and the yield and duration of discharge or recharge of two or more wells are known, then the combined drawdown or buildup at any point within their interfering area of influence may be estimated by adding algebraically the component of drawdown of each well" (Bureau of Reclamation 1995). This concept is shown in Figure 3.18.

The principle of superposition implies that, within the zone of dewatering, the aquifer attributes (transmissivity, storage coefficient, thickness, etc.) and the well attributes (pumping rate, pumping period, etc.) do not vary. For complex dewatering situations—where the attributes of the pumping wells, the aquifer, or both vary—the use of a computer program to design the system is recommended. The use of computer-aided design software for dewatering systems allows the designer to rapidly adjust the proposed number of wells, the yield, and the spacing to produce an optimum configuration at the lowest cost.

REFERENCES

Bureau of Reclamation, Ground Water and Drainage Group. 1995. *Ground Water Manual*, 2nd ed. Final Report PB96-207394. Denver, CO: U.S. Department of the Interior, Bureau of Reclamation, Technical Service Center.

Coates, D.F., ed. 1977. *Pit Slope Manual*. Report 77-13. Ottawa, ON: Canada Centre for Mineral and Energy Technology.

Cooper, H.H., Jr., and C.E. Jacob. 1946. A generalized graphical method for evaluating formation constants and summarizing well-field history. *Transactions of the American Geophysical Union* 27:526–534.

Driscoll, F.G. 1986. *Groundwater and Wells*, 2nd ed. St. Paul, MN: Johnson Division.

Freeze, R.A., and J.A. Cherry. 1979. *Groundwater*. Englewood Cliffs, NJ: Prentice-Hall.

Hunt, R.E. 1984. *Geotechnical Engineering Investigation Manual*. New York: McGraw-Hill.

Morgenstern, N.R. 1971. The influence of groundwater on stability. In *Proceedings of the First International Conference on Stability in Open Pit Mining*. Edited by C.O. Brawner and V. Milligan. New York: SME-AIME.

Todd, D.K. 1980. *Groundwater Hydrology*, 2nd ed. New York: John Wiley and Sons.

CHAPTER 4

The Rockfall Hazard Rating System*†

The rockfall hazard rating system (RHRS) is included in this text because it is a useful tool for rating a rock slope for potential instability due to rockfalls. It is primarily used by various departments of transportation (DOTs) in the United States to rate and list highway rock-cut sections that may need stabilization. It has potential for adaptation by other entities (e.g., mine operations, quarry operations) for rating rock highwall slopes. The discussion in this chapter is based primarily on the works of Ritchie (1963), Pierson et al. (1990), and Pierson and Van Vickle (1993), as well as Buss et al. (1995).

SIGNIFICANCE OF THE ROCKFALL PROBLEM

Rockfall is the movement of rock from a slope that is so steep that rock continues to move down the slope (Pierson et al. 2001). Rockfalls pose a significant hazard for mining operations and in road cuts for highways and railways in mountainous terrain. Whereas rockfalls are often neither as dramatic nor pose the same magnitude of economic risk as large-scale failures, they do, however, tend to kill a significant number of people each year. An example of a Mine Safety and Health Administration (MSHA) Fatalgram is shown in Figure 4.1 for an individual killed by falling rock.

Other examples of massive rockfalls covering roads are shown in Figures 1.9 and 4.2. Rockfall can be initiated by any number of factors, including unfavorable rock discontinuity orientation; freeze–thaw processes, especially in the spring time; blasting practices, especially overbreak or backbreak; water—rainfall, runoff, pore pressure changes,

---

* Permission to present information in this chapter has been granted from the Federal Highway Administration for SA-93-057, *Rockfall Hazard Rating System: Participant's Manual* (FHWA 1993), and from L.A. Pierson, S.A. Davis, and R. Van Vickle (1990).

† The units used in this chapter follow the U.S. customary standard as opposed to the SI (International System of Units) standard, given that the rockfall hazard rating system is inherently based on U.S. customary units.

METAL/NONMETAL MINE FATALITY - On January 7, 2013, a 49-year old assistant plant manager with 30 years of experience was injured at a crushed stone operation. The victim was working on a lift, taking samples from a highwall, when a large rock fell and struck him. He was hospitalized and died on January 19, 2013.

Courtesy of Mine Safety and Health Administration

FIGURE 4.1 MSHA Fatalgram of person killed by rockfall at a quarry

groundwater elevation changes, and so on; chemical degradation or weathering of rock or clay material; and root growth and tree levering (Brawner 1994).

In 1963, Arthur M. Ritchie, chief geologist with the Washington State Department of Highways, published his sentinel study on rockfall titled *Evaluation of Rockfall and Its Control.* The purpose of Ritchie's study was to identify the characteristics of rockfall motion relative to a slope's configuration and height, and to determine the expected impact distance of a rockfall from the base of the slope. He also investigated how to effectively stop a falling rock that had considerable angular momentum once it landed in the catchment area (Ritchie 1963; Pierson et al. 2001). Ritchie determined that the movement of a rock down a slope may be free falling, bouncing, rolling, or sliding, depending on the slope angle (Figure 4.3).

In addition, Ritchie prepared an empirical design table of maximum rock catchment area width and depth, based on the slope height and angle, which was later adapted into a design chart (Figure 4.4) and published in *Rock Slopes: Design, Excavation, Stabilization* (Hoek and Bray 1989). More than 40 years later, the chart (or modified version) is still used by many state and local transportation agencies to dimension ditch-catchment areas (Pierson et al. 2001), although design standards for catchment ditches have changed.

ORIGINS OF THE ROCKFALL HAZARD RATING SYSTEM

The RHRS originated in Oregon in 1984. At that time, the Oregon DOT began to discuss the need for a system for rating rockfall hazards. As in many other states, the method the Oregon DOT used to identify rockfall projects prior to development of the RHRS relied

FIGURE 4.2 Rockfall from a steep hillside

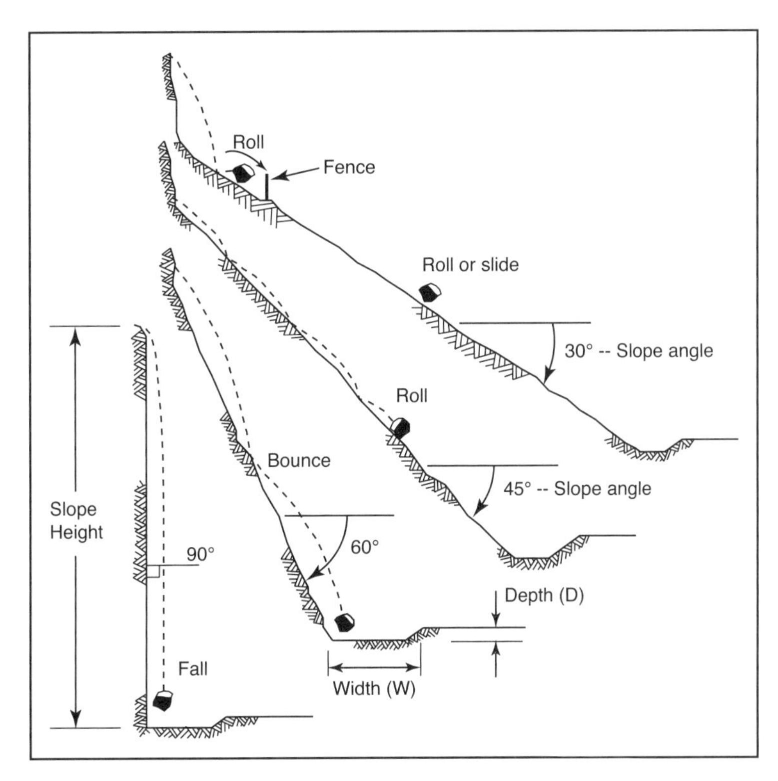

Source: Ritchie 1963

FIGURE 4.3 Rockfall travel modes

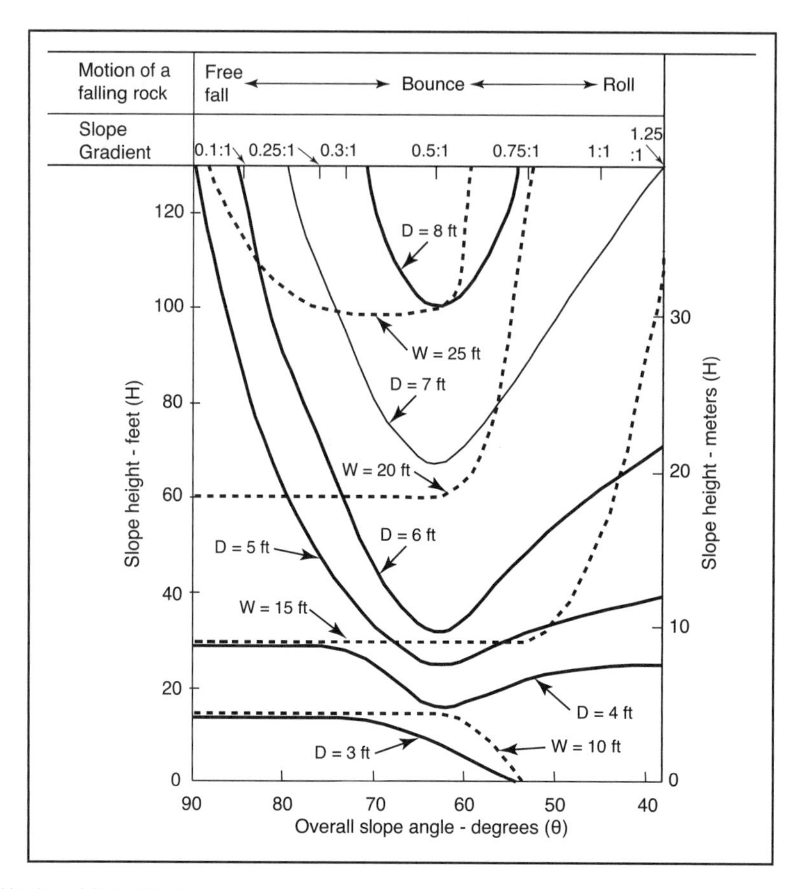

Source: Hoek and Bray 1989

FIGURE 4.4 Ritchie's rockfall catch ditch design chart

principally on the relative ratio of (1) the costs associated with the accident history plus the annual maintenance costs to (2) the estimated cost of reconstructing the roadway section. By its nature, this cost-comparison approach is a reactive prioritization technique. It places the emphasis on those roadway sections where an accident due to rockfall has already taken place—which may not reflect the potential for future rockfall events. The annual maintenance cost along a section where rockfall occurs generally represents the cost to clean out the catch ditch and to patrol the highway for rock debris on the road. However, if an adequately designed catch ditch performs well (no rock on the roadway) but needs regular cleaning, the maintenance cost may be high while the hazard to motorists is low. A detailed analysis of rockfall hazard potential for the section, in conjunction with a review of the various rockfall mitigation techniques available, may justify an improved roadway design to lower the maintenance cost. This situation would indicate that these two items—maintenance cost and hazard level—are not sufficient by themselves to develop a rockfall priority list. In addition, this technique relies on information reported by highway maintenance crews, law enforcement personnel, emergency response personnel, or the general public.

This group as a whole is not adequately trained to systematically document or evaluate rockfall events.

The Oregon DOT's management, as well as its legal counsel, believed that a proactive method should be adopted to provide a more reasonable way to prioritize rockfall projects and allocate available repair funds. To be effective, the program would include an inspection of all rock slopes along the state's highway system to identify where rockfall would most likely affect the roadway. Once identified, these sections would be rated by trained personnel to determine which were the most hazardous. To accomplish this task, an RHRS was needed.

To develop a relative rating system, the Oregon DOT initially conducted a review of available literature on the subject. As part of this search, a paper written by C.O. Brawner and D. Wyllie (1975) was reviewed. It contained rating criteria and a scoring method that grouped rockfall sections into categories A, B, C, D, or E based on the potential for a rockfall event and the expected effect. In a subsequent paper, Wyllie (1987) outlined a more detailed procedure for prioritizing rockfall sites. Wyllie's approach included specific categories for evaluation that were assessed on the basis of an exponential scoring system. This system became the prototype for Oregon's RHRS. The state adopted the rating sheet format and the exponential scoring system.

OVERVIEW

The RHRS uses a process that allows a state transportation agency to rationally evaluate and manage the rock slopes along its highway system. However, the system requires greater commitment and focus on the rock slope issue on the agency's part than is generally the case. This commitment is in the form of time and money to complete the initial survey and to update it periodically, then to develop remedial programs aimed at minimizing the risk at the worst sites.

The system contains six main features:

1. A uniform method for slope inventory: creating a geographic database of rockfall locations.
2. A preliminary rating for all slopes: grouping the rockfall sites into three broad, manageably sized categories (A, B, and C).
3. The detailed rating of all hazardous slopes: prioritizing the identified rockfall sites from least to most hazardous.
4. A preliminary design and cost estimate for the most serious sections: adding remediation information to the rockfall database.
5. Project identification and development: advancing rockfall correction projects toward construction.
6. Annual review and update: maintaining the rockfall database.

The RHRS contains two phases of inspection: the initial assessment phase (preliminary rating) as part of a statewide slope survey and the detailed rating phase. This format is the most efficient way to implement the RHRS in situations where an agency has responsibility for a large number of slopes with a broad range of rockfall potential.

SLOPE SURVEY AND PRELIMINARY RATING

The slope survey and preliminary rating represent the initial assessment phase of the RHRS.

Slope Survey

The purpose of the slope survey is to gather specific information on where rockfall sites are located. This is an important first step because only by going through the process can an agency accurately determine the number and location of its rockfall sites.

Accurate delineation of rockfall sections is important. For the purpose of the RHRS, a *rockfall section* is defined as "any uninterrupted slope along a highway where the level and occurring mode of rockfall are the same" (Pierson et al. 1990). This definition raises an important issue: Even though an extended section along a roadway may be uninterrupted, the level (i.e., frequency and/or amount) and mode (i.e., reason for the rockfall) for that section may vary. Therefore, it is not always a good idea to group separate cut slopes into one long section. Instead, because of these variations, the delineating of additional sections—each with its own level and mode—may be necessary.

Two people should conduct the slope survey: (1) a rater, who will perform the preliminary rating and, if needed, the detailed rating and (2) an agency person familiar with the rockfall history and associated maintenance along the rockfall section. The upper portion of the RHRS field data sheet (Figure 4.5) should be completed during the slope survey and preliminary rating phase. The slope survey provides an opportunity to document the historical rockfall activity. The following information should be covered and noted in the "Comments" section of the RHRS field data sheet:

1. Location of rockfall activity.
2. Estimated frequency of rockfall activity.
3. Time of year when activity is highest.
4. Estimated size/quantity of rockfall per event.
5. Physical description of rockfall material.
6. Where rockfalls come to rest.
7. History of reported accidents (note that not all accidents are reported).
8. Opinion as to rockfall cause.
9. Frequency of ditch cleaning/road patrol.
10. Cost of maintenance response.

RHRS FIELD DATA SHEET

Highway: Region:

Highway #______ Beginning M.P._____ L/R Ending M.P.________
County #_______ Date_____________ New Rated By__________
Class: A B C ADT____________ Update Speed Limit________

CATEGORY	REMARKS	CATEGORY SCORE
Slope Height _______ ft. / /		Slope Height _______
Ditch Effectiveness: G M L N		Ditch Effect _______
Average Vehicle Risk %		AVR _______
Sight Distance ________ft Percent Decision Site Distance_____%		Sight Distance _______
RoadwayWidth __________ ft		Roadway Width _______
GEOLOGIC CHARACTER		GEOLOGIC CHARACTER
CASE 1		CASE 1
Structural Condition: D C/F R A		Struct. Cond. _______
Rock Friction: R I U P C - S		Rock Friction _______
CASE 2		CASE 2
Differential Erosion Features: F O N M		Diff. Er. Features _______
Difference in Erosion Rates: S M L E		Diff. Er. Rates _______
Block Size/Volume_______ft/yd³		Block Size _______
Climate Precipitation: L M H Freezing Period: N S L Water on Slope N I C		Climate _______
Rockfall History: F O M C		Rockfall History _______
COMMENTS		**Total Score:**

Source: FHWA 1993

FIGURE 4.5 Sample RHRS field data sheet

Preliminary Rating

The purpose of the preliminary rating is to group the rockfall sections inspected during the slope inventory into three broad, more manageably sized categories. Without this step, many additional hours would be spent applying the detailed rating at sites with only a low-to-moderate chance of ever producing a hazardous condition.

The preliminary rating is a subjective evaluation of rockfall potential based on past rockfall activity. It requires experienced, insightful personnel to make valid judgments. The criteria used in the preliminary rating are shown in Table 4.1.

Estimated potential for rockfall on roadway. The criterion of "estimated potential for rockfall on roadway" is the controlling element of the preliminary rating. If there is any doubt as to whether a slope should be rated A or B, the *potential for rockfall should be*

TABLE 4.1 Preliminary rating system

Criteria	Class A	Class B	Class C
Estimated potential for rockfall on roadway	High	Moderate	Low
Historical rockfall activity	High	Moderate	Low

Source: FHWA 1993

considered first, supplemented by the historical activity. When the potential for rockfall on the roadway is being rated, the following factors should be considered:

- Estimated size of material
- Estimated quantity of material per event
- Amount of material available
- Ditch effectiveness

Historical rockfall activity. When historical rockfall activity is being rated, the following factors should be considered:

- Frequency of rockfall on highway
- Quantity of material
- Size of material
- Frequency of ditch/catch-bench cleanout

The ratings. A C rating can mean one of two things: Either a rockfall at this site is unlikely or, if a rock should fall, it is unlikely to reach the roadway. In other words, the risk of a hazardous situation occurring is low to nonexistent. If a section receives a B rating, the risk ranges from moderate to low. For A-rated sections, the risk ranges from high to moderate.

All rockfall sections that receive an A rating should be photographed and evaluated with the detailed rating system. This approach will economize the effort and ensure that such effort is directed toward the most critical areas. The B-rated sections should be evaluated as time and funding allow. The C-rated sections will receive no further attention and therefore are not included in the rockfall hazard rating database.

DETAILED RATING PHASE

The detailed rating phase represents the second main feature of the RHRS inspection. It encompasses 12 categories that—when evaluated, scored, and totaled—allow the rock slopes to be rated from least to most hazardous. Slopes with higher total scores represent a greater hazard.

The categories represent the significant elements of a rockfall section that contribute to the overall hazard. An exponential scoring system, from 1 to 100 points, is used to allow the rater greater flexibility in evaluating the relative impact of conditions that are extremely

TABLE 4.2 Exponent formulas for detailed rating parameters

Parameter	Formula for Exponent Value, x
Slope height	$x = \frac{\text{slope height (ft)}}{25}$
Average vehicle risk	$x = \frac{\text{\% time}}{25}$
Sight distance	$x = \frac{120 - (\text{\% decision sight distance})}{20}$
Roadway width	$x = \frac{52 - \text{roadway width (ft)}}{8}$
Block size	$x = \text{block size (ft)}$
Volume	$x = \frac{\text{volume (ft}^3\text{)}}{3}$

Source: FHWA 1993

variable. For some categories, exact scores can be determined by calculating the value of the exponent x in the function $y = 3^x$. The formulas that yield the exponent values are included in Table 4.2.

The 12 categories are discussed in detail in the following sections. For each of the categories, a simplified set of ratings and corresponding descriptions is given, with scores increasing exponentially from 3 to 81 points. For simplicity, Tables 4.2 and 4.3 can be used (with appropriate interpolation) instead of calculating the value of the function y. Where desired, for certain measurable categories, the exact value can be calculated by determining the exponent from Table 4.2 and then calculating the function $y = 3^x$. If the calculated value of y is greater than 100, then it should be rounded to a score of 100. To estimate scores from the tables greater than 81 but less than 100, a "worst case" for the category under consideration should be estimated and set to a point score of 100. Next, the situation under consideration should be positioned with respect to the 81-point score and the 100-point score, and then the appropriate point score can be estimated. For some categories, a score greater than 81 cannot be estimated (e.g., for the "average vehicle risk" category—because 81 points signifies that the vehicle is at risk 100% of the time). The criteria for receiving various scores for the 12 detailed rating categories are summarized in Table 4.3.

Slope Height

The "slope height" category evaluates the risk associated with the height of a slope. Rocks on high slopes have more potential energy than rocks on lower slopes; thus, they present a greater hazard and receive a higher rating. Slope height is measured up to the highest point

TABLE 4.3 Criteria for receiving listed scores for the 12 detailed rating categories

Category	Rating Score 3 Points	9 Points	27 Points	81 Points
Slope height	25 ft	50 ft	75 ft	100 ft
Ditch effectiveness	Good catchment	Moderate catchment	Limited catchment	No catchment
Average vehicle risk	25% of the time	50% of the time	75% of the time	100% of the time
Percent of decision sight distance	Adequate sight distance; 100% of low design value	Moderate sight distance; 80% of low design value	Limited sight distance; 60% of low design value	Very limited sight distance; 40% of low design value
Roadway width, including paved shoulders	44 ft	36 ft	28 ft	20 ft
Geologic character, case 1: structural condition	Discontinuous joints; favorable orientation	Discontinuous joints; random orientation	Discontinuous joints; adverse orientation	Continuous joints; adverse orientation
Geologic character, case 1: rock friction	Rough, irregular	Undulating	Planar	Clay infilling or slickensided
Geologic character, case 2: differential erosion features	Few differential erosion features	Occasional differential erosion features	Many differential erosion features	Major differential erosion features
Geologic character, case 2: difference in erosion rates	Small difference	Moderate difference	Large difference	Extreme difference
Block size or quantity of rockfall per event	1 ft or 3 yd^3	2 ft or 6 yd^3	3 ft or 9 yd^3	4 ft or 12 yd^3
Climate and presence of water on slope	Low to moderate precipitation; no freezing periods; no water on slope	Moderate precipitation or short freezing periods, or intermittent water on slope	High precipitation or long freezing periods, or continual water on slope	High precipitation and long freezing periods, or continual water on slope and long freezing periods
Rockfall history	Few falls	Occasional falls	Many falls	Constant falls

Source: FHWA 1993

from which rockfall is expected. If rocks are coming from the natural slope above the cut, the cut height plus the additional slope height (vertical distance) should be used.

The slope height can be obtained by using the following relationship, shown graphically in Figure 4.6:

$$\text{total slope height} = \frac{X \sin\delta \cdot \sin\varepsilon}{\sin(\delta - \varepsilon)} + \text{HI}$$

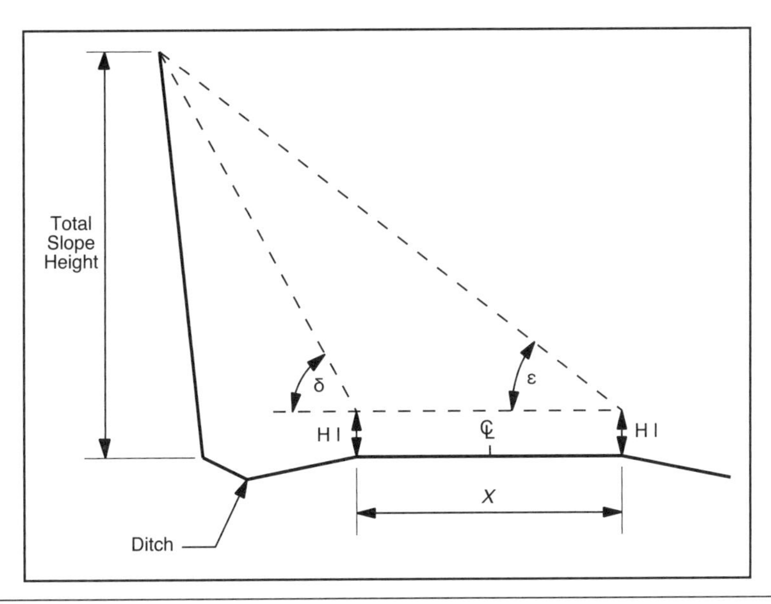

FIGURE 4.6 Diagram for determining slope height

where

X = distance between points where angle is measured

δ = vertical angle at inner measuring station to the highest point from which rockfall is expected

ε = vertical angle at outer measuring station to the highest point from which rockfall is expected

HI = height of the instrument

The distance X should be from edge of road to edge of road or from edge of pavement to edge of pavement; the angles δ and ε can be measured by using an appropriate instrument, such as a transit or geologic compass. Significant error can be introduced by careless measurement of δ and/or ε. The rating criteria and associated scores for the "slope height" category are shown in Table 4.3.

Ditch Effectiveness

The effectiveness of a ditch is measured by its ability to restrict rockfall from reaching the roadway. Factors such as the following should be considered when the ditch effectiveness is being estimated:

- Slope height and angle
- Ditch width, depth, and shape
- Anticipated block size and quantity of rockfall
- Impact of slope irregularities (features from which rocks may be launched out onto the roadway [i.e., ledges, overhangs, and protruding rock faces]) on falling rocks

TABLE 4.4 Decision sight distance values for various speed limits

Posted Speed Limit, mph	Decision Sight Distance, ft
25	375
30	450
35	525
40	600
45	675
50	750
55	875
60	1,000
65	1,050

Adapted from AASHTO 2011

Evaluating the effect of slope irregularities is especially important because these abnormalities can completely offset the benefits that the ditch may offer. It is important to observe whether any slope irregularities exist that may launch falling rocks onto or across the roadway area. The rating criteria and associated scores for the "ditch effectiveness" category are shown in Table 4.3.

Average Vehicle Risk

The "average vehicle risk" (AVR) category measures the risk associated with the percentage of time a vehicle is present in the rockfall hazard zone. This percentage is obtained by using a formula based on slope length, average daily traffic (ADT), and the posted speed limit through the hazard zone:

$$\text{AVR\%} = \frac{\text{ADT (cars/d)} \times \text{slope length (mi)} \times 100\%}{\text{posted speed limit (mph)}}$$

The preceding equation calculates the average percentage of time that a vehicle is present within the rockfall section. A rating of 100% means that, on average, a vehicle can be expected to be within the hazard section 100% of the time. Care should be taken to measure only the length of a slope where rockfall is a problem—overestimated lengths will strongly skew the formula results. It is possible to obtain values greater than 100% with the formula. When this happens, it means that more than one vehicle is present within the rockfall section at any one time. An AVR% score greater than 100% yields an AVR score of 100. The rating criteria and associated scores for the AVR category are shown in Table 4.3.

Percent of Decision Sight Distance

The decision sight distance (DSD) is used to determine the length of roadway, in feet, that a driver must have to make a complex or instantaneous decision. Sight distance, as prescribed by the American Association of State Highway and Transportation Officials (AASHTO), is the shortest distance at which a 6-in.-high object on the road is continuously visible to a driver. The DSD is the length of roadway, in feet, required by a driver to perceive a problem and then bring the vehicle to a stop.

Throughout a rockfall section, the sight distance can change appreciably. Horizontal and vertical highway curves, along with obstructions such as rock outcrops and roadside vegetation, can severely limit a driver's ability to see an object on the road.

The percent of DSD is calculated from the following formula:

$$\%\text{DSD} = \frac{\text{actual sight distance}}{\text{decision sight distance}} \times 100\%$$

where

actual sight distance = shortest distance at which a 6-in.-high object lying on the roadway will be visible for an eye height of 3.5 ft above the road surface when movement is taking place toward the object

decision sight distance = the AASHTO required sight distance at the posted speed limit (see Table 4.4)

Once the %DSD has been calculated, the RHRS score can be determined from Table 4.3. Table 3.3 in AASHTO (2011), from which the DSD values herein were derived, contains five categories for the appropriate DSD depending on whether the road is a rural highway, urban highway, suburban highway, or a combination thereof. The "low design values" of Table 4.3 are approximately equal to the lowest value in category C (rural roads) of the AASHTO (2011) categories.

Roadway Width

The roadway width is measured perpendicular to the highway centerline from edge of pavement to edge of pavement or, if the road is unpaved, from edge of road to edge of road. The unpaved shoulders are not included in the measurement. On divided highways, only the portion available to the driver is measured. The rating criteria and associated scores for the "roadway width" category are shown in Table 4.3.

Geologic Character

Four of the RHRS detailed rating categories are based on geologic character. There are two categories for each of the two cases considered. Case 1 involves rockfalls that result from discontinuities in the rock mass. Case 2 involves rockfalls due to erosion.

Case 1, structural condition. The "case 1, structural condition" category considers discontinuity orientation and type. The term *discontinuity*, as used here, implies a length greater than 10 ft. Types of discontinuities may include joints, faults, bedding planes, and shear structures. The presence of filling in the discontinuity (e.g., fault gouge) and water pressure should also be considered.

Rocks with numerous discontinuities are more prone to rockfall than is massive rock. Discontinuities oriented adversely with respect to the slope may make one or more types of rock failure modes kinematically possible. The rating criteria and associated scores for this category are shown in Table 4.3.

Case 1, rock friction. The rock friction parameter directly affects the potential for one block to move relative to another. Friction along a joint, bedding plane, fault, or other discontinuity is governed by the first- and second-order roughness of the discontinuity surface. First-order (macro) roughness is the degree of undulation of the surface; second-order (micro) roughness is the texture of the surface. Rockfall potential is greater in areas where discontinuities contain highly weathered or hydrothermally altered products; where movement has occurred, causing slickensides or fault gouge to form; where open joints dominate the slope; or where joints are water filled. The rating criteria and associated scores for this category are shown in Table 4.3.

Case 2, differential erosion features. The "differential erosion features" category is used for slopes where differential erosion or oversteepening is the dominant condition that leads to rockfall. Erosion features include oversteepened slopes, unsupported rock units (overhangs), or exposed resistant rocks on a slope that may eventually lead to a rockfall event.

For this category, rockfall is commonly caused by erosion that leads to a loss of support, either locally or throughout a slope. The types of slopes that may be susceptible to this condition are

- Layered units containing more easily erodible units that undermine more durable rock;
- Talus slopes;
- Highly variable units, such as conglomerates and mudflows that weather differentially, allowing resistant rocks and blocks to fall; and
- Rock/soil slopes that weather, allowing rocks to fall as the soil matrix material is eroded.

The rating criteria and associated scores for this category are shown in Table 4.3.

Case 2, difference in erosion rates. The rate of erosion on a slope directly relates to the potential for a future rockfall event. As erosion progresses, unsupported or oversteepened slope conditions develop. The impact of the common physical and chemical erosion processes, as well as the effects of human actions, should be considered. The degree of hazard and the score for this category should reflect how quickly erosion is occurring; the size of rocks, blocks, or units being exposed; the frequency of rockfall events; and the amount of material being released during an event. The rating criteria and associated scores for this category are shown in Table 4.3.

Block Size or Quantity of Rockfall per Event

The measurement in the "block size or quantity of rockfall per event" category should be representative of whichever type of rockfall event is most likely to occur. If individual blocks are typical of the rockfall, then block size should be used for scoring. If a mass of broken rock tends to be the dominant type of rockfall, then the quantity per event should be used. The rating criteria and associated scores for this category are shown in Table 4.3.

Climate and Presence of Water on Slope

Water and freeze–thaw cycles both contribute to the weathering and movement of rock materials. As Table 4.3 shows, the exact ratings specified for the "climate and presence of water on slope" category range from "low to moderate precipitation; no freezing periods" (3 points) to "high precipitation and long freezing periods" (81 points). The criteria *between* these two extremes contain the word *or*. Areas receiving less than 20 in. per year are low-precipitation areas; areas receiving more than 50 in. per year are high-precipitation areas.

Rockfall History

The "rockfall history" category considers the historical rockfall activity at a site as an indicator of future rockfall events. The historical information is best obtained from maintenance records. There may be no history available at newly constructed sites or where poor documentation practices have been followed. If this is the case, an estimate of the "history" based on the condition of the rockfall section should be made, along with a note to review the rating at a later date. The rating criteria and associated scores for this category are shown in Table 4.3.

REFERENCES

AASHTO (American Association of State Highway and Transportation Officials). 2011. *A Policy on Geometric Design of Highways and Streets*. Washington, D.C.: AASHTO Standing Committee on Highways.

Brawner, C.O. 1994. *Rockfall Hazard Mitigation Methods: Participant's Notebook*. FHWA-SA-93-085. Washington, DC: U.S. Department of Transportation, Federal Highway Administration.

Brawner, C.O., and D.C. Wylie. 1975. *Rock Slope Stability on Railway Projects*. Vol. 77. Chicago: American Railway Engineering Association.

Buss, K., R. Prellwitz, and M.A. Reinhart. 1995. *Highway Rock Slope Reclamation and Stabilization Black Hills Region, South Dakota, Parts I & II*. SD94-09-F and -G. Pierre, SD: South Dakota Department of Transportation.

FHWA (Federal Highway Administration). 1993. *Rockfall Hazard Rating System: Participant's Manual*. FHWA-SA-93-057. Washington, DC: U.S. Department of Transportation, FHWA. https://www.fhwa.dot.gov/engineering/geotech/pubs/009767.pdf

Hoek, E., and J. Bray. 1989. *Rock Slopes: Design, Excavation, Stabilization*. McLean, VA: Federal Highway Administration.

Pierson, L.A., and R. Van Vickle. 1993. *Rockfall Hazard Rating System: Participants' Manual.* FWHA-SA-93-057. Washington, DC: U.S. Department of Transportation, Federal Highway Administration.

Pierson, L.A., S.A. Davis, and R. Van Vickle. 1990. *Rockfall Hazard Rating System Implementation Manual.* FWHA-OR-EG-90-01. Washington, DC: U.S. Department of Transportation, Federal Highway Administration.

Pierson, L.A., C.F. Gullixson, and R.G. Chassie. 2001. *Rockfall Catchment Area Design Guide.* Final Report SPR-3(032). Salem, OR: Oregon Department of Transportation; and Washington, DC: U.S. Department of Transportation, Federal Highway Administration.

Ritchie, A.M. 1963. *An Evaluation of Rockfall and Its Control.* Highway Research Record, No. 17. Olympia, WA: Washington State Highway Commission. pp. 13–28.

Wyllie, D. 1987. Rock slope inventory system. In *Proceedings of the Federal Highway Administration Rockfall Mitigation Seminar.* Portland, OR: Federal Highway Administration, Region 10.

CHAPTER 5

Kinematic Slope Stability Analysis

Kinematic analysis, which is purely geometric, examines which modes of slope failure are possible in a rock mass with respect to an existing or proposed rock slope. In a kinematic analysis, it is the *orientation* of the combination of discontinuities, the slope face, the upper slope surface, and any other slope surface of interest—together with friction—that is examined to determine if certain modes of failure can possibly occur. The analysis is normally conducted with the aid of a stereographic representation of the planes and/or lines of interest. This chapter begins by discussing a method of kinematic analysis developed by Markland (1972). The Markland test is an extremely valuable tool for identifying those discontinuities that could lead to planar-, wedge-, or toppling-type failure in the rock mass and for eliminating other discontinuities from consideration. It is also valuable for conducting a sector analysis of a mine highwall or road cut, since each different alignment of the highwall or road cut can be analyzed independently.

This chapter also describes a technique called the friction cone method, which may be used for conducting an analysis of plane shear failure. This method is not purely a kinematic approach; it combines elements of kinematics and kinetics.

MARKLAND TEST FOR PLANE SHEAR FAILURE

The Markland test for plane shear failure (i.e., plane failure or wedge failure) uses a stereographic projection of the great circle representing the slope face together with a circle representing the friction angle, ϕ, of the discontinuity (Figure 5.1). The friction circle, as plotted on the stereographic projection, has a radius equal to $90° - \phi$. (In other words, ϕ is measured inward from the north, south, east, and west [N, S, E, and W] points of the stereograph, and a circle—whose center is the center of the stereograph—is drawn containing the points.) The zone between the great circle representing the slope face and the friction circle (i.e., the shaded area on Figure 5.1) represents a critical zone within which

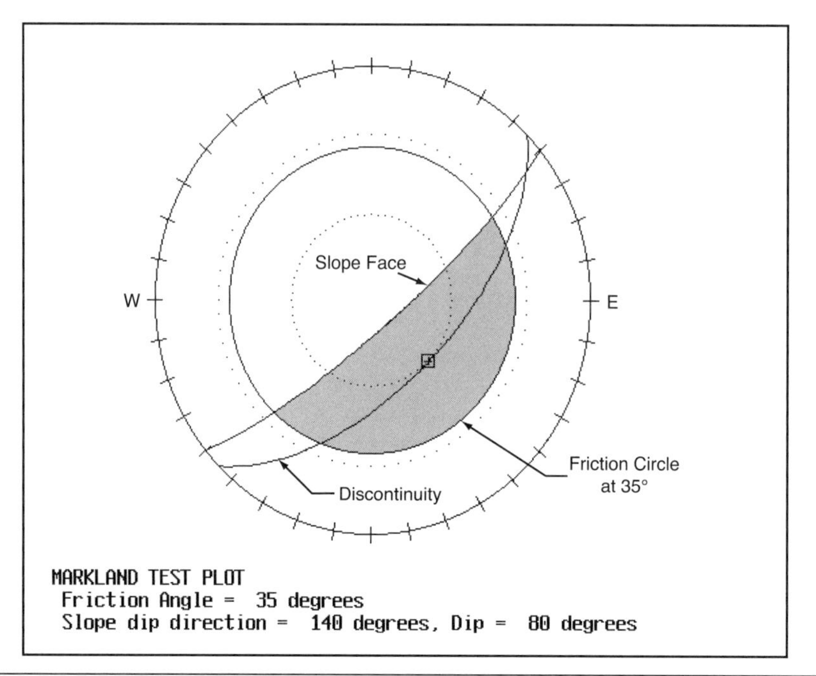

FIGURE 5.1 The Markland test for plane shear

plane shear failure is kinematically possible if three additional conditions are met (Hoek and Bray 1981; Piteau and Martin 1982):

1. The dip of the discontinuity (or plunge of the line of intersection of two discontinuities) must exceed the angle of friction for the rock surface, ϕ.
2. The discontinuity (or line of intersection of two discontinuities) must daylight in the slope face.
3. The dip of the discontinuity (or plunge of the line of intersection of two discontinuities) must be less than the dip of the slope face, ψ.

These conditions can be expressed by the following relationship:

$$\psi > \beta > \phi$$

where

ψ = dip of the slope face

β = dip of the discontinuity or plunge of the line of intersection

ϕ = angle of friction for the rock surface

One additional condition must be met for the possibility of plane failure to exist; namely, the strike (or dip direction) of the discontinuity must be within ±20° (some experts say

±15°) of the strike (or dip direction) of the slope face (Hoek and Bray 1981). Outside of that 20° zone, discontinuities disappear into the slope such that they often lock themselves in.

The Markland test stereographic analysis is conservative because it makes two very important assumptions:

1. All of the discontinuities are assumed to be continuous and through-going, though in reality many of them are not. Even a small percentage of intact rock along a discontinuity can generate enough strength to make the rock safe against sliding.
2. The stereonet procedure assumes that the cohesion, *c*, equals zero; in other words, the effects of cohesion are ignored. When this assumption is made, the fundamental limiting equilibrium equation for the factor of safety (see Equation 1.1) reduces to $FS = \tan\phi/\tan\beta$.

Therefore, whenever the dip of the discontinuity is greater than the friction angle, the factor of safety is less than 1.0 and the dip vector will plot between the friction circle and the great circle representing the slope face on the stereonet. Whenever the dip of the discontinuity is less than the friction angle, the factor of safety is greater than 1.0 and the dip vector will plot outside the friction circle. For both of the preceding situations, the strike of the discontinuity must be within about ±20° of the strike of the slope face.

Figure 5.2 shows two potentially unstable wedges formed by (1) the intersection of great circles representing planes A and B and (2) the intersection of great circles representing planes A and C. The points representing the plunges of the lines of intersection of these two wedges fall in the critical zone. Figure 5.2 shows that wedge A-B has a plunge of the line of intersection approximately midway between the friction angle, ϕ, and the dip of the slope face, ψ. On the other hand, wedge A-C has a plunge of the line of intersection slightly less than the dip of the slope face. Nevertheless, both wedges should be further analyzed kinetically for potential instability (as discussed in Chapter 8). Figure 5.2 also shows a wedge formed by the intersection of two discontinuities—planes B and C—that does not meet the conditions of the Markland test (i.e., the intersection is not in the critical zone); hence, the wedge is stable.

Figure 5.3 shows a Markland test stereographic plot of potential plane failure. Shown on the figure are the great circle representations of two planes, both of which pass through the critical zone. Both planes, therefore, meet the Markland test conditions of $\psi > \beta > \phi$. However, plane B has a dip direction greater than ±20° from the dip direction of the slope face; plane A has a dip direction very close to that of the slope face. Plane A, therefore, would be the most critical for further analysis.

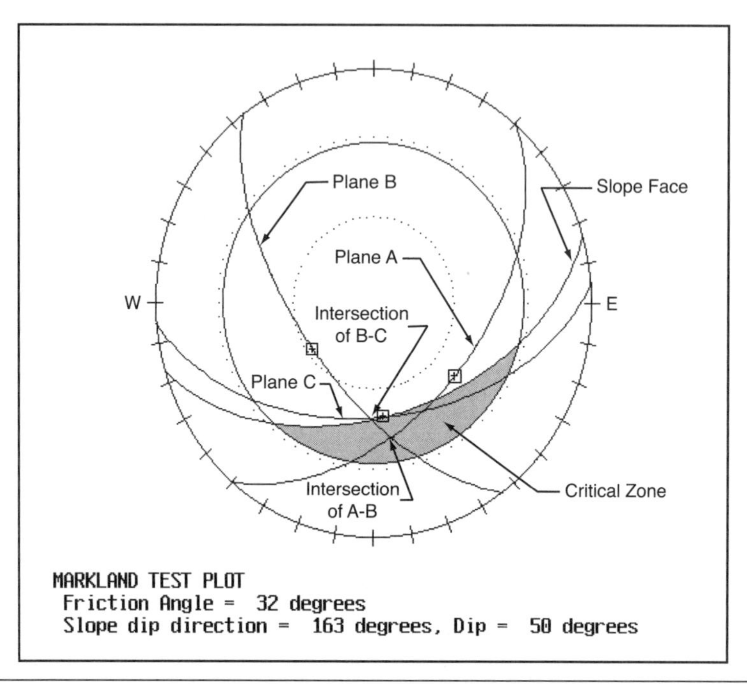

FIGURE 5.2 Potential wedge failure (planes A and B) and stable wedge (planes B and C)

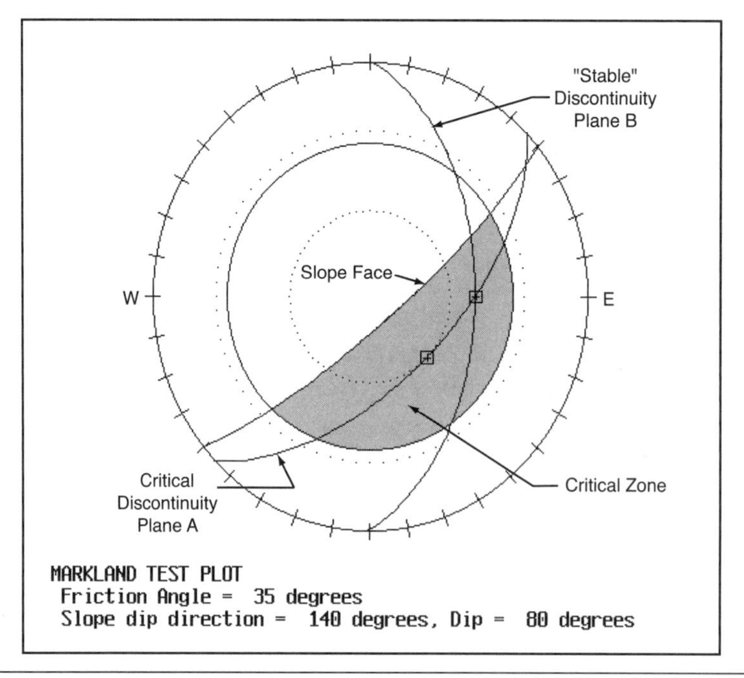

FIGURE 5.3 Potential plane failure and stable plane

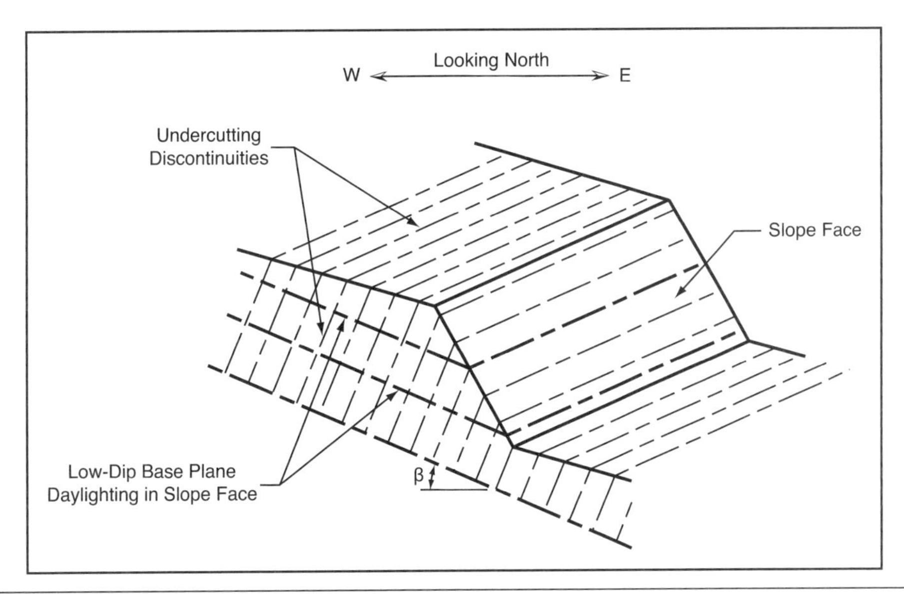

FIGURE 5.4 The toppling failure mode showing low-dip base plane and undercutting discontinuities

MARKLAND TEST FOR TOPPLING FAILURE

The stereographic projection of the slope face and the friction circle can also be used to determine a kinematic potential for the toppling failure mode. For toppling failure to be kinematically possible, the following conditions must exist:

1. There is an inclined base plane dipping in a direction approximately equal to the slope face and inclined at a low angle $\beta < \phi$.
2. Undercutting discontinuities (i.e., a joint set approximately perpendicular to the inclined surface) are present.

Figure 5.4 depicts a case in which these two conditions are present, and Figure 5.5 shows a stereographic plot of the slope shown in Figure 5.4. The undercutting discontinuities and the low-dip base planes combine to form blocks that, if they are of the correct geometric proportions, can topple.

COMPUTER-AIDED KINEMATIC ANALYSIS

Modern computer software packages such as Rocscience's Dips for the analysis of orientation-based geological data (Rocscience 2015) allows the user to conduct a kinematic analysis for planar sliding, wedge sliding, and toppling.

For planar sliding, the kinematic analysis can be conducted with either discontinuity poles using the critical pole vector zone (Figure 5.6) or dip vectors using the critical dip vector zone (Figure 5.7). As shown in Figures 5.6 and 5.7, lateral limits (of ±20° to 30°, depending on

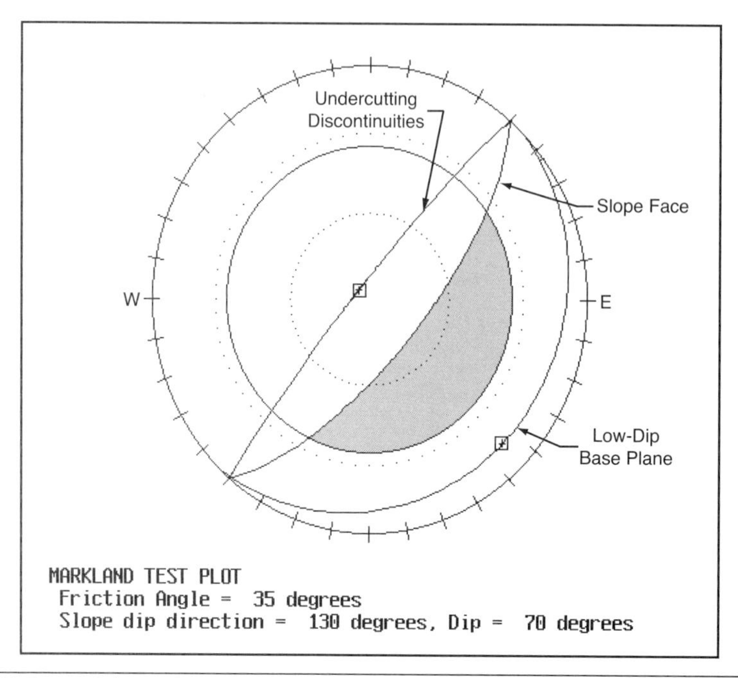

FIGURE 5.5 Potential toppling failure

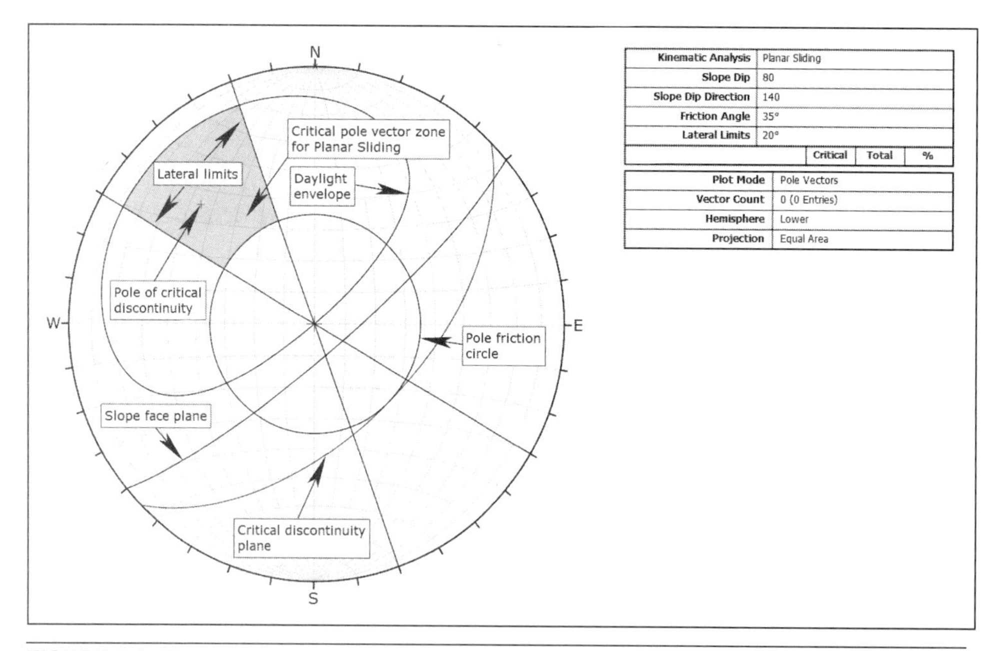

FIGURE 5.6 Kinematic analysis for planar sliding using the critical pole vector zone

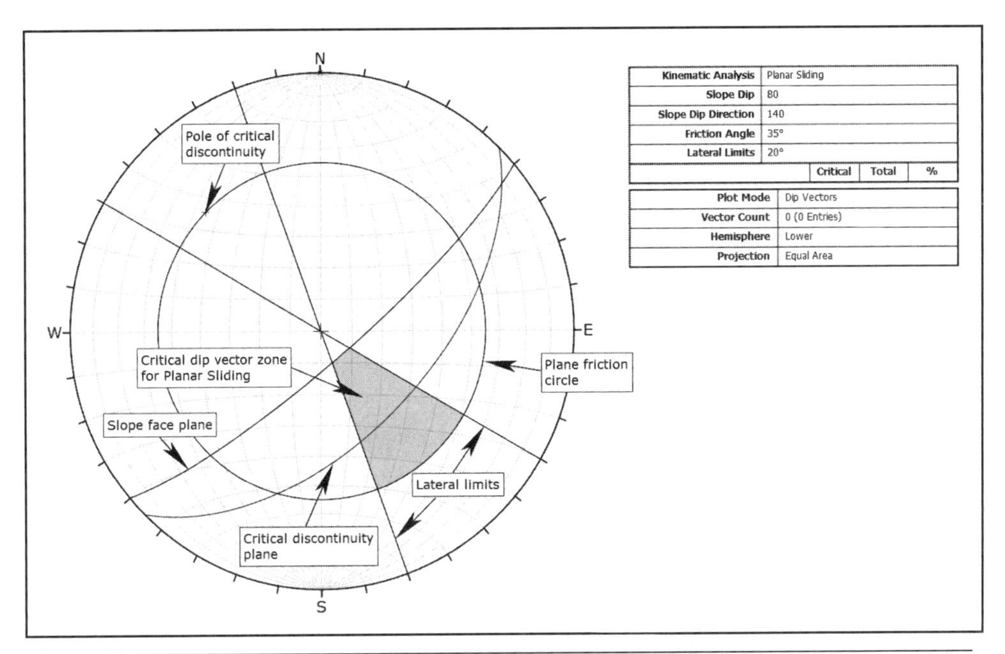

FIGURE 5.7 Kinematic analysis for planar sliding using the critical dip vector zone

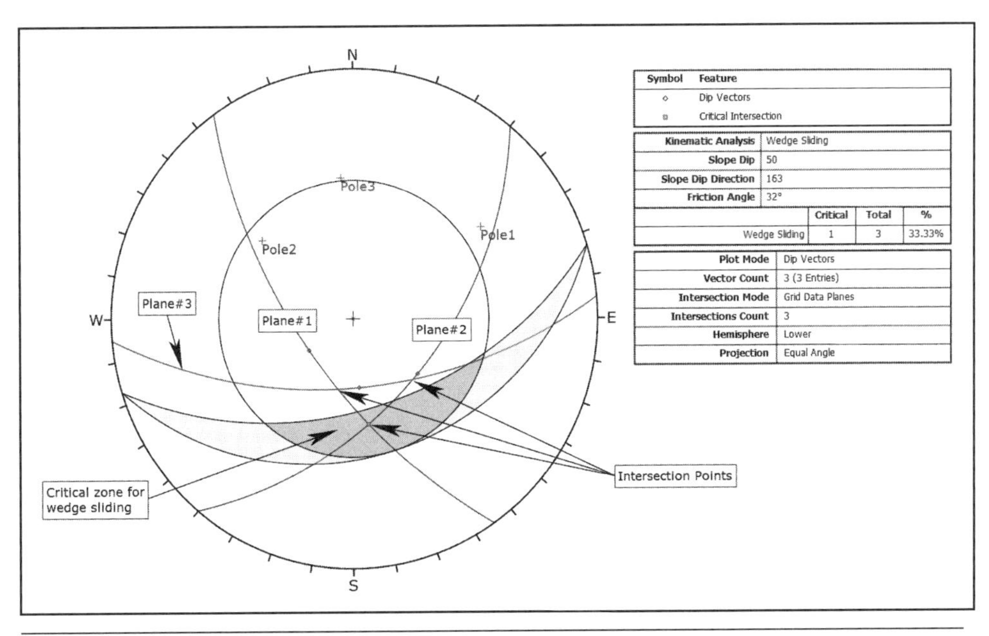

FIGURE 5.8 Wedge sliding kinematic analysis showing intersection points

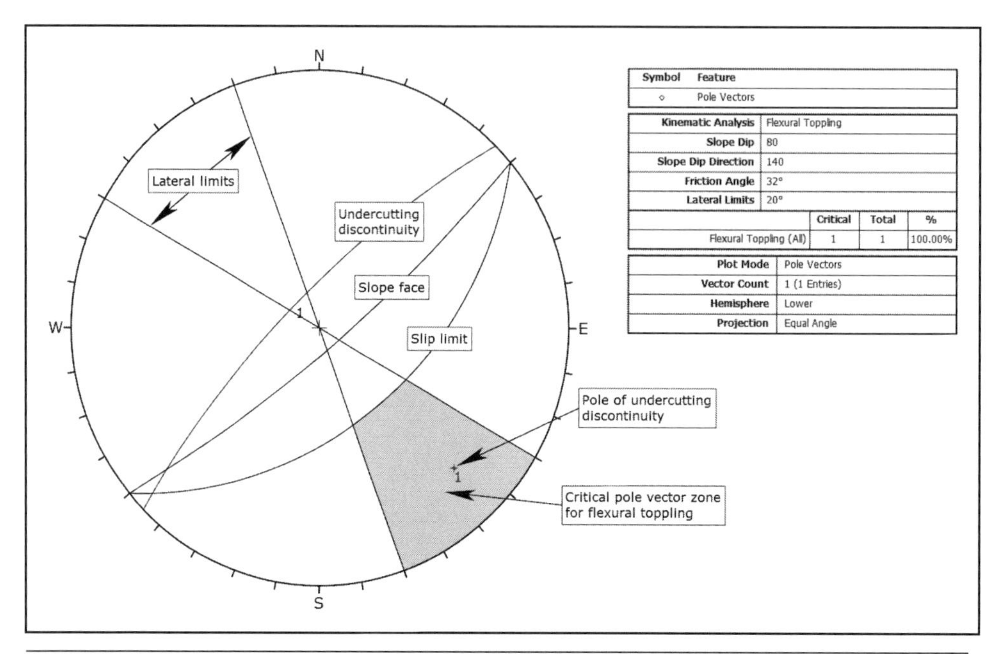

FIGURE 5.9 Flexural toppling kinematic analysis using the pole vector mode

the user's preference) can be included to restrict the critical zone to a certain limit on either side of a single plane's dip vector. As mentioned previously, and discussed in more detail in Chapters 6 and 7, one of the conditions for plane failure or wedge failure to possibly occur is that the dip direction of the failure plane (or the plunge of the line of intersection for wedge-type failure) must be within about ±20° of the dip direction of the slope face.

For wedge sliding, the *critical zone* is as defined in Figure 5.2. However, points representing planes that intersect within the critical zone can be plotted and analyzed, as shown in Figure 5.8. Other options exist (Rocscience 2015), such as plotting the intersections of *all* planes in the file, contouring of the intersections, and plotting the intersections of just the major planes (i.e., mean set planes and/or user planes).

The analysis for flexural toppling can be carried out using either the pole vector mode or the dip vector mode (Rocscience 2015). In either case, key elements of the flexural toppling kinematic analysis are

- The slope plane,
- The slip limit plane (based on slope angle and friction angle), and
- Lateral limits.

For flexural toppling using pole vectors, the critical zone for toppling is defined by the region:

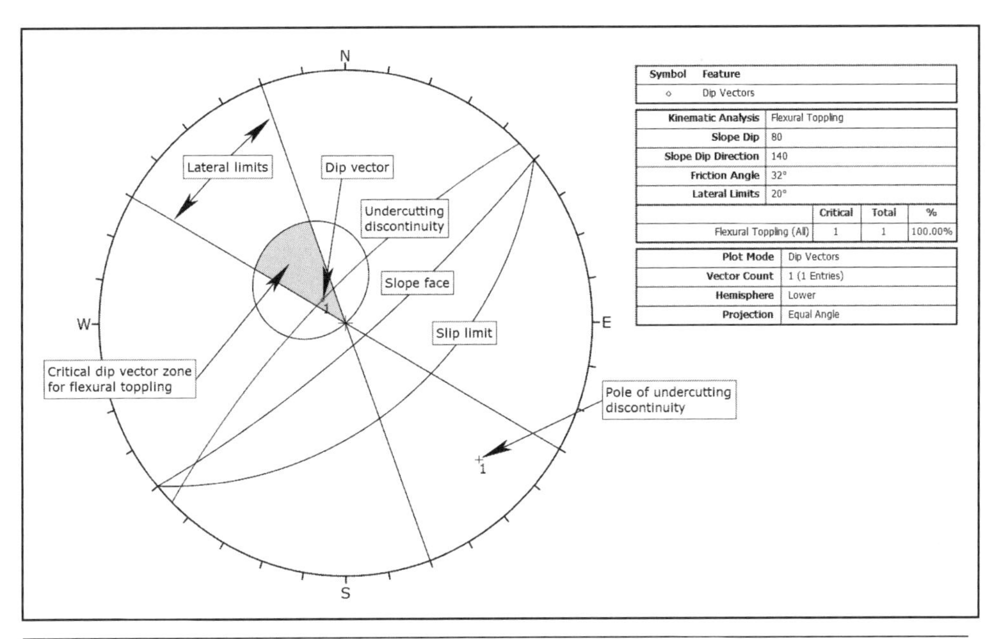

FIGURE 5.10 Flexural toppling kinematic analysis using the dip vector mode

- **Outside** what Rocscience designates as the *slip limit plane* (Goodman [1980] states that for slip to occur, the bedding normal must be inclined less steeply than a line inclined at an angle equivalent to the friction angle above the slope. This results in a "slip limit" plane that defines the critical zone for flexural toppling. The dip angle of the slip limit plane is derived from the slope angle – friction angle. The dip direction of the slip limit plane is equal to that of the slope face).
- **Inside** the lateral limits (The lateral limits for flexural toppling have the same purpose as described for planar sliding. They define the lateral extents of the critical zone with respect to the dip direction of the slope).

All poles that plot in the critical pole vector zone for flexural toppling (Figure 5.9) represent discontinuity orientations that pose a toppling risk.

The flexural toppling kinematic analysis using the dip vector mode is shown in Figure 5.10. In this analysis, flexural toppling is kinematically possible if the dip vectors of the discontinuities plot within the *critical dip vector zone for flexural toppling*, defined as the region between the lateral limits and the daylight envelope of the slip limit plane (the *daylight envelope* is that circle tangent to the intersection point of the lateral limits and with diameter equal to the dip angle [in degrees] of the slope plane).

FRICTION CONE CONCEPT

Another technique that can be employed for stability analysis of plane shear failure is known as the friction cone method. This method is a combination of kinematic and kinetic

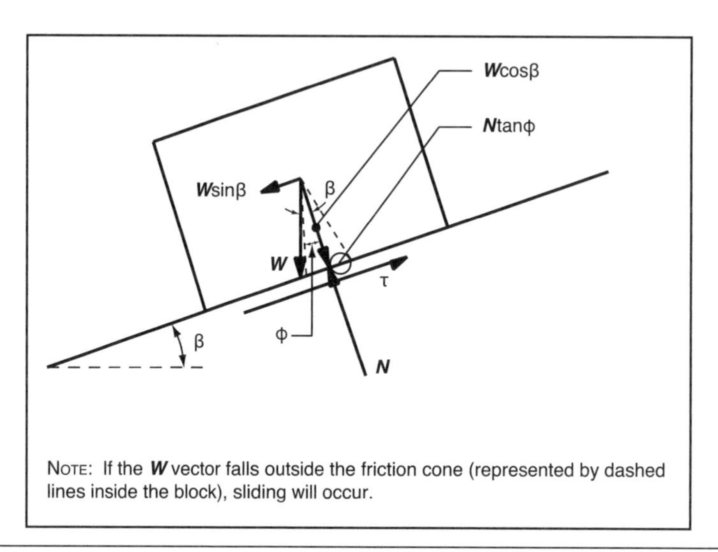

NOTE: If the *W* vector falls outside the friction cone (represented by dashed lines inside the block), sliding will occur.

FIGURE 5.11 The friction cone concept for a block resting on an inclined plane

analysis. The friction cone method is a convenient way to resolve the forces normal to and acting down a potential failure plane into a graphical representation plotted on a stereographic projection. After resolution of the forces onto the stereonet in the form of a friction cone, the position of the mass weight, W, with respect to the edge of the friction cone indicates whether or not the mass is potentially unstable. This method, in conjunction with a purely kinematic analysis, is especially useful for wedge-type stability analyses (see Chapter 8).

Zero-Cohesion Assumption

To visualize the theory of the friction cone, we must look again at a block resting on an inclined plane, as well as the corresponding stereographic representation (Figures 5.11 and 5.12). Note that the block in Figure 5.11 is not at limiting equilibrium; it is stable with respect to sliding on the inclined surface given that $\beta > \phi$ (as discussed in more detail later). Plotted on the stereographic projection is the great circle representing the discontinuity and the normal, N, to the discontinuity, represented by the pole, **P**, of the discontinuity plane (see Chapter 2). For the zero-cohesion (i.e., $c = 0$) assumption, the resisting forces are

$$N \tan\phi = (W \cos\beta)\tan\phi$$

If N is the normal to the plane, then the friction angle, ϕ, forms a cone around N of radius $N\tan\phi$ (see Figure 5.11). On the stereographic projection, a "circle" is formed around **P** of radius ϕ (see Figure 5.12). Note, however, that the friction "circle" may not be truly circular in shape, depending on the type of stereonet used (as discussed in Chapter 2). For friction circles plotted with the equal-angle stereonet, the friction circle will be circular in shape, but with the center of the circle offset from the pole, **P**. For friction circles plotted with the

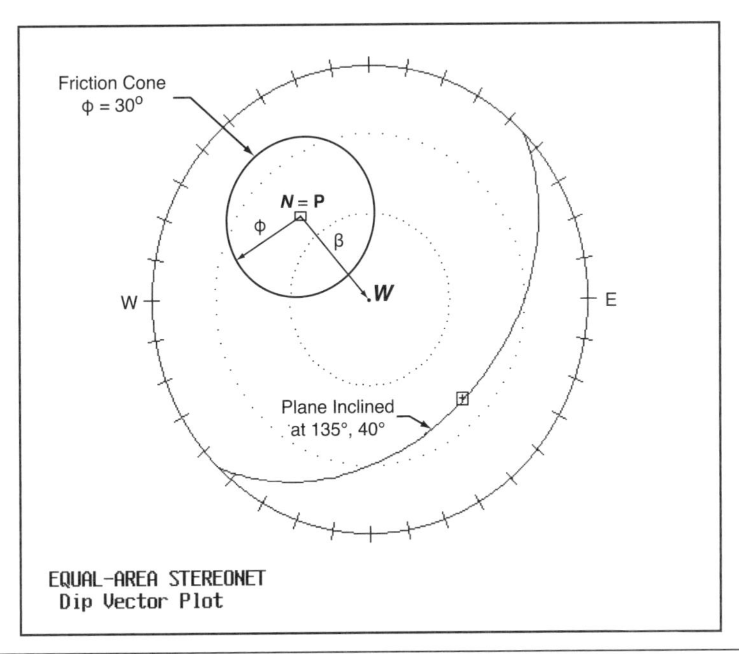

FIGURE 5.12 Stereographic representation of the friction cone concept from Figure 5.11

equal-area net (the stereonet used for Figure 5.12), the friction circle will be centered at **P** but will not be circular in shape.

The driving force for sliding on the plane is $W\sin\beta$. If we assume W acts vertically downward, then W is represented by the center pivot point on the stereographic projection. The driving force, from vector analysis, is represented by angle **P**–W on the stereographic projection.

At limiting equilibrium, the resisting force equals the driving force, or

$$N \tan\phi = W \sin\beta$$

which may also be expressed as

$$W \cos\beta \cdot \tan\phi = W \sin\beta$$

On the stereographic projection, this is represented by

$$\mathbf{P} - W = \phi$$

at limiting equilibrium. As was discussed in Chapter 2, the pole, **P**, is plotted on the stereonet by measuring the dip angle, β, from the center point of the stereonet, W, in the opposite direction of the dip direction of the great circle representing the discontinuity. That is, angle **P**–W represents β on the stereographic projection. Thus, at limiting equilibrium,

$$\beta = \phi$$

By vector analysis, at limiting equilibrium, this becomes

$$W \cos\beta \cdot \tan\phi = W \sin\beta$$

or

$$\tan\phi = W \sin\beta / W \cos\beta$$

or

$$\tan\phi = \tan\beta$$

for a factor of safety of 1.0 at limiting equilibrium.

The general factor-of-safety equation then becomes

$$\text{FS} = \frac{\tan\phi}{\tan\beta} \qquad \text{(EQ 5.1)}$$

Equation 5.1 is a simplified form of the more general Equation 1.1, with the assumption that $c = 0$. Remember, Equation 1.1 may be written as

$$\text{FS} = \frac{cA + W\cos\beta\tan\phi}{W\sin\beta} \qquad \text{(EQ 1.1)}$$

If $c = 0$, then $cA = 0$ and Equation 1.1 becomes Equation 5.1.

Nonzero Cohesion Assumption

With the addition of cohesion, the applicable equation for the resisting force of the inclined block in Figure 5.11 is

$$\text{magnitude of resisting force} = cA + W\cos\beta \cdot \tan\phi$$

where

A = block area resting on the surface
W = magnitude of weight

This expression rearranges to

$$\text{magnitude of resisting force} = \frac{cA}{W\cos\beta} + \tan\phi$$

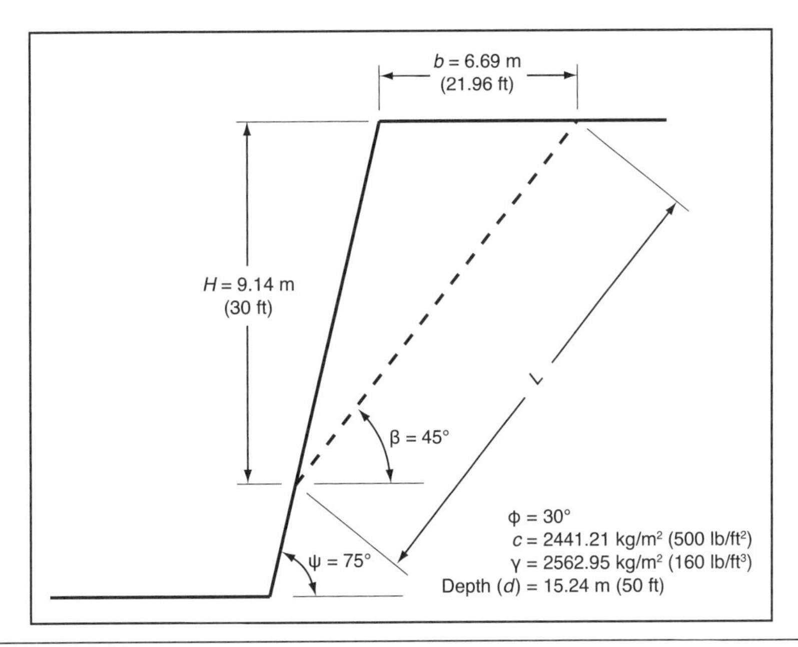

FIGURE 5.13 Slope for friction cone method example

Referring to the failure plane (Figure 5.11, or see Figure 1.16 for general details without the friction cone), we note that cA is the component of shear resistance along the plane of sliding (failure plane) and that $W\cos\beta$ is the normal component, so $cA/(W\cos\beta)$ becomes an equivalent tangent angle. Adding this equivalent tangent angle to the tangent of the internal friction angle, we get an apparent friction angle, ϕ_a, given by

$$\tan\phi_a = \tan\phi + \frac{cA}{W\cos\beta} \qquad \text{(EQ 5.2)}$$

This apparent friction angle—incorporating friction, ϕ, and cohesion, c—can be plotted on the stereonet as a new friction cone around the normal, **P**. The new factor of safety can be determined by the following expression:

$$\text{FS} = \frac{\tan\phi_a}{\tan\beta} \qquad \text{(EQ 5.3)}$$

Example: Friction Cone Method for Planar Failure

As an example of applying the friction cone method to assess the potential for planar failure, consider the slope shown in Figure 5.13. The slope shown has an orientation of the slope face of 150°, 75° (dip direction, dip). The critical discontinuity has an orientation of 143°, 45°. Assume that the cross section shown in Figure 5.13 has been drawn through the center of the width of the potential failure mass; therefore, the weight of the mass can be calculated by using the volume of a right, three-sided pyramid (i.e., the difference in

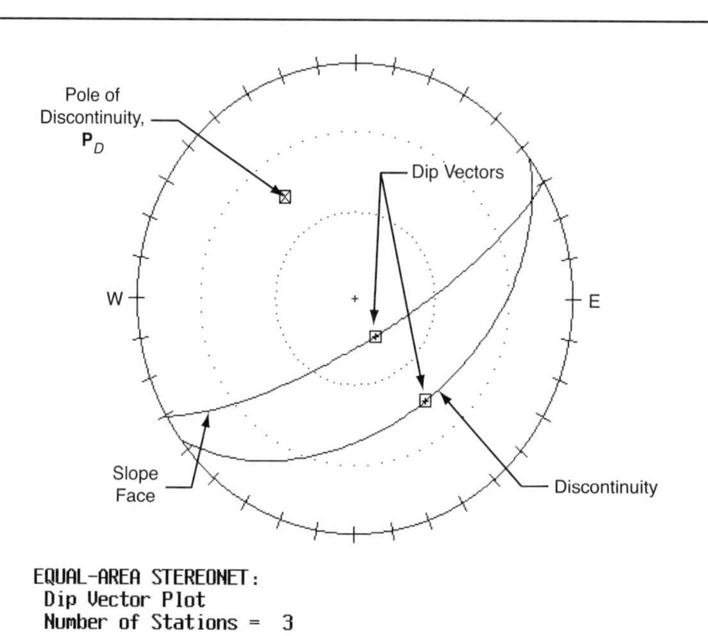

Step 1. Plot great circles representing the slope face and discontinuity and the pole of the discontinuity ($\mathbf{P}_D$).

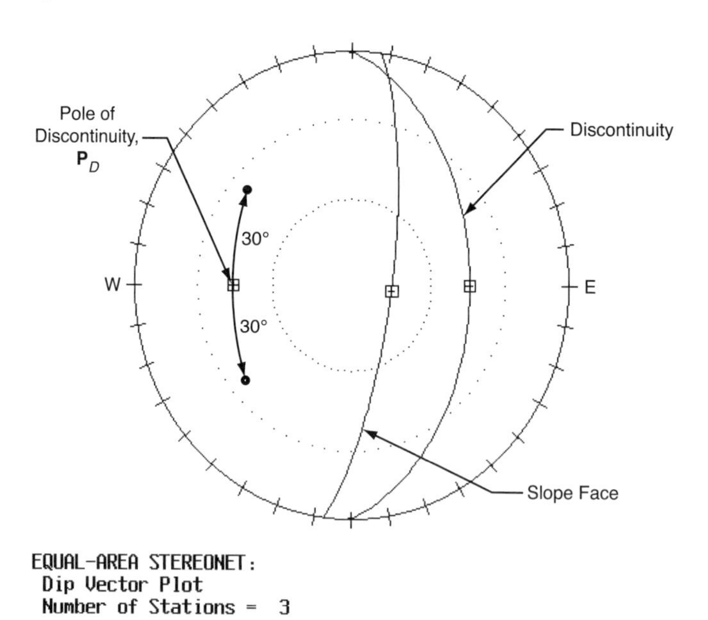

Step 2. Rotate $\mathbf{P}_D$ to a great circle on the stereonet, then measure the plot ϕ along the great circle on both sides of $\mathbf{P}_D$.

FIGURE 5.14 Plotting the friction cone for the example of Figure 5.13

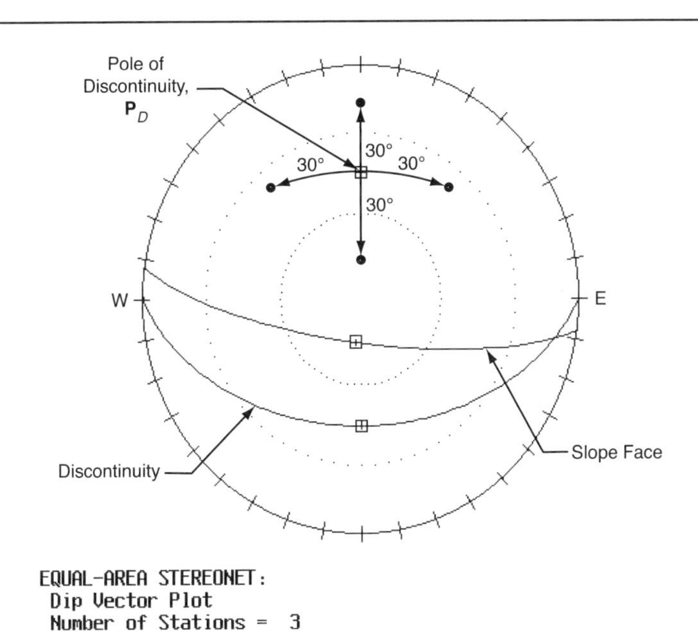

Step 3. Rotate P_D to another great circle, then measure and plot ϕ.

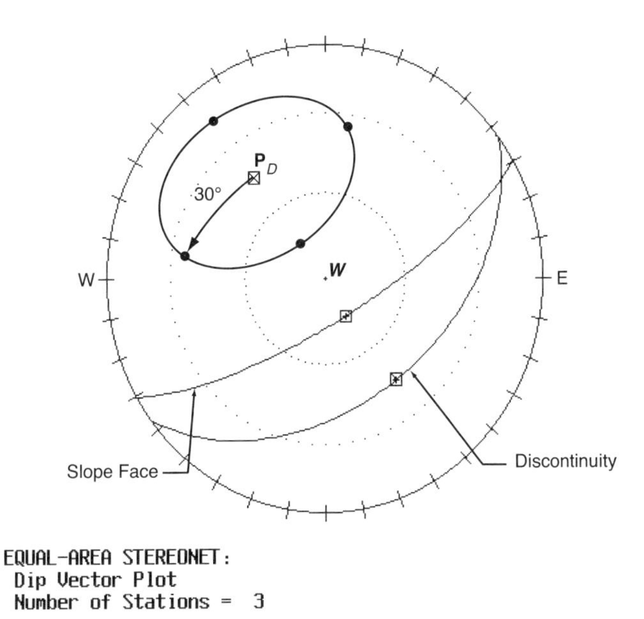

Friction circle of 30° around the pole of the discontinuity oriented at 143°, 45°

FIGURE 5.14 Plotting the friction cone for the example of Figure 5.13 (continued)

orientations of the slope face and the discontinuity are accounted for). The first task at hand is to plot the friction cone for this situation by taking the following steps, which are also shown in Figure 5.14:

1. Plot great circles of the discontinuity and the slope face. Also plot the pole of the discontinuity ($\mathbf{P}_D$).
2. Rotate $\mathbf{P}_D$ to a convenient great circle on the stereonet. Measure and plot ϕ along the great circle on both sides of $\mathbf{P}_D$.
3. Repeat step 2 at other meridians until the friction circle is formed.

Once this task is completed, calculating the factor of safety assuming zero cohesion is simply (by Equation 5.1)

$$\text{FS} = \frac{\tan\phi}{\tan\beta} = \frac{\tan(30°)}{\tan(45°)} = 0.5774$$

If cohesion is accounted for (i.e., is nonzero), the factor-of-safety analysis is more involved (see Figure 5.13), requiring the following calculations to determine the apparent friction angle:

$$\begin{aligned} b &= H \cdot \tan(\beta) - H \cdot \tan(90 - \psi) \\ &= 9.14 \cdot \tan(45°) - 9.14 \cdot \tan(15°) \\ &= 6.69 \text{ m } (21.96 \text{ ft}) \end{aligned}$$

$$L = \sqrt{H^2 + \left(b + \left(H \cdot \tan(90 - \psi)\right)\right)^2}$$

or

$$\begin{aligned} L &= \sqrt{(2H)^2} \text{ for } 45° \text{ triangle} \\ &= \sqrt{(9.14)^2 + (9.14)^2} = 12.93 \text{ m } (42.43 \text{ ft}) \end{aligned}$$

$$\begin{aligned} W &= \frac{1}{2} b \cdot H \cdot d \cdot \gamma_r \\ &= ½(6.69)(9.14)(15.24)(2{,}562.95) = 1{,}195{,}000 \text{ kg } (2{,}635{,}200 \text{ lb}) \end{aligned}$$

$$\begin{aligned} \tan\phi_a &= \tan\phi + \frac{cA}{W\cos\beta} \\ &= \tan(30°) + \frac{(2{,}441.21)(12.93)(15.24)}{1.195 \times 10^6 \cos(45°)} = 1.15 \end{aligned}$$

Plot the friction cone at 49° by following preceding steps 1 through 3 (Figure 5.15). Then we can determine the factor of safety in two ways: by calculation and from the friction cone. The calculation yields

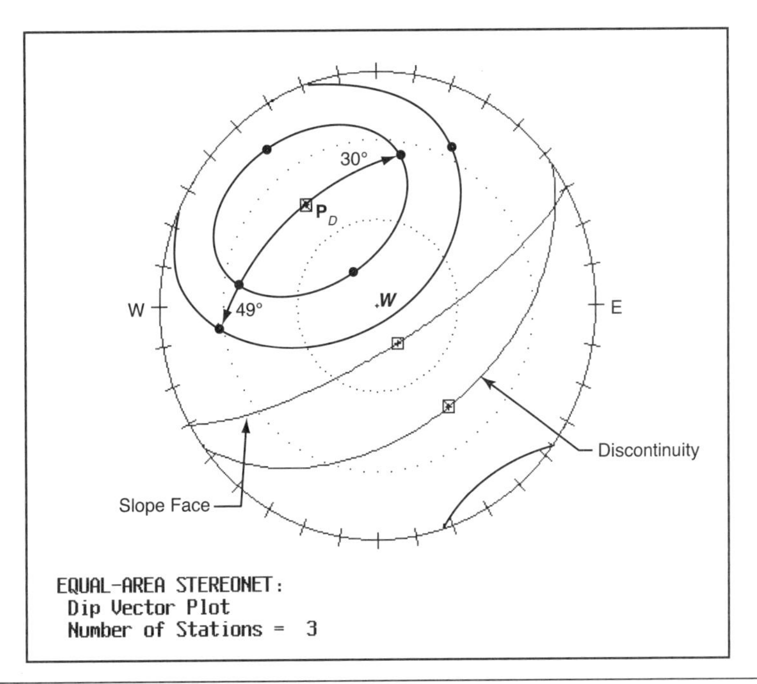

FIGURE 5.15 The friction cones for $\phi = 30°$ and $\phi_a = 49°$ for the discontinuity oriented at 143°, 45°

$$\text{FS (calculated)} = \frac{cL + W\cos\beta \cdot \tan}{W\sin\beta}$$

$$= \frac{(2{,}441.21)(12.93) + \frac{1}{2}(6.69)(9.14)(2{,}562.95)\cos(45°)\tan(30°)}{\frac{1}{2}(6.69)(9.14)(2{,}562.95)\tan(45°)}$$

$$= 1.147$$

whereas the friction cone yields

$$\text{FS (from the friction cone)} = \frac{\tan(49°)}{\tan(45°)} = 1.150$$

REFERENCES

Goodman, R.E. 1980. *Introduction to Rock Mechanics*. Toronto: John Wiley.

Hoek, E., and J.W. Bray. 1981. *Rock Slope Engineering*, 3rd ed. London: Institution of Mining and Metallurgy.

Markland, J.T. 1972. *A Useful Technique for Estimating the Stability of Rock Slopes When the Rigid Wedge Sliding Type of Failure Is Expected*. Imperial College Rock Mechanics Research Report No. 19. London: Imperial College Press.

Piteau, D.R., and D.C. Martin. 1982. Mechanics of rock slope failure. In *Stability in Surface Mining*, Vol. 3. Edited by C.O. Brawner. New York: SME-AIME.

Rocscience. 2015. *Dips 6.0 Documentation*. Toronto: Rocscience.

CHAPTER 6

Kinetic Slope Stability: Analysis of Shear Failure—Plane Shear and Rotational Shear

PLANE SHEAR

Planar failure is a special case of the more general wedge type of failure; in planar failure, the rock mass slides on a single surface, as shown in Figures 6.1 and 6.2. Because only one surface is involved, we may carry out a two-dimensional (2-D) analysis by using the concept of a block resting on an inclined plane at limiting equilibrium. The 2-D plane failure analysis is easy to understand; it is especially useful for demonstrating the sensitivity of the slope to changes in shear strength, groundwater, and applied forces (both resisting and driving forces).

As discussed in Chapter 5, the Markland test for plane shear states that, for failure to occur by sliding on a single plane, the following geometrical conditions must be satisfied (Hoek and Bray 1981; Piteau and Martin 1982):

- A continuous plane on which sliding occurs must strike parallel or nearly parallel (within approximately ±20°) to the slope face.
- The failure plane must daylight in the slope face; in other words, the failure plane's dip must be *less than* the dip of the slope face, or $\beta < \psi$.
- The dip of the failure plane must be *greater than* the angle of friction of this plane; that is, $\beta > \phi$.
- Surfaces of separation (relief surfaces) that provide negligible resistance to sliding must be present in the rock mass to define the lateral boundaries of the failure block.

The first three conditions can be expressed by the following relationship:

$$\psi > \beta > \phi$$

FIGURE 6.1 A mass of rock that slid along a single plane

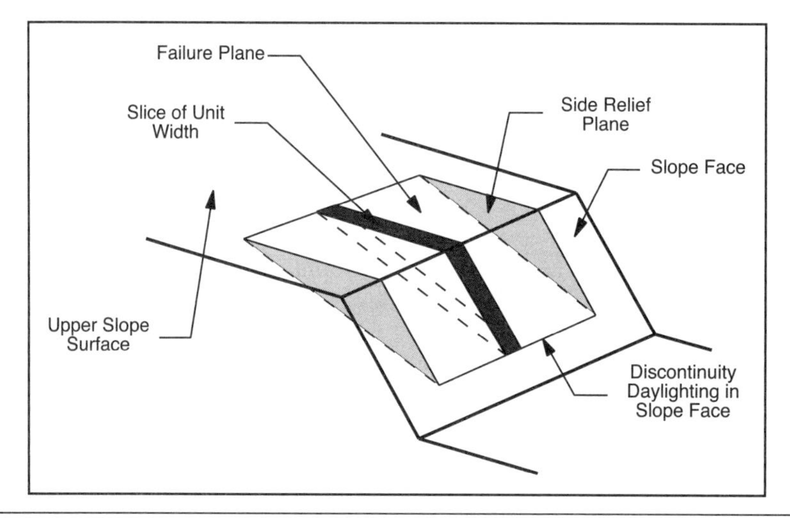

FIGURE 6.2 The planar failure mode

where

ψ = dip of the slope face

β = dip of the discontinuity

ϕ = angle of friction for the rock surface

The presence of surfaces of separation (i.e., the fourth condition) that outline the lateral boundaries of the failure block is an important element because the side planes of the potential failure mass offer shearing resistance to movement of the failure mass (see Figure 6.1). Without the relief planes, the net positive driving force of the failure mass applied over the two side plane areas must be greater than the shearing resistance of the rock mass. With

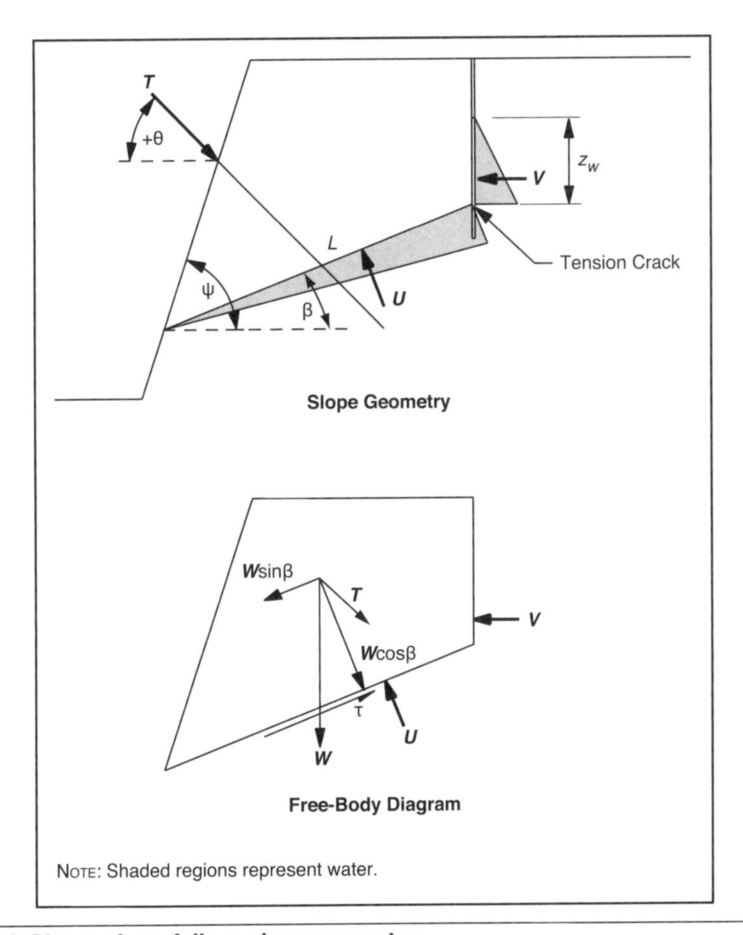

FIGURE 6.3 Planar shear failure plane geometry

the side relief planes, as is generally the assumed case, the general situation exists where the driving forces down the failure plane must be greater than the resisting forces along the failure plane for failure to be possible.

Method of Analysis for Plane Shear Failure

Plane shear failure analysis is carried out by using a vertical cross section of the slope that illustrates the slope geometry, failure plane geometry, and forces acting on the failure plane (Figure 6.3) for a unit width as shown on Figure 6.2 (Major et al. 1977). A unit width is used so that the area of the sliding plane can be represented by the length of the failure surface and the volume of the sliding block can be represented by the cross-sectional area of the block.

Simple analysis of a block resting on an inclined plane at limiting equilibrium is used to resolve the forces acting about the failure plane (Figure 6.4). Other forces acting on the

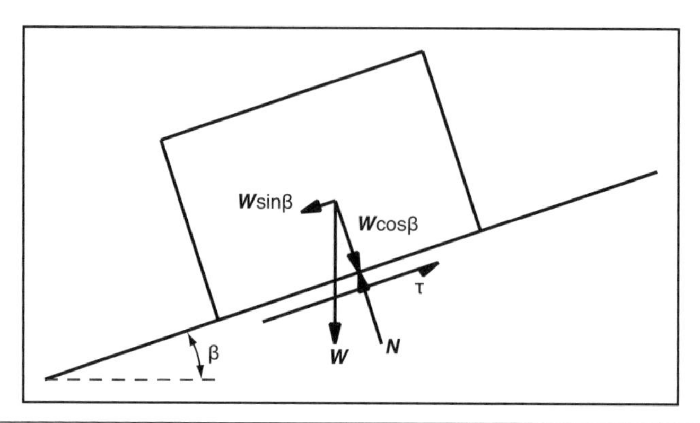

FIGURE 6.4 Block on an inclined plane at limiting equilibrium

rock block (i.e., water, surcharge, earthquake, and rock bolts) can be resolved by use of statics force diagrams, as discussed later in this chapter.

Summary of forces. Forces that may be considered to be acting on the failure block are as follows:

1. **Weight of the block.** The weight of the block, W, is calculated from the slope geometry and failure geometry.
2. **Normal water pressure.** Normal water pressure, or uplift pressure acting on the failure plane, is obtained by assuming a triangular or uniform curve of pressure distribution that acts normal to the failure plane. The resultant force, U, is assumed to act through the centroid of the mass, thus causing no rotational moments of the block.
3. **Tension crack water pressure.** Hydrostatic force in a tension crack is calculated from the measured or assumed height of water in the crack. This water pressure acts horizontally, and its resultant force, V, is assumed to also act through the centroid of the mass.
4. **Surcharge.** Additional weight on the ground surface at the crest of the slope can be caused by placement of a waste dump or tailings basin or by the weight of water from a lake or pond. This surcharge is usually assumed to act vertically downward along the same line as the weight vector.
5. **Dynamic forces.** Dynamic forces due to earthquakes or blasting may be of extreme importance for permanent slopes. The possible dynamic forces, as determined from seismic risk analysis, are generally expressed as a percentage of the gravitational force (g); they are presumed to be a combination of a horizontal force due to ground acceleration and an oscillating upward and downward vertical force due to the ground motion.
6. **Forces from artificial support.** If artificial support in the form of rock bolts or anchors is to be used, the total support force (T) required on the failure plane can be determined from the proposed inclination ($\pm\theta$) of the anchors with respect to the horizontal.

Resolution of forces and determination of factor of safety. Consider once again the block resting on an inclined plane at limiting equilibrium (Figure 6.4), with rock-bolt forces, water pressure, and so on, being neglected for the time being. The forces acting on the block are as follows:

$$\text{resisting force} = \text{shear force}$$
$$= \tau A = (c + \sigma \cdot \tan\phi)A = \left[c + (N/A)\tan\phi\right]A$$
$$\text{driving force} = \boldsymbol{W}\sin\beta$$

where

τ = shear stress along the failure plane
A = area of the base of the plane
c = cohesion along the failure plane
σ = normal stress on the failure plane
ϕ = angle of internal friction for the failure plane
$\boldsymbol{N}$ = normal force magnitude across the failure plane
$\boldsymbol{W}$ = weight of the failure mass
β = dip angle of the failure plane

By equating the driving forces and the resisting forces at limiting equilibrium, we get an equation for the factor of safety, FS:

$$\text{FS} = \frac{\text{resisting forces}}{\text{driving forces}} = \frac{cA + \boldsymbol{W}\cos\beta \cdot \tan\phi}{\boldsymbol{W}\sin\beta}$$

Factor of safety when surcharge is included. Surcharge on the ground surface at the crest of a slope, such as from a waste embankment or a heavy piece of machinery, creates an additional weight (of magnitude $\boldsymbol{W}_s$), which is assumed to act through the centroid of the potential failure mass. This additional weight is illustrated in Figure 6.5, which is a free-body diagram of the same slope geometry as in Figure 6.3.

Under these circumstances, the factor of safety becomes

$$\text{FS} = \frac{\text{resisting forces}}{\text{driving forces}} = \frac{cA + (\boldsymbol{W} + \boldsymbol{W}_s)\cos\beta \cdot \tan\phi}{(\boldsymbol{W} + \boldsymbol{W}_s)\sin\beta}$$

Factor of safety when water pressure is included. When groundwater is present above the potential failure surface, an uplift force ($\boldsymbol{U}$) caused by the water pressure will act on the potential failure plane in a direction opposite that of the weight's normal component. The magnitude of the water force depends on the area (length) of the portion of failure plane below the groundwater surface. This water force magnitude also depends on whether

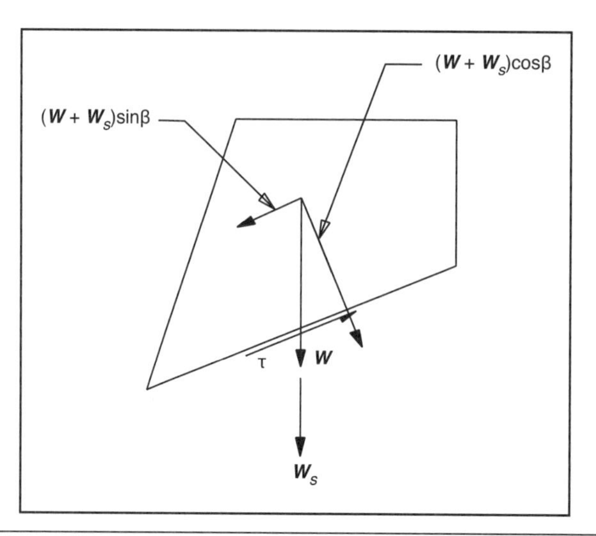

FIGURE 6.5 Free-body diagram of the failure surface of Figure 6.3 when surcharge force is included

or not a tension crack is present and, if so, whether or not the tension crack intersects the potential failure plane below the groundwater table.

If a tension crack does intersect the potential failure plane below the groundwater table, a horizontal water pressure is exerted on the face of the tension crack. The horizontal water pressure can be resolved into a horizontal force ($\boldsymbol{V}$) with components that tend to increase the driving force of the potential failure mass and reduce the normal force, $\boldsymbol{N}$, across the potential failure plane.

For the failure plane geometry of Figure 6.3, the magnitudes of $\boldsymbol{V}$ and $\boldsymbol{U}$ are expressed as follows:

$$\boldsymbol{V} = \tfrac{1}{2}\gamma_w z_w^{\ 2}$$

$$\boldsymbol{U} = \tfrac{1}{2}\gamma_w L z_w$$

where

γ_w = unit weight of water

z_w = vertical height of the tension crack from the water table surface (phreatic surface) down to where it intersects the potential failure plane

L = length of the potential failure plane (the length of the portion of the potential failure plane from where it daylights in the slope face to where it intersects the tension crack is denoted L')

Figure 6.6 shows a free-body diagram of the failure plane geometry of Figure 6.3, with $\boldsymbol{U}$ and $\boldsymbol{V}$ and the resulting resisting and driving forces accounted for. The factor of safety for this case then becomes

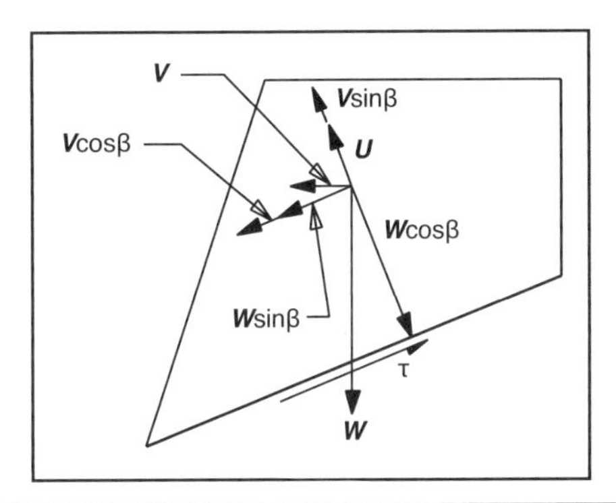

FIGURE 6.6 Free-body diagram of the failure surface of Figure 6.3 when water forces are included

Courtesy of Peabody Energy

FIGURE 6.7 Planar-type failures in sedimentary rock at a coal mine associated with water seeping from the highwall

$$FS = \frac{\text{resisting forces}}{\text{driving forces}} = \frac{cA + (W\cos\beta - V\sin\beta - U)\cdot\tan\phi}{W\sin\beta + V\cos\beta}$$

Figure 6.7 shows a series of planar-type failures associated with steeply dipping joints in sedimentary rock that intersect a weak layer saturated with water located above a major coal seam. Water can be seen seeping from the weak layer (light gray streaks). The failures are most likely caused by a combination of high-pressure gases from blasting venting into the weak discontinuities, thus weakening them further (backbreak), plus the water pressure.

Factor of safety when vibrational force is included. Ground vibrations from earthquakes, blasting, the operation of heavy equipment, vibrating compactors, pile drivers, and

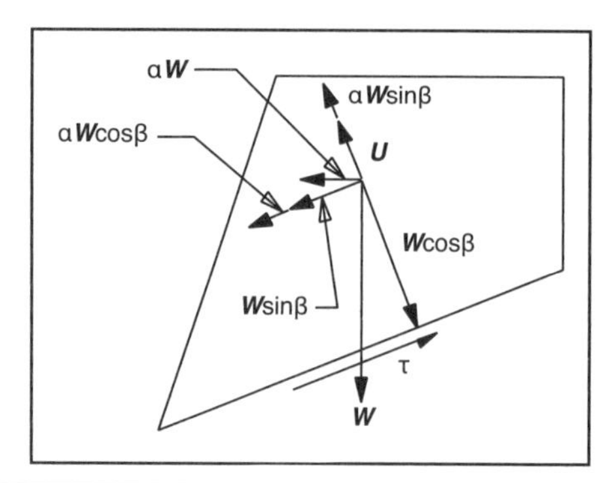

FIGURE 6.8 Free-body diagram of the failure surface of Figure 6.3 when vibrational force is included

so on, can induce dynamic forces to a rock slope. In the analysis of a slope, these forces are generally expressed as a decimal percentage of the gravitational force (e.g., $0.1g$) and are assumed to act in a horizontal direction. As a result, vibrational forces will tend to decrease the normal forces and increase the driving forces on a potential failure mass. Figure 6.8 shows a free-body diagram of the failure plane geometry of Figure 6.3 along with the resulting resisting and driving forces from vibration. The factor of safety for this case becomes

$$\text{FS} = \frac{\text{resisting forces}}{\text{driving forces}} = \frac{cA + (W\cos\beta - \alpha W\sin\beta)\cdot\tan\phi}{W\sin\beta + (\alpha W)\cos\beta}$$

where α is the percentage of gravitational force (expressed as a decimal).

As an example, consider a blasting seismograph that measured $0.50g$ as the longitudinal (compressive) ground vibration in the vicinity of a slope. In this case, the factor α in the preceding equation would be 0.50.

Factor of safety when rock-bolt force is included. Two types of rock-bolt force may be applied to stabilize a rock mass (Das and Stimpson 1986):

1. **Passive rock-bolt force.** One that is placed in the rock mass and does not play a role in ground support until the mass moves and subsequently loads the fixture (e.g., untensioned, grouted resin bolt or split set).
2. **Active rock-bolt force.** One that loads the rock masses in which it is placed immediately upon installation (e.g., point anchor or tensioned roof bolt).

For the purpose of this discussion, it is assumed that the rock-bolt force (T) will be applied to the face of the potential failure mass at an angle ($+\theta$ upward or $-\theta$ downward) from the horizontal (Figure 6.9). Figure 6.10 shows a free-body diagram of the failure plane geometry of Figure 6.3, with the resolved rock-bolt forces shown normal to and parallel to the potential failure plane.

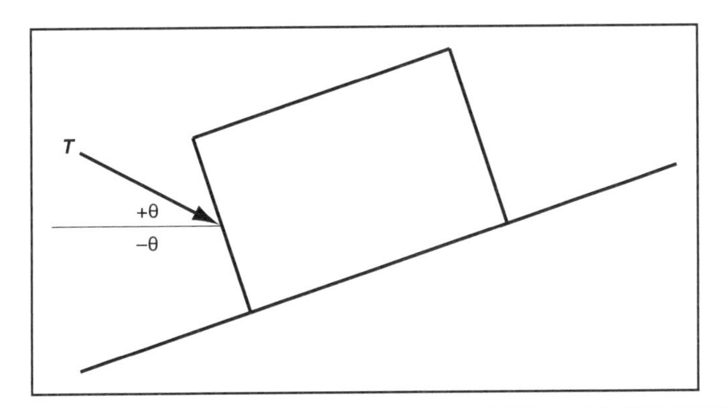

FIGURE 6.9 Rock-bolt angle θ, and rock-bolt force, *T*

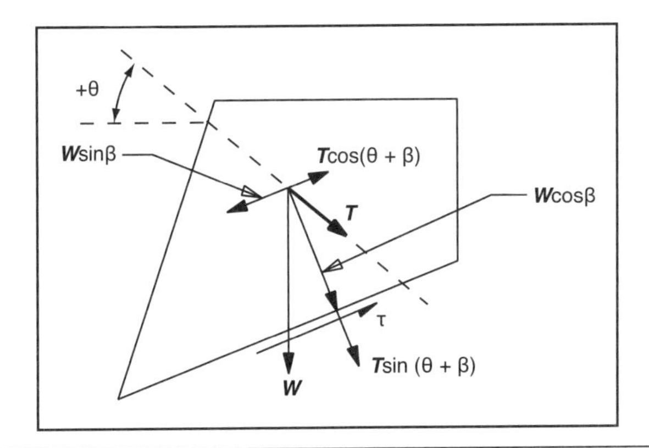

FIGURE 6.10 Free-body diagram of the failure surface of Figure 6.3 when rock-bolt forces are included

In the case of *passive rock-bolt force*, the assumption is that the resulting force acting parallel to the failure plane tends to decrease the driving forces of the potential failure mass. The factor of safety becomes

$$FS = \frac{cA + \left[W\cos\beta + T\sin(\theta+\beta)\right]\tan\phi}{W\sin\beta - T\cos(\theta+\beta)}$$

where T is the magnitude of the rock-bolt force.

In the case of *active rock-bolt force*, the assumption is that the resulting force parallel to the failure plane tends to increase the resisting forces along the potential failure plane. The factor of safety becomes

$$FS = \frac{cA + \left[W\cos\beta + T\sin(\theta+\beta)\right]\tan\phi + T\cos(\theta+\beta)}{W\sin\beta}$$

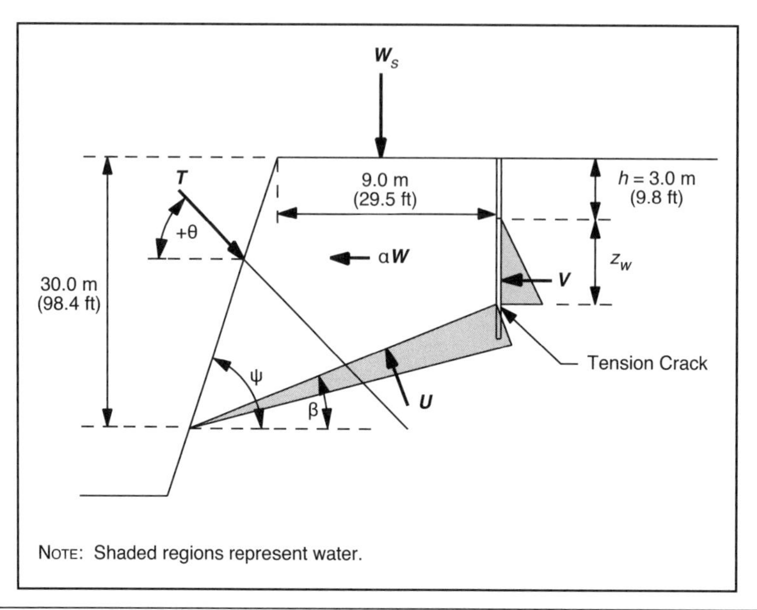

FIGURE 6.11 Slope geometry for example problem

Solving Plane Shear Problems: An Example

A rock slope in a mine 42 m (138.7 ft) high and with an overall slope angle of 50° is intersected by a fault that dips at 35° in the same direction as the slope face (Figure 6.11). The fault daylights in the slope face 30 m (98.4 ft) below the crest. A tension crack has developed 9 m (29.5 ft) behind the crest of the slope.

Water in the slope has always been a problem. Seepage of groundwater is occurring where the fault intersects the slope face. Also, static water was detected 3 m (9.8 ft) deep in the tension crack.

Laboratory testing of the rock and the fault zone has been conducted to determine the unit weight of rock, as well as the cohesion and friction angle along the fault. A fill weighing approximately 37,000 kg/m (24,865.9 lb/ft) of slope width was placed between the crest and the tension crack before the fault was known to exist. The mine is located in a seismically inactive area; however, blasting in the vicinity has produced ground accelerations of around 0.10*g*. Table 6.1 summarizes the values of the slope geometry parameters for this example.

A proposal has been made to stabilize the slope, if needed, to an FS of 1.50 by using no. 14, grade 60, epoxy-coated rock dowels. The rock dowels have a yield strength of 413.69 MPa (60,000 psi) and a cross-sectional area of 14.516 cm^2 (2.25 in.^{2}) (DYWIDAG 1994). The rock dowels will be fully grouted (FOSROC 1994) with polyester resin and nominally tensioned (*passive condition*). The proposal calls for the rock dowels to be inserted at an inclination angle of 30° from the horizontal into the face of the slope and through the fault plane. The question of interest is what rock-bolt force magnitude, ***T***, will be required

TABLE 6.1 Slope geometry parameters for the example problem

Slope Geometry Parameter	Value
Slope height (H)	30 m (98.4 ft)
Slope angle (ψ)	50°
Dip of fault plane (β)	35°
Cohesion on fault plane (c)	7,000 kg/m^2 (9.97 psi)
Friction angle (ϕ)	30°
Surcharge weight per unit width ($\boldsymbol{W}_s$)	37,000 kg/m (24,866 lb/ft)
Horizontal distance between tension crack and slope crest	9 m (29.5 ft)
Depth of water in tension crack	3 m (9.8 ft)
Seismic acceleration	0.10g
Rock-bolt inclination angle (θ)	30°
Unit weight of rock (γ_r)	2.790 kg/m^3 (171.7 lb/ft^3)

to achieve the desired factor of safety for these conditions and the proposed stabilization technique.

Factor-of-safety analysis. To determine the factor of safety versus sliding, it is necessary to take the following steps:

1. Resolve the slope geometry.
2. Determine the magnitudes of the forces $\boldsymbol{W}$, $\boldsymbol{U}$, and $\boldsymbol{V}$.
3. Resolve the forces into components resisting sliding and components driving the failure; equate; and write the equation for factor of safety.
4. Determine the required rock-bolt force to stabilize the slope.

To resolve the slope geometry, the following values must be calculated:

- Length along the top of slope to the intersection of the fault (b):

$$b = \frac{H}{\tan\beta} - \frac{H}{\tan\psi} = \frac{30}{\tan 35°} - \frac{30}{\tan 50°} = 17.7 \text{ m (58.0 ft)}$$

- Length of the fault from where it daylights in the slope face to where it also daylights in the upper slope surface (L):

$$L = \sqrt{H^2 + \left(\frac{H}{\tan\beta}\right)^2} = \sqrt{30.0^2 + \left(\frac{30.0}{\tan 35°}\right)^2} = 52.3 \text{ m (171.6 ft)}$$

- Length along the upper slope surface from the intersection of the fault to the tension crack (b'):

$$b' = b - 9.0$$

(where 9.0 m is the distance of the tension crack from the crest of the slope.)

$$=17.7-9.0=8.7\text{ m }(28.4\text{ ft})$$

- Depth of the tension crack from the top of the slope surface to the fault (TC):

$$\text{TC}=b'\tan\beta$$
$$=(8.7)\tan 35°=6.1\text{ m }(19.9\text{ ft})$$

- Depth of water in the tension crack (z_w):

$$z_w=\text{TC}-3.0$$

(where 3.0 m is the depth to water in the tension crack.)

$$=6.1-3.0=3.1\text{ m }(10.1\text{ ft})$$

- Length of the fault from where it daylights in the slope face to the tension crack (L'):

$$L'=L-\sqrt{(b')^2+(\text{TC})^2}=52.3-\sqrt{8.7^2+6.1^2}=41.7\text{ m }(136.9\text{ ft})$$

- Magnitude of the horizontal water force, $\boldsymbol{V}$:

$$\boldsymbol{V}=\tfrac{1}{2}\gamma_w(\text{width}=1\text{ m})(z_w)^2$$
$$=\tfrac{1}{2}\gamma_w(1)(3.1)^2=4{,}805\text{ kg per meter of width }(3{,}230\text{ lb per foot of width})$$

- Magnitude of the uplift water force, $\boldsymbol{U}$:

$$\boldsymbol{U}=\tfrac{1}{2}\gamma_w(\text{width}=1\text{ m})(L')(z_w)$$
$$=\tfrac{1}{2}\gamma_w(1)(41.7)(3.1)$$
$$=64{,}635\text{ kg per meter of width }(43{,}440\text{ lb per foot of width})$$

- Weight magnitude of potential sliding mass, $\boldsymbol{W}$:

$$\boldsymbol{W}=\left[\tfrac{1}{2}(b)(H)-\tfrac{1}{2}(b')(\text{TC})\right](1\text{ m width})\gamma_r$$
$$=\left[\tfrac{1}{2}(17.7)(30)-\tfrac{1}{2}(8.7)(6.1)\right](1)\gamma_r$$
$$=666{,}700\text{ kg/m of width }(448{,}065\text{ lb/ft of width})$$

Given the values just calculated, we can determine the factor of safety for this example before a rock-bolt force is applied to help stabilize the slope:

$$\text{FS}=\frac{cL'+\left[(\boldsymbol{W}+\boldsymbol{W}_s)(\cos\beta-\alpha\sin\beta)-\boldsymbol{V}\sin\beta-\boldsymbol{U}\right]\tan\phi}{(\boldsymbol{W}+\boldsymbol{W}_s)\sin\beta+\boldsymbol{V}\cos\beta+\alpha(\boldsymbol{W}+\boldsymbol{W}_s)\cos\beta}=1.22$$

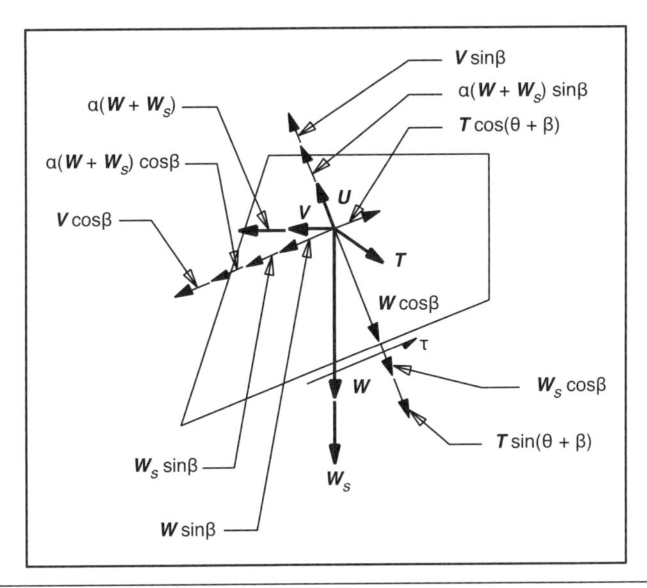

FIGURE 6.12 Free-body diagram of example problem

Required rock-bolt force. As shown in Figure 6.10, a rock-bolt force inclined from the horizontal at angle θ can be resolved into a component parallel to the potential failure plane, $T\cos(\theta + \beta)$, and a component normal to the potential failure plane, $T\sin(\theta + \beta)$. As stated previously, in the case of *passive rock-bolt force*, the component $T\cos(\theta + \beta)$ tends to decrease the driving force of the potential failure mass. The resolved rock-bolt forces, as well as the other resolved forces, are shown on the free-body diagram for this problem (Figure 6.12). The factor-of-safety equation for the slope geometry of Figure 6.11 becomes

$$\mathrm{FS}=\frac{cL'+\left[(W+W_s)(\cos\beta-\alpha\sin\beta)-V\sin\beta-U+T\sin(\theta+\beta)\right]\tan\phi}{(W+W_s)\sin\beta-T\cos(\theta+\beta)+V\cos\beta+\alpha(W+W_s)\cos\beta}$$

Rewriting this equation, letting FS = 1.5 (the required value), letting θ = 30°, and solving for T, we get the following (breaking a very long equation into three parts):

$$\begin{aligned}T=&\frac{(W+W_s)\{\mathrm{FS}\cdot\sin\beta-[\cos\beta-\alpha(\sin\beta)]\tan\phi\}}{\sin(\theta+\beta)\tan\phi+\mathrm{FS}\cdot\cos(\theta+\beta)}\\&+\frac{(V\sin\beta+U)\tan\phi-cL'}{\sin(\theta+\beta)\tan\phi+\mathrm{FS}\cdot\cos(\theta+\beta)}\\&+\frac{\mathrm{FS}[V\cos\beta+\alpha(W+W_s)\cos\beta]}{\sin(\theta+\beta)\tan\phi+\mathrm{FS}\cdot\cos(\theta+\beta)}\\=&\,116{,}900\ \text{kg/m}\ (78{,}550\ \text{lb/ft})\ \text{of slope width}\end{aligned}$$

Conducting the same derivation of an equation for the *active rock-bolt force* required to stabilize the slope geometry of Figure 6.11 to a factor of safety of 1.5, we get the following:

$$\begin{aligned} \boldsymbol{T} &= \frac{\mathrm{FS}\cdot\left[\left(\boldsymbol{W}+\boldsymbol{W}_s\right)\cdot\sin\beta+\boldsymbol{V}\cos\beta+\alpha\cdot\left(\boldsymbol{W}+\boldsymbol{W}_s\right)\cos\beta\right]}{\sin(\theta+\beta)\tan\phi+\cos(\theta+\beta)} \\ &- \frac{cL'}{\sin(\theta+\beta)\tan\phi+\cos(\theta+\beta)} \\ &+ \frac{\tan\phi\cdot\left[\boldsymbol{V}\sin\beta+\boldsymbol{U}-\left(\boldsymbol{W}+\boldsymbol{W}_s\right)\left(\cos\beta-\alpha\cdot\sin\beta\right)\right]}{\sin(\theta+\beta)\tan\phi+\cos(\theta+\beta)} \\ &= 143{,}100 \text{ kg/m } (96{,}160 \text{ lb/ft}) \text{ of slope width} \end{aligned}$$

If we calculate the active and passive rock-bolt force ($\boldsymbol{T}_{\text{Active}}$ and $\boldsymbol{T}_{\text{Passive}}$) to stabilize the slope of Figure 6.11 to a factor of safety of 1.5 at a range of inclination angles from $\theta = +70°$ to $\theta = -30°$, we get results as shown in Table 6.2 and Figure 6.13.

ROTATIONAL SHEAR

The most common examples of rotational slides are little-deformed slumps (Figures 6.14 and 6.15), which are slides along a surface of roughness that is curved concavely upward. In many cases, the rupture surface of a rotational slide has the general appearance as if a great tablespoon had taken a scoop out of the earth.

Rotational-type slides are also often referred to as "circular failures," although the failure surface is rarely a purely circular segment. Most often the failure surface is partially to wholly controlled by structure—which may be a weak clay layer, a fault, a discontinuity or set of discontinuities, or various soil layer contacts—that dictates its shape. The influence of such discontinuities or other weak layers must be properly taken into account when conducting a stability analysis that assumes a certain configuration for the surface of rupture.

In slumps and other rotational-type failures, the movement is more or less rotational about an axis that is parallel to the slope (Varnes 1978). In the head area, the movement may be almost wholly downward and have little apparent rotation; however, the top surface of each unit commonly tilts backward toward the slope, but some blocks may tilt forward. More than one scarp may exist at the head, a main scarp plus additional minor scarps. The flanks, or sides, of the rotational failure are areas of shear where the moving mass has sheared through the intact mass with little or no rotation involved. At the toe area is a zone of bulging where the sliding mass has either piled up and over the intact ground mass or it has slid into the intact ground mass in a plowing manner causing it to bulge upward, or both. Longitudinal and transverse cracks are often visible in the zone of bulging. Springs or seeps may occur in the failed mass as a result of the release of water pressure, which may have contributed to the failure of the mass (Figures 6.16 and 6.17).

TABLE 6.2 T_{Active} and $T_{Passive}$ values versus θ angle for the slope of Figure 6.11

Angle θ (+/–)	T_{Active} (kg/m of width)	$T_{Passive}$ (kg/m of width)
70	452,768	798,551
60	277,283	304,473
50	204,306	191,693
40	165,725	143,051
30	143,056	116,933
20	129,299	101,488
10	121,319	92,118
0	117,633	86,746
–10	117,633	84,388
–20	121,319	84,659
–30	129,299	87,604

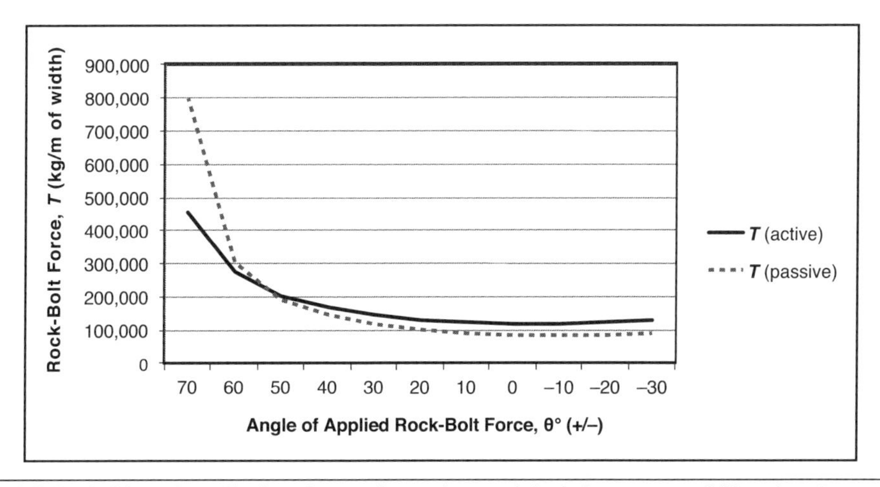

FIGURE 6.13 Plot of applied active and passive rock-bolt force to stabilize the slope of Figure 6.11 to an FS of 1.5

Numerous methods have been devised over the years for rotational shear stability analysis, with most consideration being given to 2-D analysis techniques, although much recent work has been done with three-dimensional analysis tools.

The numerous approaches available for slope stability analysis can be categorized as either (1) limit (plasticity-type) formulation or (2) displacement formulation, such as the finite element or finite difference methods (Fredlund 1984). Limit formulations provide a theoretical context for understanding the range of answers that can be expected from a slope stability analysis (Mendelson 1968). These formulations are referred to as *upper-* and *lower-bound* solutions. The method of characteristics for stresses is an example of a lower-bound solution (Fredlund 1984). The method of characteristics for displacements is an example of an upper-bound solution (Fredlund 1984).

FIGURE 6.15 Small rotational failure (slump) in the slope of a road cut

Courtesy of Jonathan Brinson

FIGURE 6.14 Slumps in dragline spoil heaps at a surface coal mine

A *complete* limit analysis of a soil or soil-like material requires that three conditions be satisfied, namely, the stress equilibrium equations, the stress–strain relations, and the compatibility equations relating strain and displacement (Fredlund 1984). In the absence of satisfying all of these equations, it is possible to satisfy only certain conditions. This gives rise to the lower- and upper-bound solutions. The upper-bound solution results in a higher factor of safety than for the complete solution, and vice versa for the lower-bound solution (Hoek and Bray 1977).

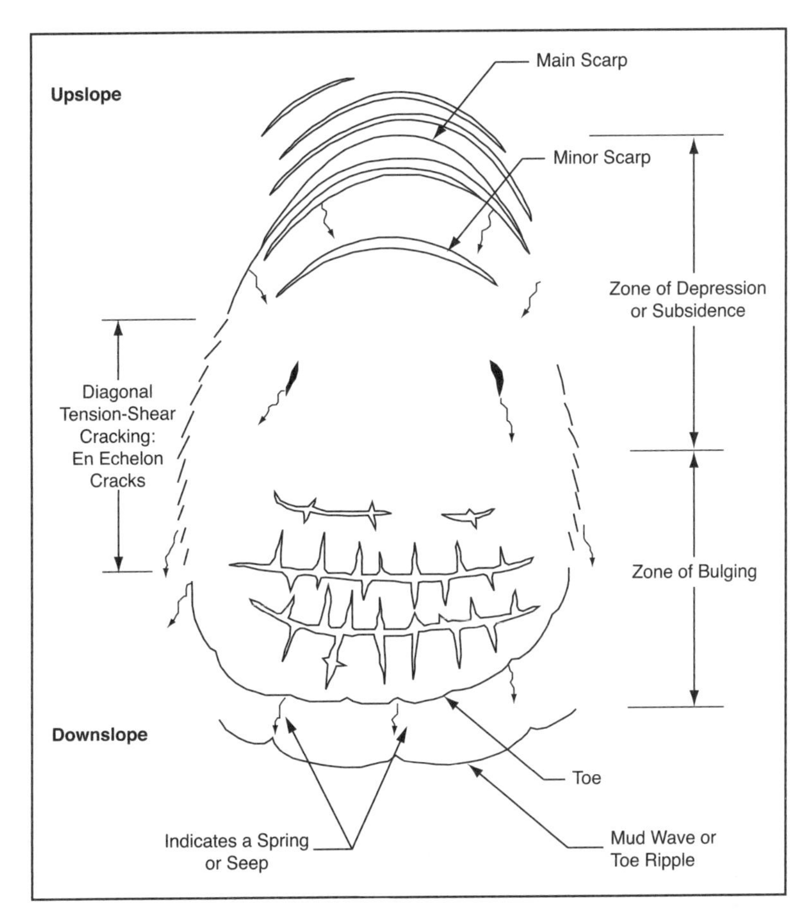

Source: Varnes 1-978

FIGURE 6.16 Typical rotational-type failure topography illustrating cracks, bulges, scarps, and springs

FIGURE 6.17 Evidence of rotation: (1) material from head of slide tilts backward toward the slide; (2) other blocks are tilting forward

According to Fredlund (1984), the lower-bound solution satisfies the equilibrium equations, the stress boundary conditions, and the failure criteria for the soil. The resulting stresses yield a statically admissible stress field for the problem. The lower-bound solution normally gives no consideration to kinematics. The soil approaches the failure condition as a result of applying a load. Fredlund (1984) states that it is possible, however, to factor down the strength of the soil and thereby reduce the applied load to zero for slope stability problems. Classically, this method has been used mainly for bearing capacity–type problems.

The upper-bound solution satisfies velocity boundary conditions, strain and velocity compatibility conditions, and the failure criterion of the soil (Fredlund 1984). This results in a kinematically admissible velocity field, which can also be thought of as a displacement field if all processes are assumed to be independent of time (Fredlund 1984). As the stress conditions are defined only in the region that is deforming, they need not be in equilibrium. Interfaces within the deforming mass must also be at the yield conditions, or at limiting equilibrium. The shear strength parameters can be reduced by a constant—the factor of safety. The upper-bound solution can be obtained by either the energy approach or the equilibrium approach, both developing the same results. Therefore, the equilibrium approach is of the most interest in solving slope stability problems, and the "limiting equilibrium" method common to slope stability analysis is an upper-bound-type of solution because it specifies a simplified mode of failure and applies statics (Fredlund 1984).

Although limit equilibrium methods adopt the general philosophy of an upper-bound solution, they do not meet all the precise requirements (Fredlund 1984). Limit equilibrium methods fall short of a complete solution in that no consideration is given to kinematics and that equilibrium conditions are only satisfied in a limited case (Fredlund 1984). Izbicki (1981) called the results from limit equilibrium methods a "reduced" upper-bound solution in that the limit equilibrium solutions are somewhat less than the upper-bound factor of safety.

Lambe and Whitman (1969), Taylor (1948), and Hoek and Bray (1977) all have examined whether the upper-bound solution or the lower-bound solution is a meaningful practical solution to the factor of safety for slope stability problems. All suggest that the actual factor of safety for simple circular failure of a homogeneous slope lies reasonably close to the lower-bound limit. Both Taylor (1948) and Hoek and Bray (1977) developed stability charts to aid in the solution of simplified rotational failure problems. Taylor's stability charts are illustrated in Figure 6.18.

Given that the solution to a slope stability problem is statically indeterminate, several methods of analysis are available to the engineer, including the limit equilibrium method (LEM), the limit analysis method, the finite element method, or the finite difference method. By far, most engineers still use the LEM with which they are most familiar.

Computerized analysis techniques are the preferred method of solution for slope stability problems, including rotational failure. Some of the early LEMs, however, are fairly simple

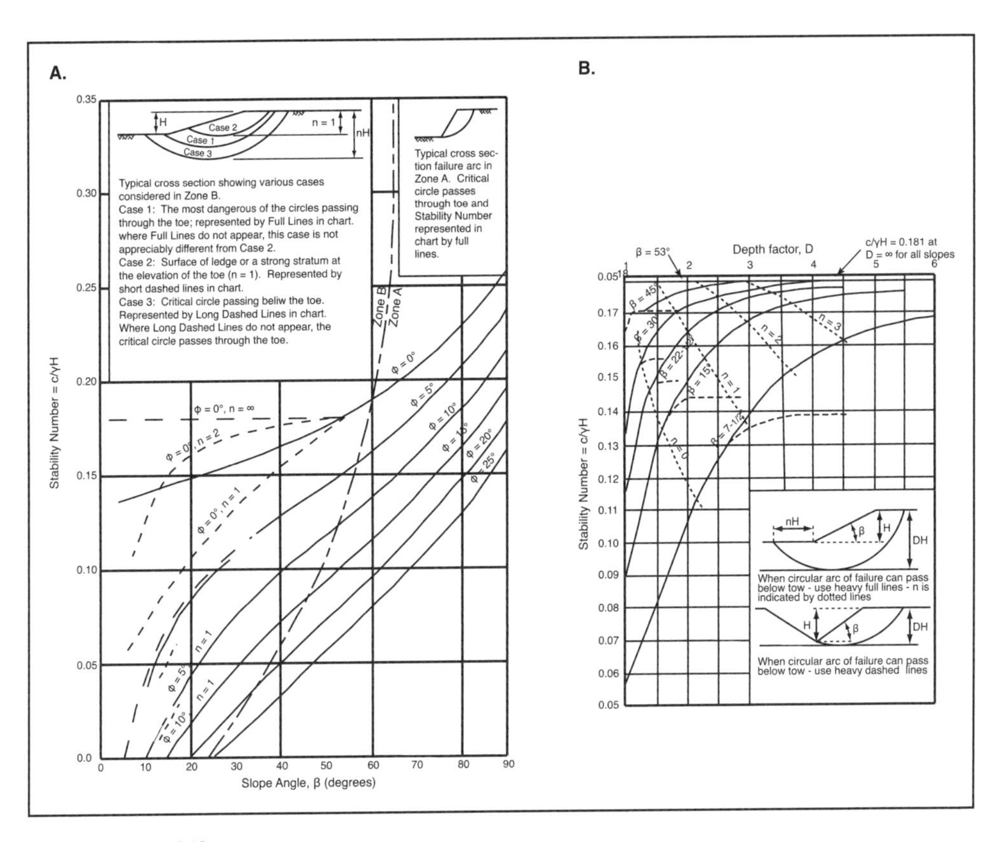

Source: Taylor 1948

FIGURE 6.18 (A) Stability chart for slopes in homogeneous material; (B) effect of depth limit on Taylor's curves

and can be computed by hand methods with a calculator or by using the various stability charts.

The greatest number, though, of stability analyses carried out is for cases in which the materials involved are stratified, or sufficiently heterogeneous, so that the use of stability charts provides only a very rough guide (Perloff and Baron 1976). In such cases, it is necessary to consider conditions within the slide mass, to take into account the material variability. The method by which this is done is called the *method of slices*. Most of the computerized limit equilibrium techniques in use are based on the method of slices, which can be vertical, horizontal, or inclined. The first slice technique (Fellenius 1936) was based more on engineering intuition than on a rigorous mechanics principle. There was a rapid development of the slice methods in the 1950s and 1960s by Bishop (1955), Janbu et al. (1956), Lowe and Karafiath (1960), Morgenstern and Price (1965), and Spencer (1967).

The Limit Equilibrium Method of Slices

LEMs of slices have been widely used for assessing stability of natural or human-made slopes. Numerous (more than 10) methods of slices have been developed dealing with circular or arbitrarily shaped slip surfaces, the major difference between the methods being the assumptions made to render the problem statically determinate. Common features of the LEM of slices are as follows (Zhu et al. 2003):

1. The sliding body over the failure surface is divided into a finite number of slices (generally vertical).
2. The strength of the slip surface is mobilized to the same degree to bring the sliding body into a limiting state.
3. Assumptions regarding interslice forces are employed to render the problem determinate.
4. The factor of safety is computed from force and/or moment equilibrium equations.

The LEM can be broadly classified into two main categories: "simplified" methods and "rigorous" methods. For the simplified methods, either force or moment equilibrium can be satisfied but not both at the same time. For the rigorous methods, both force and moment equilibrium can be satisfied, but usually the analysis is more tedious and may sometimes experience non-convergence problems.

The Generalized Method of Slices*

For a slope that may be composed of many materials that may vary in both horizontal and vertical directions, and which may be subjected to additional forces such as water pressure, seismic forces, surcharge or reinforcing, it is worthwhile to consider a generalized failure surface as shown in Figure 6.19A. This figure shows a slope with some loading acting on the surface and a failure surface of a general shape. It may also be subjected to pore pressures that may vary with position.

To investigate conditions at a point along the failure surface, the slope has been divided into several thin *slices*. A typical one, the *i*th slice, is shown in Figure 6.19B. The slice is of thickness Δx_i, with one of its faces at position x and the other face at position $x + \Delta x_i$. The forces that act on the slice, shown in Figure 6.19B, are the horizontal and vertical components of the resultant external loading, $\boldsymbol{Q}_{ih}$ and $\boldsymbol{Q}_{iv}$, respectively, and the weight of the slice, $\boldsymbol{W}_i$. The horizontal forces $\boldsymbol{E}_x$ and $\boldsymbol{E}_{(x+\Delta x)}$ and the shear forces $\boldsymbol{T}_x$ and $\boldsymbol{T}_{(x+\Delta x)}$ act on the vertical faces and represent the interaction between the slice and the slice next to it. The normal force on the bottom of the slice $\boldsymbol{N}_i$, and the shear force, $\boldsymbol{S}_i$, are the forces on the potential failure surface. If Δx is assumed to be small enough, the segment of arc on the base can be considered as a straight line (NOTE: This does not have to be the assumption).

* Much of this section has been adapted from Perloff and Baron (1976).

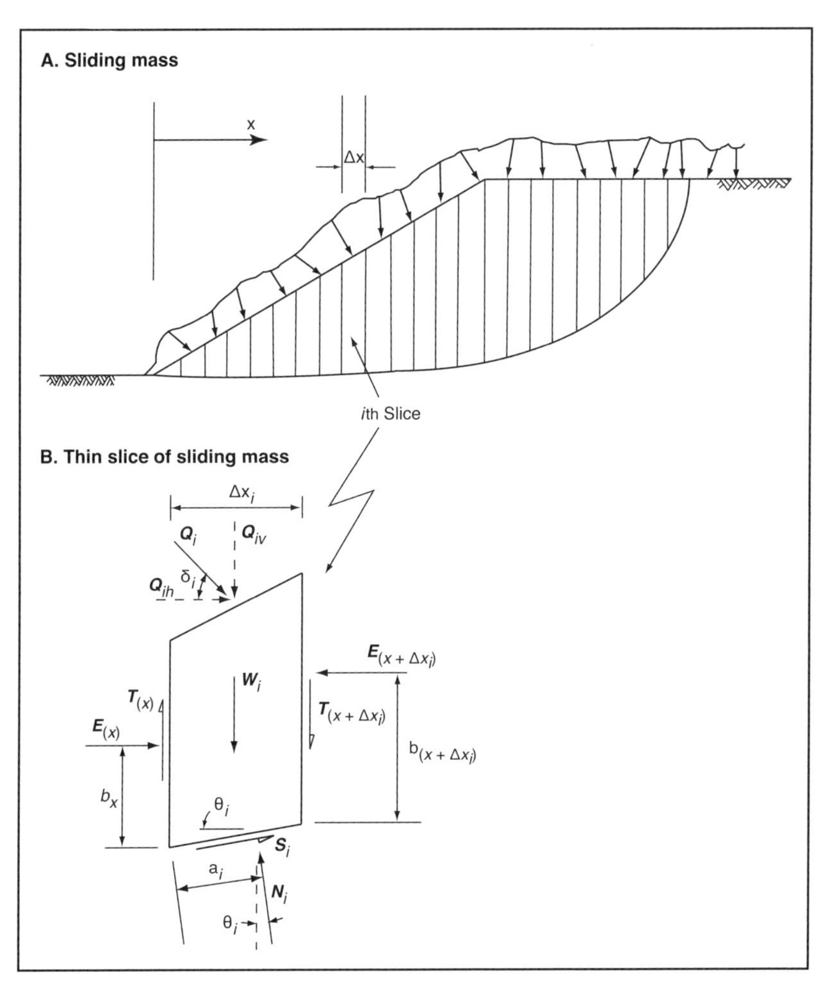

Source: Perloff and Baron 1976

FIGURE 6.19 Division of the general slide mass into slices

Commonly, water pressure on the base of a slice may be determined by one of two ways (Figure 6.20). The most correct way is to measure the difference in elevation *along the equipotential* that passes through the center of the base of the slice. This requires constructing the vertical flow net beneath the phreatic surface. The equipotential, or line of equal potential, is constructed perpendicular to each of the vertically spaced flow lines passing through the slice. The second, most common, way is to simply determine the vertical head at the center of the slice. In either case, the water pressure is the water head times the unit weight of water (in units of meters or feet times metric tons per cubic meter or pounds per cubic foot).

Going back to the preliminary discussion of rotational failure and Figure 1.17, we see that for n slices we find a total of $6n - 3$ unknowns, and $3n$ equations to work with, as shown

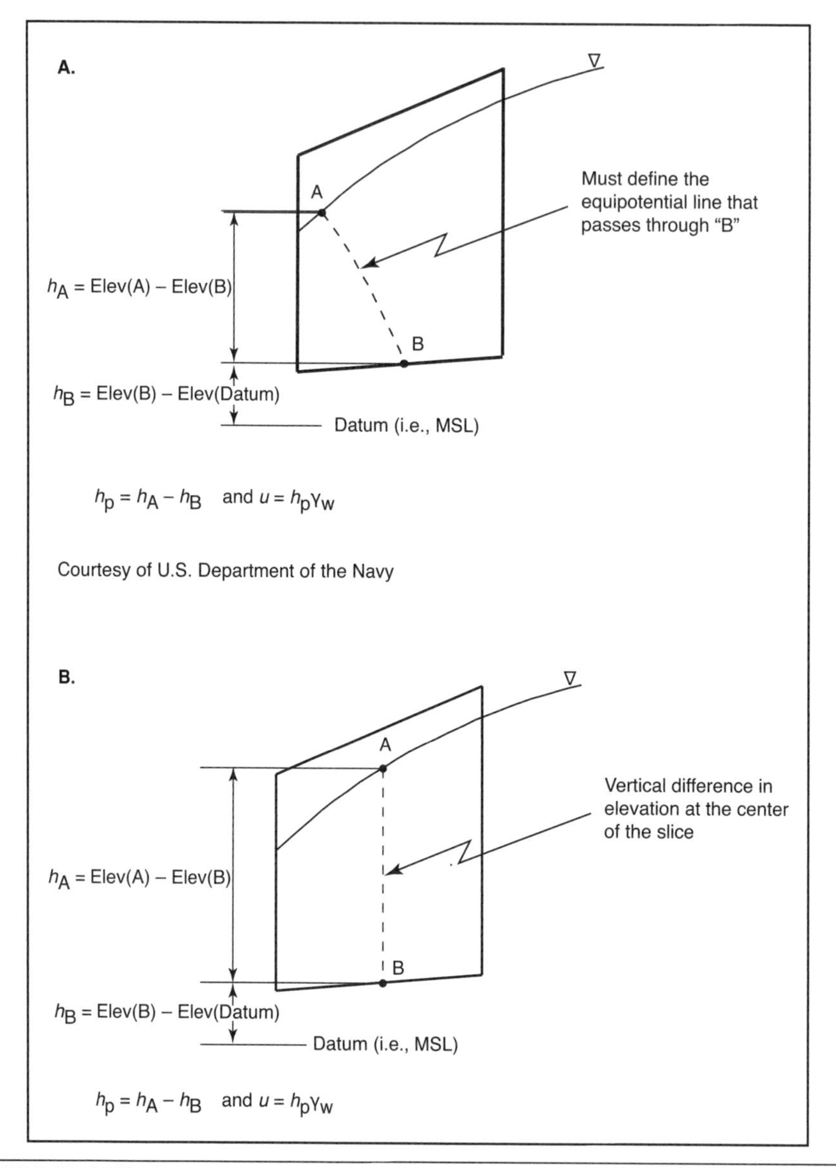

FIGURE 6.20 Definition of water pressure by (A) difference in elevation along an equipotential, and (B) difference in vertical elevation at the center of the slice

in Table 6.3, assuming the slope is in equilibrium. Thus, we need to make some additional assumptions that will give us a sufficient number of conditions to make the available equations equal the number of unknowns. The assumptions we will make are as follows (Perloff and Baron 1976):

1. There is a fixed relationship between $\boldsymbol{N}_i$ and $\boldsymbol{S}_i$ that is

$$\boldsymbol{S}_i = \frac{1}{\text{FS}}\left(c_i \Delta x_i \sec\theta_i + \boldsymbol{N}_i \tan\phi_i\right) \qquad \text{(EQ 6.1)}$$

TABLE 6.3 Factors in equilibrium formulation of slope stability for *n* slices

Unknown	Number
$\boldsymbol{E}_x$	$n - 1$
$\boldsymbol{T}_x$	$n - 1$
b_x	$n - 1$
$\boldsymbol{N}_i$	n
$\boldsymbol{S}_i$	n
a_i	n
Total unknowns	$6n - 3$

in which FS, the factor of safety, is assumed to be constant at every point along the failure surface, c_i is the cohesive component of strength, and $\Delta x_i \cdot \sec\theta_i$ is the length of the failure surface within the slice.

Because Equation 6.1 applies to each slice, it is equivalent to adding n additional equations so that there are now $4n$ equations available to work with. However, the FS is one additional unknown, so that the number of unknowns now becomes $6n - 2$.

2. We shall assume that at any point x, the relationship between $\boldsymbol{E}_x$ and $\boldsymbol{T}_x$ is only a function of x, that is,

$$\frac{\boldsymbol{T}_x}{\boldsymbol{E}_x} = \lambda f(x) \qquad \text{(EQ 6.2)}$$

in which λ is a constant, and $f(x)$ is some function of x that we shall assume. For n slices, Equation 6.2 corresponds to $n - 1$ additional relationships, increasing the number of equations available to $5n - 1$. However, λ is an additional unknown, bringing the total of unknowns to $6n - 1$.

3. We also require that Δx approaches zero, or at least becomes so small that α_i becomes negligible, that is, $\alpha i \cong 0$. This eliminates n unknowns, bringing the total to $5n - 1$, which is now the same as the number of equations.

Thus we see that for any assumed function $f(x)$, we can, in principle, determine the forces acting on each slice and thereby evaluate the stability of the slope.

The ratio $\boldsymbol{T}_x/\boldsymbol{E}_x$ really defines the orientation of the resultant force on the side of each slice. So $f(x)$ describes the way in which the direction of the resultant force on a vertical slice changes within the sliding mass. The choice, therefore, of an appropriate $f(x)$ is a matter of engineering judgment. The primary criteria applied to make this choice are that the location of the line of thrust, b_x, is "reasonable," and the required $\boldsymbol{T}_x$ on the sides of the slices be smaller than the maximum permitted by the shear strength of the soil (Morgenstern and Price 1965).

Because of the difficulty of choosing the appropriate $f(x)$ for a given situation, the generalized method of slices is often simplified by using various assumptions regarding the interslice force along with various combinations of equilibrium conditions (force and/or moment) considered, giving different values of factor of safety for the same slip surface (Zhu et al. 2003). For a detailed discussion of the calculus of the generalized method of slices formulation, see Zhu et al. (2003).

Taylor's Stability Charts Method

For two similar slopes made from two different soils, it can be shown that the ratio

$$\frac{c_m}{\gamma \mathbf{H}} \tag{EQ 6.3}$$

is the same for each slope, provided that the two soils have the same angle of friction (Smith 1982).

Equation 6.3 is known as the stability number and is given the symbol $\boldsymbol{N}_s$, where

c_m = mobilized cohesion with respect to total stress
γ = unit weight of the soil
$\mathbf{H}$ = vertical height of the embankment

Taylor (1948) prepared two sets of curves that relate the stability number to the angle of slope. The first (Figure 6.18A) is for the general case of $c - \phi$ soil with a slope angle less than 53°, and the second one (Figure 6.18B) is for a soil with $\phi = 0$ and a layer of stiff material or rock at depth **DH** below the top of the embankment. **D** is known as the *depth factor*, and depending on its value, the slip circle will either emerge at a distance $n\mathbf{H}$ in front of the toe or pass through the toe (the value of n can be determined from the set of curves).

Figure 6.21 illustrates some general comments on circular failure using Taylor's charts. Taylor's method using his stability charts is as follows.

1. The factor of safety with respect to cohesion, FS_c:

$$FS_c = \frac{c}{c_d} = \frac{\text{actual cohesion}}{\text{cohesion for } FS_c = 1}$$

2. The factor of safety with respect to friction, FS_ϕ:

$$FS_\phi = \frac{\tan\phi}{\tan\phi_d} = \frac{\text{actual friction}}{\text{friction required for } FS_\phi = 1}$$

3. For stability: $FS = FS_c = FS_\phi$

The solution procedure for Taylor's method using his stability charts is as follows:

1. Estimate (guess at) FS_ϕ.

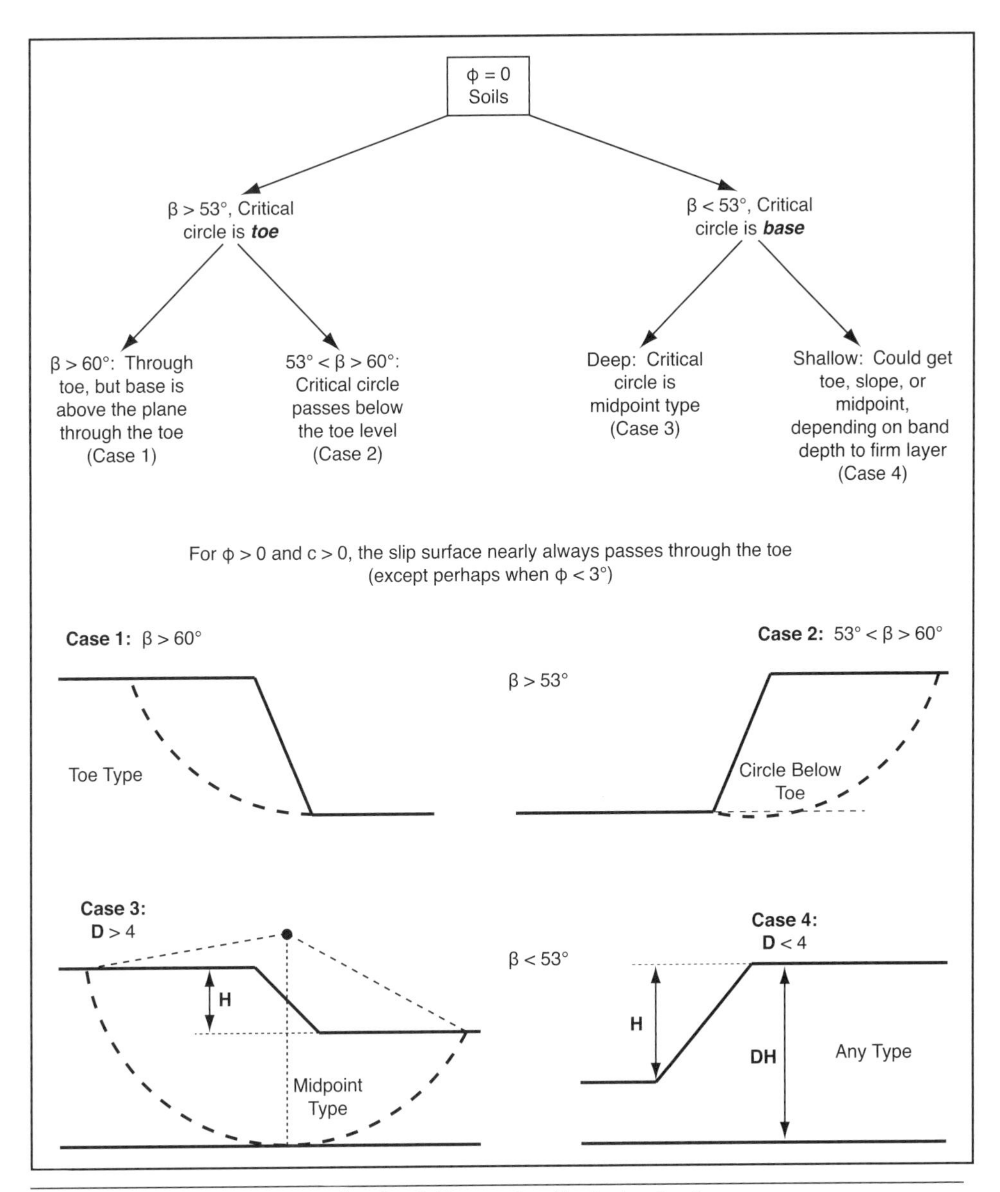

FIGURE 6.21 Comments on circular failure using Taylor's charts

2. Calculate $\phi_d = \tan^{-1}\left[\dfrac{\tan\phi}{FS_\phi}\right]$.

3. Determine the *stability number*, N_s, from Taylor's charts (Figure 6.18), using ϕ_d as ϕ on the chart.

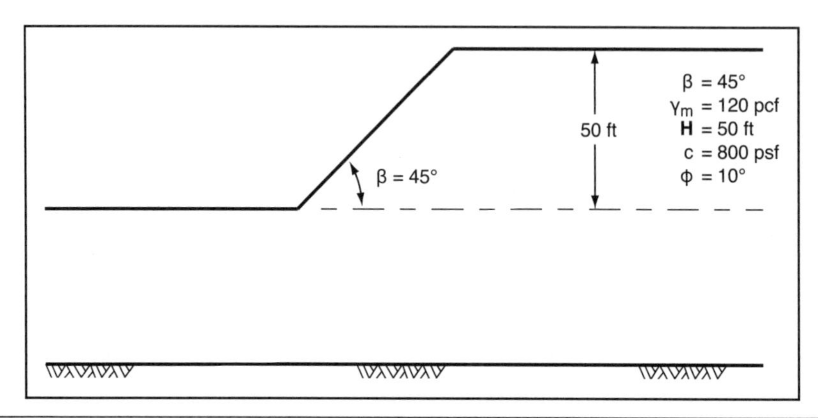

FIGURE 6.22 Simple slope for example using Taylor's charts

4. Calculate $c_d = N_s{\cdot}\gamma{\cdot}\mathbf{H}$.
5. Calculate $FS_c = \dfrac{c}{c_d}$, where c = the known cohesion.
6. If $FS_c = FS_\phi$, then stop. $FS = FS_c = FS_\phi$.
7. If $FS_c \neq FS_\phi$, then do steps 1–5 again for another assumed FS_ϕ.

   NOTE: A minimum of two trials is usually necessary. The factor of safety (FS) is determined by interpolation.

An Example Using Taylor's Stability Charts Method

For the given slope and attributes of Figure 6.22, determine the factor of safety using Taylor's stability charts method.

Use Taylor chart of Figure 6.23.

1. Assume $FS_\phi = 1.0 = \dfrac{\tan\phi}{\tan\phi_d}$

$$\phi_d = \tan^{-1}\left[\frac{\tan\phi}{FS_\phi}\right] = \tan^{-1}\left[\frac{\tan(1.0)}{1.0}\right] = 10°$$

2. For $\beta = 45°$ and $\phi_d = 10°$, we get

$$N_s = 0.108 \text{ (on Figure 6.23, line — —)}$$

$$N_s = 0.108 = \frac{c_d}{\gamma H} = \frac{c_d}{(120)(50)},\ c_d = 648.0$$

$$FS_c = \frac{c}{c_d} = \frac{800}{648} = 1.23 \text{ (plotted as ● on Figure 6.24)}$$

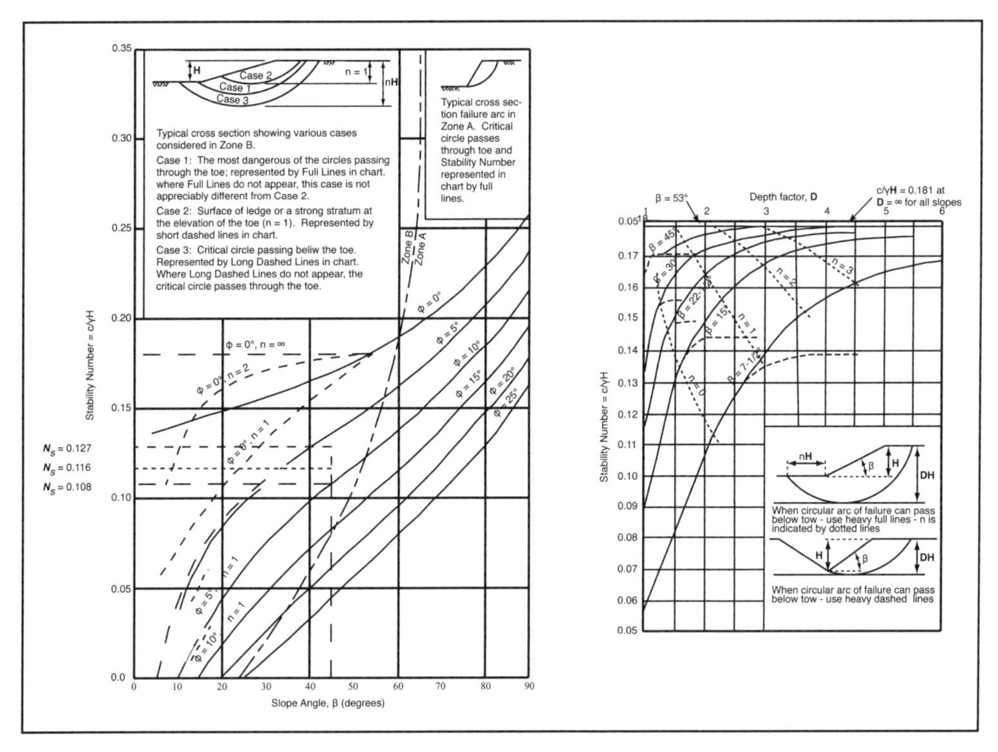

Source: Taylor 1948

FIGURE 6.23 Determination of N_s for example with ß = 45° and ϕ_d assumed using Taylor's charts

3. Assume $FS_\phi = 1.50 = \dfrac{\tan\phi}{\tan\phi_d}$

$$\phi_d = \tan^{-1}\left[\frac{\tan\phi}{F_\phi}\right] = \tan^{-1}\left[\frac{\tan(1.0)}{1.5}\right] = 6.7044°$$

4. For $\beta = 45°$ and $\phi_d = 6.7044°$, we get

$$N_s = 0.127 \text{ (on Figure 6.23, line — — —)}$$

$$N_s = 0.127 = \frac{c_d}{\gamma H} = \frac{c_d}{(120)(50)},\ c_d = 762.0$$

$$FS_c = \frac{c}{c_d} = \frac{800}{762} = 1.05 \text{ (plotted as ■ on Figure 6.24)}$$

5. Try $FS_\phi = 1.20 = \dfrac{\tan\phi}{\tan\phi_d}$

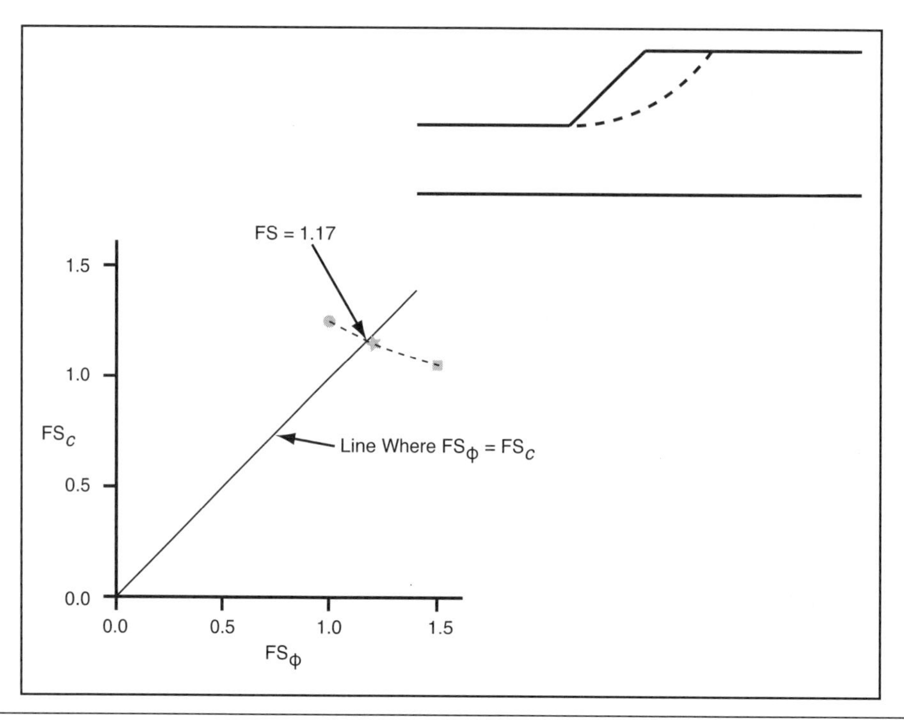

FIGURE 6.24 Determination of FS by interpolation for Taylor's method

$$\phi_d = \tan^{-1}\left[\frac{\tan\phi}{F_\phi}\right] = \tan^{-1}\left[\frac{\tan(1.0)}{1.2}\right] = 8.3592°$$

6. For $\beta = 45°$ and $\phi_d = 8.3592°$, we get

$$N_s = 0.116 \text{ (on Figure 6.23, line - - - - - -)}$$

$$N_s = 0.116 = \frac{c_d}{\gamma H} = \frac{c_d}{(120)(50)},\ c_d = 696.0$$

$$FS_c = \frac{c}{c_d} = \frac{800}{696} = 1.15 \text{ (plotted as ★ on Figure 6.24)}$$

From Figure 6.24, the factor of safety (FS), by interpolation, is approximately 1.17.

Factor of Safety by Bishop's Modified Method

Tabular forms for the determination of the factor of safety by Bishop's modified method (BMM) have been developed and are presented in Figure 6.25 and Table 6.4. Figure 6.25 can be used to determine the weights of the slices; Table 6.4 is used for calculating the safety factor of the trial slip surface.

b

h_i

Layer i

γ_i = unit weight of layer i
h_i = height of center of slice
$\boldsymbol{W}_i$ = partial weight = $(b)(h_i)(\gamma_i)$
$\Sigma \boldsymbol{W}_i$ = total weight of slice

Slice No.	b	h_i	γ_i	W_i	ΣW_i

FIGURE 6.25 Tabular form for computing weights of slices

Regardless of the specific procedure for carrying out the computations, the following principles are common to all methods of limit equilibrium analysis:

- A slip mechanism is postulated. This is done without any major kinematic restriction except that the mechanism be feasible and sensible.
- The shearing resistance required to equilibrate the assumed slip mechanism is calculated by means of statics. The physical concepts used here are that the potential slip mass is in a state of limiting equilibrium and the failure criterion of the soil or rock is satisfied everywhere along the proposed surfaces.
- The calculated shearing resistance required for equilibrium is compared with the available shear strength. This comparison is made in terms of the *factor of safety*.
- The mechanism and/or the trial slip surface with the lowest factor of safety is found by iteration. This is the *critical surface*.

The solution procedure for the BMM is as follows for a *single trial slip surface*:

1. Draw the slope profile to an appropriate scale.
2. Assume a trial slip surface with center at (x, y) relative to a local (0, 0) coordinate on the drawing.
3. Determine the material properties, strength attributes, and groundwater situation of any soil layer within the profile.
4. Divide the trial slip surface into n vertical slices. Usually 8 to 12 slices are sufficient. The slices should correspond to changes in slope geometry and/or moment direction, and the base should be wholly contained in a layer.
5. Determine the weight, W, of each slice using the tabular form of Figure 6.25.
6. Assume a trial value for FS.

The factor of safety of the trial surface via BMM is

$$FS = \frac{\sum_{n=1}^{n} \left[\frac{c_n b_n + \left[\left(W_n + Q_{nv} - u_n b_n \right) + \left(x_n - x_{n+1} \right) \right] \tan\phi_n}{\cos\alpha_n + \frac{\sin\alpha_n \tan\phi_n}{FS}} \right]}{\sum_{n=1}^{n} \left[\left(W_n + Q_{vh} \right) \sin\alpha_n - Q_{hn} \cdot \frac{d_n}{R} \right]} \quad \text{(EQ 6.4)}$$

where

$$\cos\alpha_n + \frac{\sin\alpha_n \tan\phi_n}{FS} = M_\alpha \quad \text{(EQ 6.5)}$$

and where

d_n = moment arm from the center of the trial surface to the center of each slice

x_n and x_{n+1} = side shear forces; and for equilibrium, $\Sigma(x_n - x_{n+1}) = 0$, so we can neglect them

α = angle of inclination from the horizontal of the base of the slice

7. For each slice, determine the following:
   - Cohesion, c, along the base of the slice
   - Friction, ϕ, along the base of the slice
   - The pore water pressure, u, at the center of the base of each slice (refer to Figure 6.20)
   - The weight of each slice from Figure 6.25
   - The thickness of the slice, b
   - The angle of inclination of the slice base with the horizontal, α
   - M_α from Equation 6.5
8. Determine the FS using Equation 6.4 (tabulating in Table 6.4) for the trial FS no. 1.
9. If the FS of step 8 = the assumed trial FS of step 6, then STOP. This is the FS for this trial slip surface of step 2.
10. If the FS of step 8 ≠ the assumed trial FS of step 6, then assume a second trial FS and redo steps 6–8.
11. Do a third or fourth trial FS, if necessary.
12. The FS for the first trial slip surface of step 2 can be determined by interpolation or by plotting the $FS_{Assumed}$ versus $FS_{Calculated}$ values similar to Figure 6.24.

Example. Figure 6.26 shows a soil embankment composed of five dipping layers with the material properties as designated. We want to determine the FS using BMM for the trial slip surface shown with center at 9.9, 75.5. Coordinate (0, 0) is located at the toe of the slope; positive X coordinate values are into the slope (left to right), and positive Y coordinate values are from the toe of the slope upward. Figure 6.27 shows the same slope with 11 vertical slices used for the stability analysis.

The table portion of Figure 6.25 filled in with appropriate calculated values (English customary units) is presented in Figure 6.28 for the given trial circular failure with center at (9.9, 75.5). Table 6.5 shows Table 6.4 filled in with the appropriate values for three iterations of FS for the same trial circular failure, and Figure 6.29 shows the *critical circle* for this sample problem determined by computerized techniques using BMM. Fifty iterations were performed via computerized techniques for this sample problem to determine the critical circle with the center shown within the contoured box.

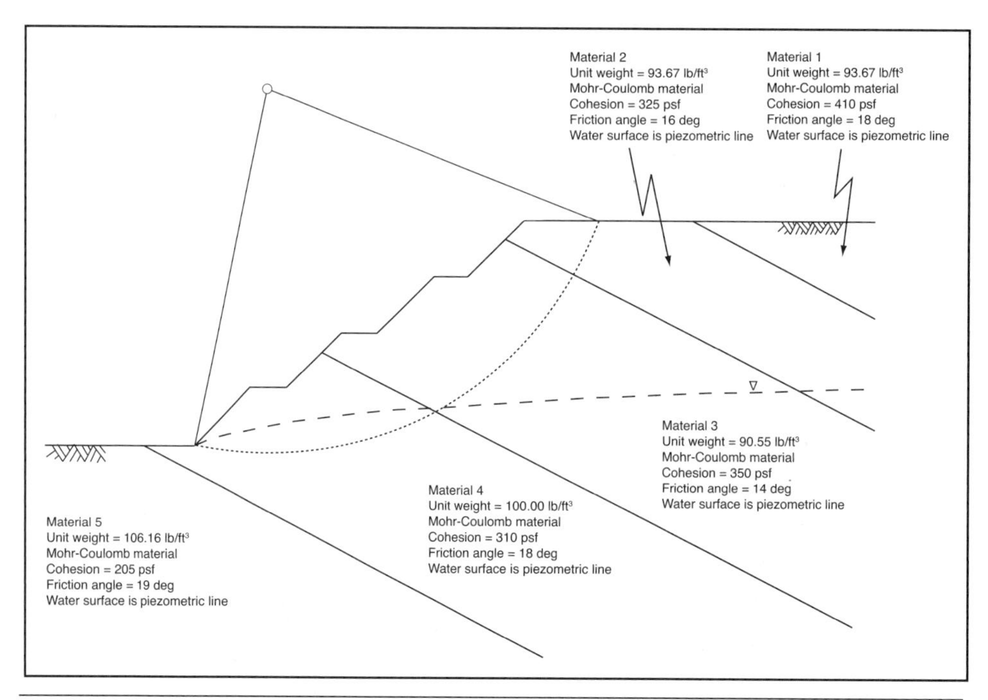

FIGURE 6.26 Trial slip surface with center at 9.9, 75.5 where 0,0 is located at the toe of the slope

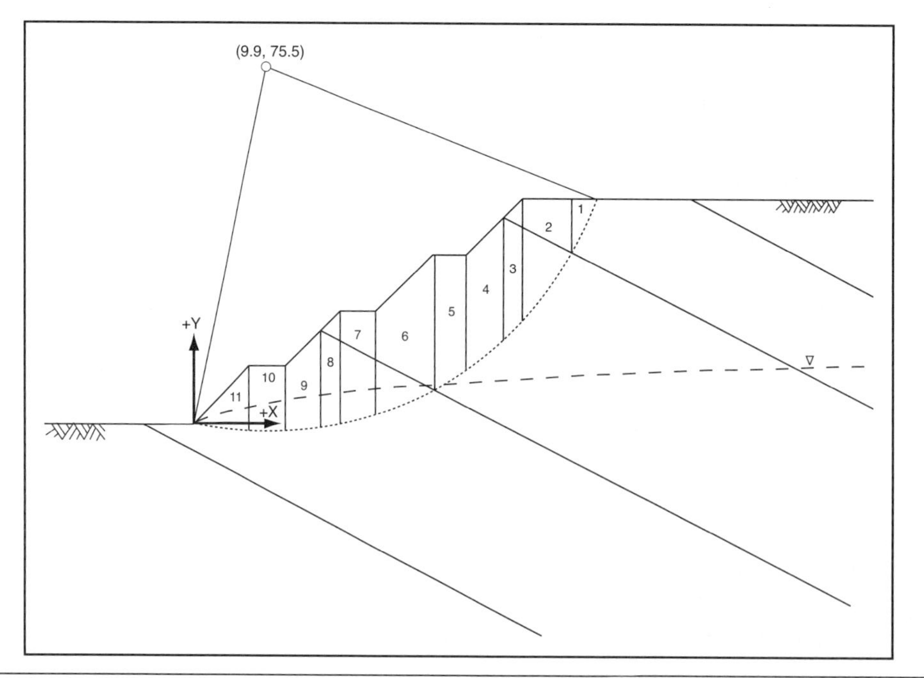

FIGURE 6.27 Trial slip surface with center at 9.9, 75.5 showing 11 vertical slices used for FS determinationby BMM

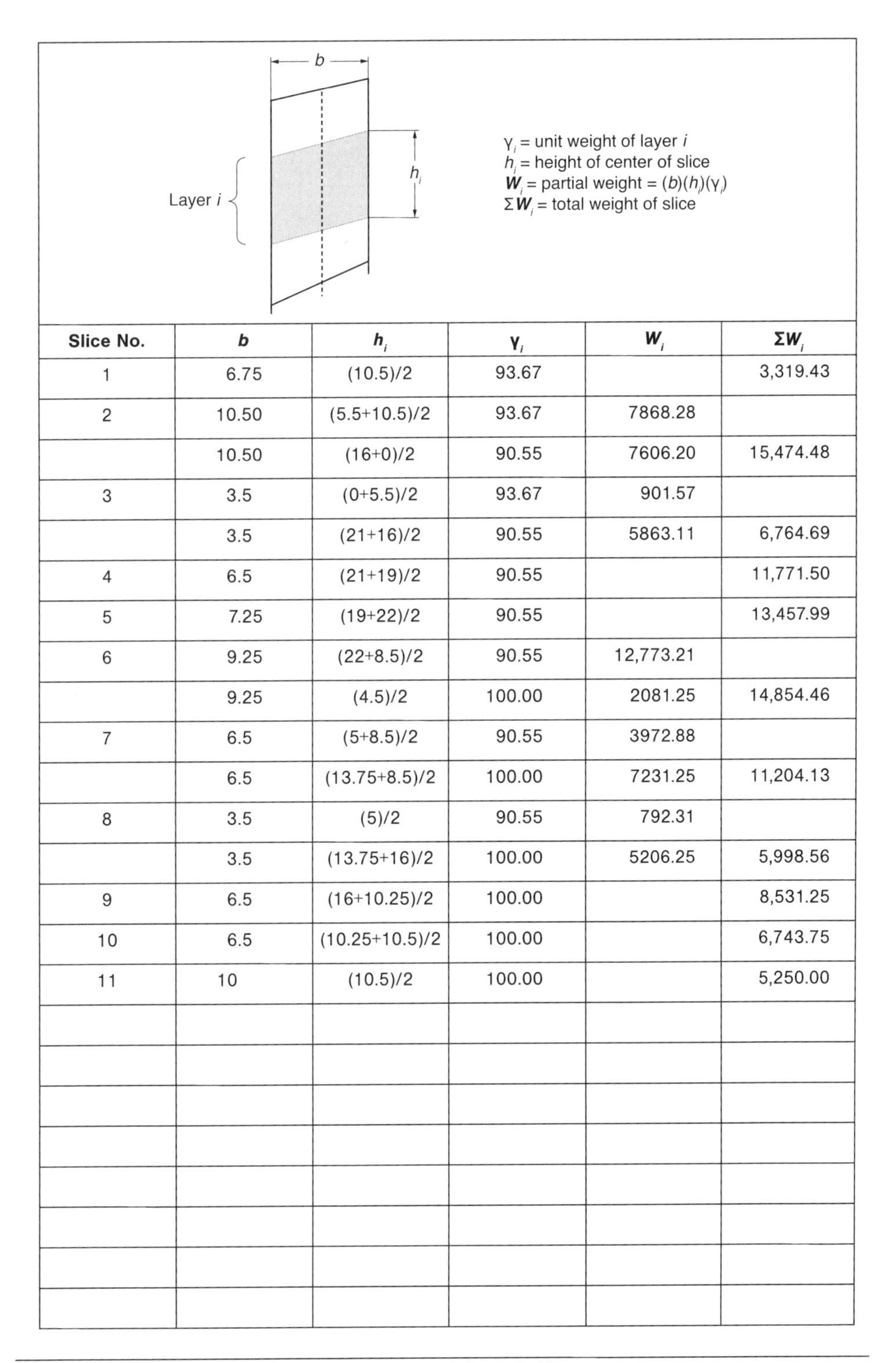

Slice No.	b	h_i	γ_i	W_i	ΣW_i
1	6.75	(10.5)/2	93.67		3,319.43
2	10.50	(5.5+10.5)/2	93.67	7868.28	
	10.50	(16+0)/2	90.55	7606.20	15,474.48
3	3.5	(0+5.5)/2	93.67	901.57	
	3.5	(21+16)/2	90.55	5863.11	6,764.69
4	6.5	(21+19)/2	90.55		11,771.50
5	7.25	(19+22)/2	90.55		13,457.99
6	9.25	(22+8.5)/2	90.55	12,773.21	
	9.25	(4.5)/2	100.00	2081.25	14,854.46
7	6.5	(5+8.5)/2	90.55	3972.88	
	6.5	(13.75+8.5)/2	100.00	7231.25	11,204.13
8	3.5	(5)/2	90.55	792.31	
	3.5	(13.75+16)/2	100.00	5206.25	5,998.56
9	6.5	(16+10.25)/2	100.00		8,531.25
10	6.5	(10.25+10.5)/2	100.00		6,743.75
11	10	(10.5)/2	100.00		5,250.00

FIGURE 6.28 Tabular form for computing weights of slices

TABLE 6.4 Tabular form for calculating the FS by Bishop's modified method

												FS =		FS =		FS =	
(1)	(2)	(3)	(4)	(5)	(6)	(7)	(8)	(9)	(10)	(11)	(12)	(13)	(14)	(13)	(14)	(13)	(14)
Slice	ϕ	c	u	$\boldsymbol{W}$	b	α	$\boldsymbol{W}\sin\alpha$	cb	ub	$(\boldsymbol{W}-ub)\tan\phi$	$cb+(\boldsymbol{W}-ub)\tan\phi$	M_α	12/13	M_α	12/13	M_α	12/13
						$\Sigma =$						$\Sigma =$					

Trial No.	1	2	3
FS = $\Sigma(14)/\Sigma(8)$			

TABLE 6.5 Tabular form for calculating the FS by Bishop's modified method

												FS =		FS =		FS =	
(1)	(2)	(3)	(4)	(5)	(6)	(7)	(8)	(9)	(10)	(11)	(12)	(13)	(14)	(13)	(14)	(13)	(14)
Slice	ϕ	c	u	W	b	α	$W\sin\alpha$	cb	ub	$(W-ub)\tan\phi$	$cb+(W-ub)\tan\phi$	M_α	12/13	M_α	12/13	M_α	12/13
1	16	325	0	3319.43	6.75	55.3	2729	2193.75	0		3145.58	0.81	3907	0.74	4228	0.74	4250
2	14	350	0	15474.48	10.50	46.53	11230	3675	0		7533.22	0.87	8670	0.82	9164	0.82	9197
3	14	350	0	6764.69	3.50	39.28	4283	1225	0		2911.63	0.93	3124	0.89	3268	0.89	3277
4	14	350	0	11771.50	6.50	34.43	6656	2275	0		5209.96	0.97	5394	0.93	5607	0.93	5620
5	14	350	0	13457.99	7.25	28.24	6368	2537.5	0		5892.95	1.00	5899	0.97	6085	0.97	6097
6	18	310	132.60	14854.46	9.25	21.38	5415	2867.5	1226.55		7295.48	1.05	6951	1.02	7160	1.02	7173
7	18	310	265.20	11204.13	6.50	15.11	2921	2015	1723.80		5095.35	1.05	4852	1.03	4956	1.06	4962
8	18	310	304.20	5998.56	3.50	11.42	1188	1085	1064.70		2688.11	1.05	2574	1.03	2615	1.05	2618
9	18	310	304.20	8531.25	6.5	7.43	1103	2015	1977.30		4144.51	1.03	4010	1.02	4052	1.02	4055
10	18	310	257.40	6743.75	6.50	2.52	297	2015	1673.10		3662.55	1.01	3614	1.01	3628	1.01	3629
11	18	310	109.20	5250.00	10	-4.70	-430	3100	1092		4451.62	0.92	4589	0.98	4556	0.98	4554
						Σ =	41760					Σ =	53584		55319		55432

Trial No.	1	2	3
FS = Σ(14)/Σ(8)	1.28	1.32	1.33

By interpolation: FS = 1.31

Computerized 2-D Limit Equilibrium

Numerous very powerful commercial computer packages are available for the general solution of slope stability problems by 2-D limiting equilibrium methods. Some examples include

- Rocscience's (2015) Slide 6.0,
- Purdue University's (2015) PCSTABL 6.0, and
- Geo-Slope International's (2015) Slope/W.

Many of the 2-D limit equilibrium slope stability analysis programs allow stability analysis using various methods, including ordinary (or Fellenius) method, Bishop simplified method, Janbu simplified or Janbu corrected method, Spencer method, Morgenstern–Price method, Corps of Engineers method 1 and 2, Lowe–Karafiath method, generalized limit equilibrium method, and finite element stress method and probabilistic analysis to account for variability and uncertainty associated with the analysis input parameters.

A probabilistic analysis allows the user to statistically quantify the probability of failure of a slope using the Monte Carlo method. The results from all Monte Carlo trials can then be used to compute the probability of failure and generate the FS probability density and distribution functions. Variability can be considered for material parameters such as unit weight, cohesion and friction angles, pore-water pressure conditions, applied line loads, and seismic coefficients (Geo-Slope International 2015).

Many of the packages also include the analysis of reinforced soil slopes with geosynthetics, nailing, active versus passive support, piles and micropiles, tiebacks, and other user-defined support models.

Most have the ability to model heterogeneous soil types, complex stratigraphic and slip surface geometry, and variable pore-water pressure conditions using a large selection of soil models. Numerous strength models can be incorporated into the models, including the Mohr–Coulomb, undrained, anisotropic strength, Hoek–Brown, Barton–Bandis, and many other strength functions.

The programs are also useful for analyzing seismic or earthquake loading of slopes and for back analysis to compute the required support force for a given factor of safety.

Figure 6.29 shows one of the output screens from a computerized 2-D analysis of the example from Figure 6.26 using BMM.

REFERENCES

Bishop, A.W. 1955. The use of the slip circle in the stability analysis of slopes. *Geotechnique*, 5:7–17.

Das, B., and B. Stimpson. 1986. Passive reinforcement of pit slopes by bolting. Paper presented at International Symposium on Geotechnical Stability in Surface Mining, November, at Calgary, Alberta, Canada.

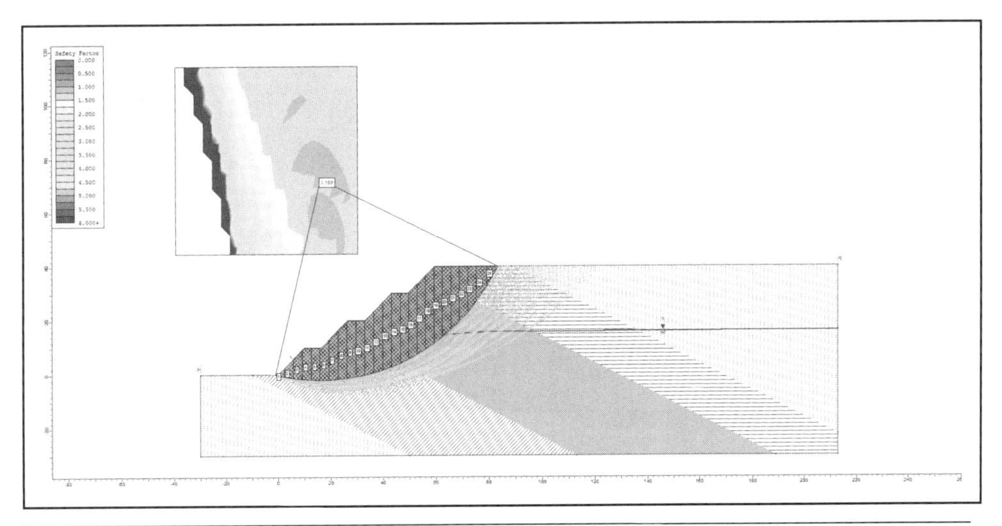

FIGURE 6.29 Output screen from computerized 2-D analysis of the slope of Figure 6.26

DYWIDAG-Systems International, USA, Inc. 1994. *DYWIDAG Threadbar Reinforcing Systems.* Bolingbrook, IL: DYWIDAG.

Fellenius, W. 1936. Calculation of the stability of earth dams. In *Transactions of the 2nd Congress on Large Dams (ICOLD)*, Paris. Vol. 4, pp. 445–462.

FOSROC Inc. 1994. *Lokset® Resin Cartridge Anchors.* Cleveland, OH: FOSROC.

Fredlund, D.G. 1984. Analytical methods for slope stability analysis. In *Proceedings of the Fourth International Symposium on Landslides, State-of-the-Art.* Toronto: Canadian Geotechnical Society. pp. 229–250.

Geo-Slope International. 2015. *Slope/W 2012 Slope Stability Analysis.* www.geo-slope.com/products/slopew.aspx (accessed Oct. 15, 2015).

Hoek, E., and J.W. Bray. 1977. *Rock Slope Engineering*, revised 2nd ed. London: Institution of Mining and Metallurgy.

Hoek, E., and J.W. Bray. 1981. *Rock Slope Engineering*, 3rd ed. London: Institution of Mining and Metallurgy.

Izbicki, R. 1981. Limit plasticity approach to slope stability problems. *American Society of Civil Engineering, Journal of Geotechnical Engineering Division*, 107(GT2):228–233.

Janbu, N., N. Bjerrum, and B. Kjaernsli. 1956. *Stabilitetsberegning for Fyllinger Skjaeringer og Naturlige Skraninger.* Norwegian Geotechnical Publication, no. 16. Oslo.

Lambe, W.T., and R.V. Whitman. 1969. *Soil Mechanics.* New York: Wiley.

Lowe, J., and L. Karafiath. 1960. Stability of earth dams upon drawdown. In *Proceedings of the 1st Pan-American Conference on Soil Mechanics and Foundation Engineering.* Mexico City. Vol. 2, pp. 537–552.

Major, G., H-S. Kim, and D. Ross-Brown. 1977. Plane shear analysis. In *Pit Slope Manual.* Report 77-16. Edited by D.F. Coates. Ottawa, ON: Canada Centre for Mineral and Energy Technology.

Mendelson, A. 1968. *Plasticity: Theory and Application.* New York: MacMillan.

Morgenstern, N.R., and V.E. Price. 1965. The analysis of the stability of general slip surfaces. *Geotechnique*, 15(1):79–93.

Perloff, W.H., and W. Baron. 1976. *Soil Mechanics Principles and Applications.* New York: John Wiley and Sons.

Piteau, D.R., and D.C. Martin. 1982. Mechanics of rock slope failure. In *Stability in Surface Mining*, Vol. 3. Edited by C.O. Brawner. New York: SME-AIME.

Purdue University. 2015. PCSTABL 6.0. West Lafayette, IN: Purdue University, School of Civil Engineering.

Rocscience. 2015. *Slide 6.0.* https://www.rocscience.com/ (accessed Oct. 15, 2015).

Smith, G.N. 1982. *Elements of Soil Mechanics for Civil and Mining Engineers*, 5th ed. London: New York: Granada Publishing.

Spencer, E. 1967. A method of analysis of the stability of embankments assuming parallel inter-slice forces. *Geotechnique*, 17(1):11–26.

Taylor, D.W. 1948. *Fundamentals of Soil Mechanics*. London: Wiley.

Varnes, D.J. 1978. Slope movement types and processes. In *Landslides Analysis and Control*. Edited by R.J. Schuster and R.J. Krizek. Special Report 176. Washington, DC: Transportation Research Board, National Academy of Sciences.

Zhu, D.Y., C.F. Lee, and H.D. Jiang. 2003. Generalized framework of limit equilibrium methods for slope stability analysis. *Geotechnique*, 53(4):377–395.

CHAPTER 7

Kinetic Slope Stability: Analysis of Toppling Failure

Toppling is possible whenever a set of well-developed or through-going discontinuities dips steeply into the slope (Piteau and Martin 1982). In this type of failure, long, thin columns of rock formed by the steeply dipping discontinuities (Figure 7.1) may rotate about a pivot point located at the lowest corner of the block (Figure 7.2).

The basic conditions for sliding and toppling of a single block on an inclined plane are shown in Figure 7.3. The figure is divided into four regions as delineated by (1) the line where the dip of the discontinuity equals the friction angle (i.e., $\beta = \phi$) and (2) the dashed curve that separates the region where the methods of limiting equilibrium can be used for analysis (to the left of the curve) from the region where methods of toppling failure analysis must be used (to the right of the curve). The terms t and h represent the block thickness

FIGURE 7.1 Toppling failure mode

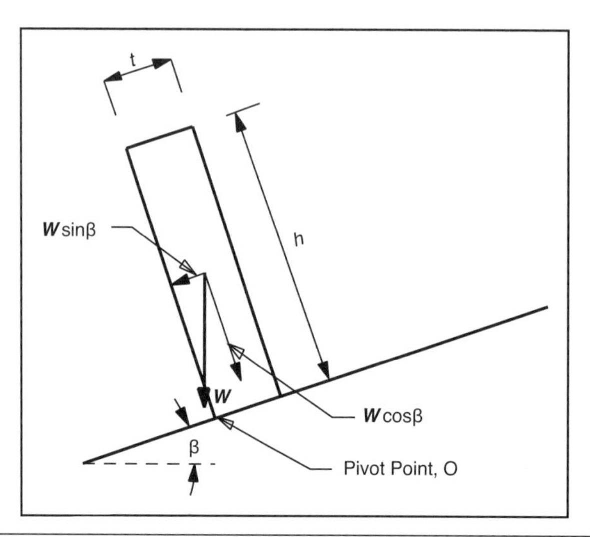

FIGURE 7.2 General model for toppling failure

and height, respectively; the ratio t/h is known as the slenderness ratio. The four regions are defined as follows.

- Region 1: $\beta < \phi$ and $t/h > \tan\beta$; the block is stable and will neither slide nor topple.
- Region 2: $\beta > \phi$ and $t/h > \tan\beta$; the block will slide but will not topple.
- Region 3: $\beta < \phi$ and $t/h < \tan\beta$; the block will topple but will not slide.
- Region 4: $\beta > \phi$ and $t/h < \tan\beta$; the block can slide and topple simultaneously.

In other words, the governing factor for toppling of a block is the location of the weight vector, $\boldsymbol{W}$, with respect to the pivot point on the block. If the block thickness-to-height ratio is less than $\tan\beta$, then the resultant force due to the weight of the block will occur outside the toe of the block, and an overturning moment will develop about the pivot point (see Figure 7.2). The governing factor for sliding, on the other hand, is the relationship between β and ϕ; if $\beta > \phi$, sliding can occur. If the condition exists where $\beta > \phi$ and $t/h < \tan\beta$, then both sliding and toppling are possible. In the analysis for toppling failure (to be developed later), the potential for a block to fail by sliding must also be considered.

GENERAL MODEL FOR TOPPLING FAILURE

For toppling failure of only an individual block or column to be kinematically and kinetically possible (not considering the potential for sliding of the block at this time), the following conditions must exist:

- There is an inclined surface upon which the block rests.
- There is a joint set approximately perpendicular to the inclined surface—that is, undercutting discontinuities are present (Figure 7.4).
- $\beta < \phi$ and $t/h < \tan\beta$.

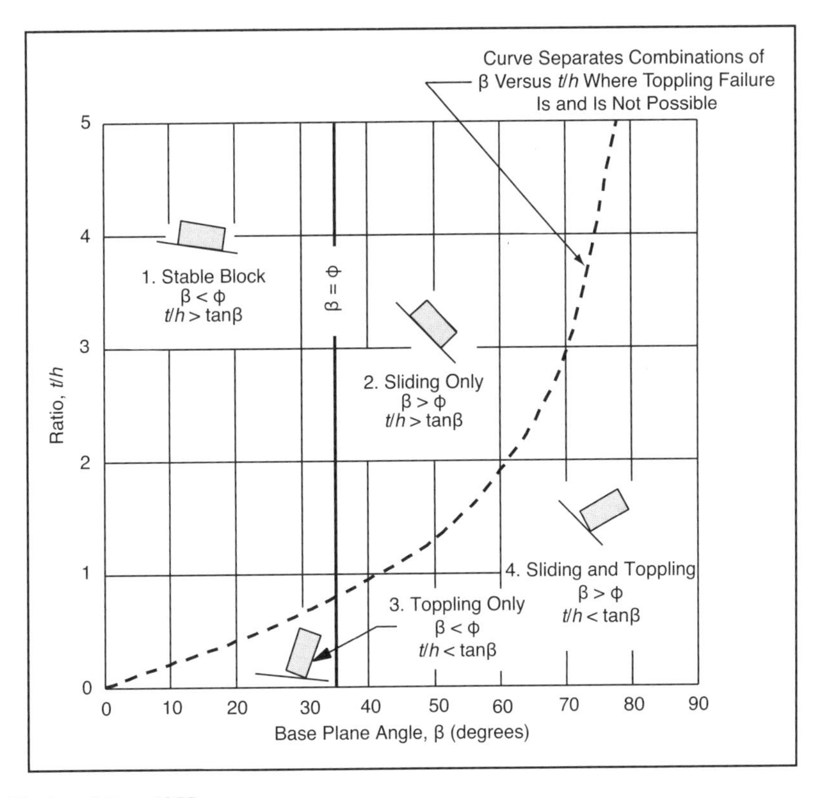

Source: Hoek and Bray 1977

FIGURE 7.3 Conditions for sliding and toppling of a block on an inclined plane

We can determine the factor of safety (FS) against toppling failure by taking the moments about the pivot point, O (see Figure 7.2), at limiting equilibrium:

$$\Sigma M_O = W\cos\beta \cdot (t/2) - W\sin\beta \cdot (h/2) = 0$$

We then calculate the factor of safety as the ratio of resisting moments to driving moments:

$$FS = \frac{\Sigma(\text{resisting moments})}{\Sigma(\text{driving moments})} = \frac{W\cos\beta \cdot (t/2)}{W\sin\beta \cdot (h/2)} = \frac{(t/h)}{\tan\beta}$$

LIMITING EQUILIBRIUM ANALYSIS OF TOPPLING ON A STEPPED BASE

The kinetic analysis of a slope for potential toppling failure is more of an academic exercise than an applicable field engineering technique. It involves making many simplifying assumptions about the slope to make the solution statically determinate. However, the theory and analysis technique presented in this section, along with a kinematic analysis, can be used as an overall indication of the stability of a slope against toppling failure.

FIGURE 7.4 Undercutting discontinuities in the highwall of a mine striking approximately parallel to the slope face

The stability analysis method presented here is based on the technique of Goodman and Bray (1976), as further developed by Zanbak (1983). The article by Zanbak presents a computer program, TOPANL, for generalized toppling analysis of dry rock slopes; this program can greatly simplify the analysis. Furthermore, newer computer programs, such as Rocscience's (2015) RocTopple 1.0, an interactive software tool for performing toppling analysis and support design, are available to help one analyze a general slope for toppling-type failure.

Toppling Forces

In jointed rock masses, where the slope is composed of rock columns (blocks) on a flat base, an unstable rock column ($i + 1$) will exert a force of magnitude $\boldsymbol{P}_{i+1}$ on the downslope adjacent column (i). Then the ith column will no longer be under gravitational forces only. The forces acting on the ith column for limiting equilibrium conditions are shown in Figure 7.5A. When the base plane of the columns is flat, overturning of the columns is not kinematically possible without displacement of the pivot points (point O) of the columns due to required dilation of the columnar structure. For the blocks on both the upslope and downslope of the block under consideration to remain in contact with the block pivoting about its rotation point O, a displacement (or dilation), Δt, is required (Figure 7.6). Such dilation is possible if the magnitude of the overturning forces exceeds the shear resistance of the rotation points of the columns. In addition, the limiting equilibrium conditions of each overturning column on a flat base are statically indeterminate because of the indefinite moment arm lengths of the $\mathbf{P}_i$ and $\boldsymbol{P}_{i+1}$ forces; these lengths are indefinite because there are planar contacts, rather than point contacts, between the columns.

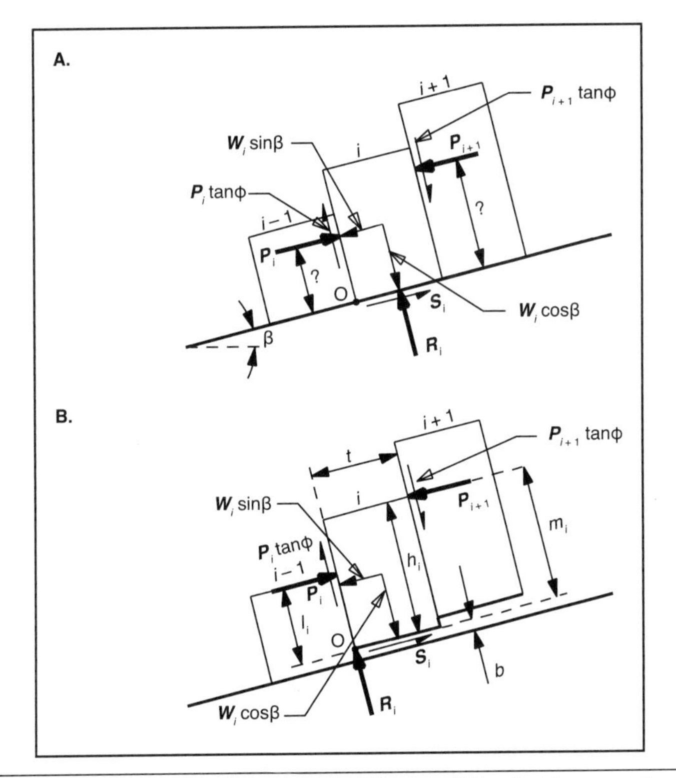

FIGURE 7.5 Forces acting on the *i*th column sitting on (A) a flat base and (B) a stepped base

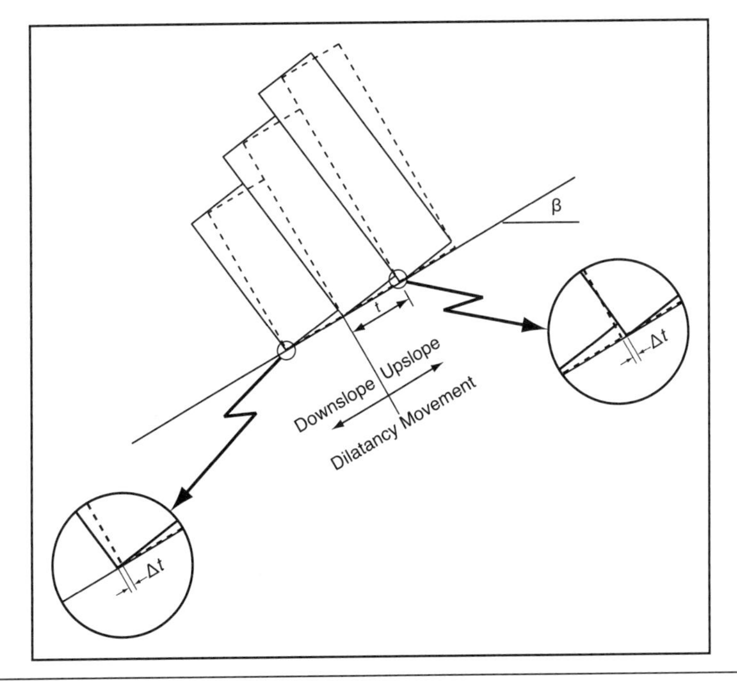

FIGURE 7.6 Dilation process in columnar slopes on a flat base

On the other hand, where the slope is composed of rock columns on a stepped base, rotation of the *i*th block about its pivot point will cause shear forces ($\boldsymbol{S}_i$) and normal forces ($\boldsymbol{N}_i$) to develop on the block, as shown in Figure 7.5B. While overturning about the pivot point, O, the *i*th block will maintain contact with blocks ($i + 1$) and ($i - 1$), as shown in Figure 7.5B. Because the locations of the shear and normal forces are known, the stability of the column against rotation and sliding is determinate.

Slope Geometry

The following are the general input parameters for static stability analysis of slopes susceptible to toppling on a stepped base (Figures 7.5 and 7.7 graphically depict many of these parameters):

- Thickness of the columns (t)
- Height of column i from the base plane (h_i)
- Dip angle of the base plane (β)
- Step angle (θ)
- Slope angle (α_1)
- Dip angle of the slope top bench (α_2)
- Friction angle at the column–column contact (ϕ_{c-c})
- Friction angle at the column–base contact (ϕ_{c-b})
- Unit weight of the rock columns (γ_r)
- Normal force at the base of the column, $\boldsymbol{N}_i = \boldsymbol{W}_i \cos\beta$
- Shear force along the column–base contact, $\boldsymbol{S}_i = \boldsymbol{N}_i \tan\phi_{c-b}$

Heights of the individual columns on the stepped base plane are controlled by the geometrical input parameters, that is,

$$h_i = f(H, t, \beta, \theta, \alpha_1, \alpha_2)$$

in which H is the slope height. The parameters β, θ, α_1, and α_2 are angular input parameters, whereas H and t are dimensional input parameters.

The slope height, dip of the base plane, dip of the slope face, and dip of the slope crest can be determined with reasonable confidence by field measurements, as can the average column thickness, t. However, to resolve the slope geometry into individual columns as shown in Figure 7.7A, a step height (b) or step angle (θ) must be assumed.

Limiting Equilibrium for Sliding and Toppling

Analysis of toppling forces is valid only if the columns are stable against sliding on their base planes. An individual column is stable under gravitational forces when ϕ_{c-b}, the friction angle between the column–base contact, is greater than β. However, if the column is

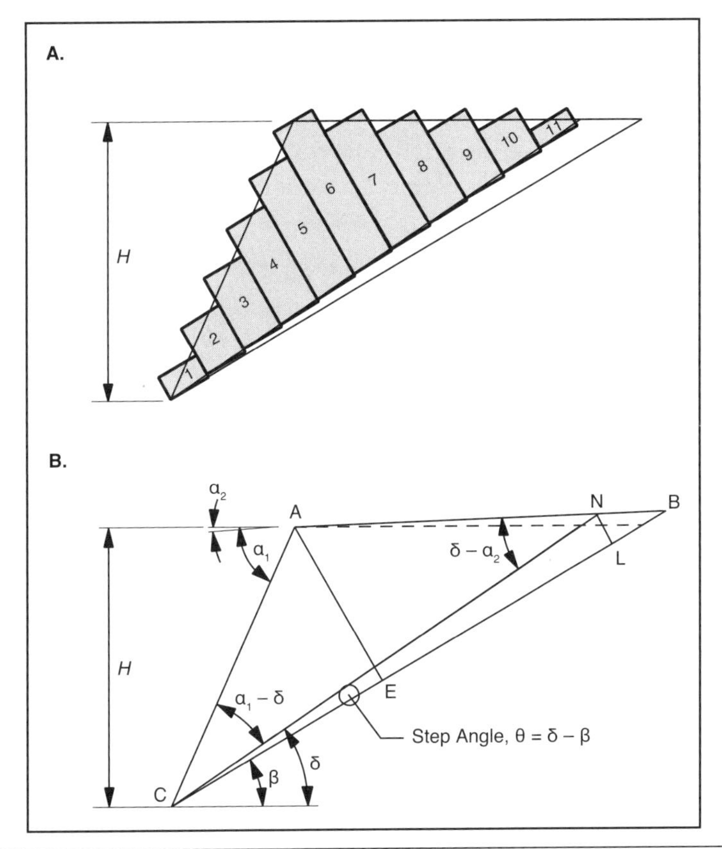

FIGURE 7.7 Slope geometry for the toppling failure mode of columns on a stepped base

under external forces applied by the adjacent columns in addition to the inherent gravity forces, then its stability against sliding should be checked before the analysis of toppling forces on the downslope columns continues. The greater of the $\boldsymbol{P}$ forces on the ith column—shear resistance or toppling force—is what will be exerted on the $(i - 1)$th column. If the magnitudes of both the toppling force and the shear force are positive, the block will tend to both topple and slide.

If the magnitude of the $\boldsymbol{P}_{is}$ force (shear resistance of the ith column) is greater than that of $\boldsymbol{P}_{it}$ (toppling force of the ith column), then the $(i - 1)$th column will be stable against toppling but will tend to slide. This case is possible when the height of the $(i - 1)$th column is smaller than the height of the ith column, as may occur at the toe region of the slope.

The overall stability of a toppling slope is defined by the magnitude of $\boldsymbol{P}_1$, the column force exerted by the first column at the toe. The slope is considered unstable if the magnitude of the $\boldsymbol{P}_1$ force at the toe column is greater than zero ($\boldsymbol{P}_1 > 0$). When $\boldsymbol{P}_1$ is greater than zero,

P_1 is the retaining force in the direction of the base plane necessary to achieve limiting equilibrium conditions.

Limiting equilibrium for sliding of the *i*th block. To determine the magnitude of P_{is}, we set the sum of the forces in both the x- and y-directions to zero (taking the x-axis to be along the base plane, at the angle β with respect to horizontal—refer to Figure 7.5). In addition, assume $\phi_{c-c} = \phi_{c-b} = \phi$. Hence, we have

$$\Sigma F_x = 0 = P_i - P_{i+1} - W_i \sin\beta + S_i$$

Rewriting and solving for S_i we get

$$S_i = W_i \sin\beta - P_i + P_{i+1}$$

$$\Sigma F_y = 0 = R_i + P_i \tan\phi - W_i \cos\beta - P_{i+1} \tan\phi \qquad \text{(EQ 7.1)}$$

Rewriting and solving for R_i we get

$$R_i = W_i(\cos\beta) + (P_{i+1} - P_i)\tan\phi \qquad \text{(EQ 7.2)}$$

where S_i is the shear force along the column–base contact. From the Mohr–Coulomb criterion, we have $\tau_i = c + \sigma_i \tan\phi$, where τ_i is the shear stress along the column–base contact and σ_i is the normal stress at the column–base contact. If we assume cohesion (c) is equal to zero and divide both sides of the equation by the column–base contact area, A, then we get $S_i = R_i \tan\phi$, where R_i is the normal force across the column–base contact ($W_i \cos\beta$). By solving the F_y equation for R_i and substituting into the F_x equation, we get

$$P_i - P_{i+1} - W_i \sin\beta + W_i \cos\beta \cdot \tan\phi - (P_i - P_{i+1})\tan^2\phi = 0$$

$$(P_i - P_{i+1})(1 - \tan^2\phi) - W_i \sin\beta + W_i \cos\beta \cdot \tan\phi = 0$$

This leads to

$$(P_i - P_{i+1})(1 - \tan^2\phi) = W_i \cos\beta(\tan\beta - \tan\phi)$$

$$(P_i - P_{i+1}) = \frac{W_i \cos\beta(\tan\beta - \tan\phi)}{1 - \tan^2\phi}$$

Therefore, we get the following equation for P_{is}:

$$P_{is} = P_{i+1} + \frac{W_i \cos\beta(\tan\beta - \tan\phi)}{1 - \tan^2\phi} \qquad \text{(EQ 7.3a)}$$

or, stated differently:

$$P_{is} = P_{i+1} - \frac{W_i(\cos\beta\tan\phi - \sin\beta)}{1-\tan^2\phi} \quad \text{(EQ 7.3b)}$$

Limiting equilibrium for toppling of the *i*th block. To determine the magnitude of P_{it}, we set the moments about the pivot point, O, to zero (refer to Figure 7.5):

$$\begin{aligned} \Sigma M_O &= 0 \\ &= (h_i/2)W_i\sin\beta - (t/2)W_i\cos\beta + m_i P_{i+1} - t(P_{i+1}\tan\phi) - l_i P_i \\ &= (W_i/2)(h_i\sin\beta - t\cos\beta) + P_{i+1}(m_i - t\cdot\tan\phi) - l_i P_i \end{aligned}$$

This can be rewritten as

$$l_i P_i = (W_i/2)(h_i\sin\beta - t\cos\beta) + P_{i+1}(m_i - t\cdot\tan\phi)$$

Therefore, the magnitude of P_{it} may be expressed as

$$P_{it} = \frac{(W_i/2)(h_i\sin\beta - t\cos\beta) + P_{i+1}(m_i - t\cdot\tan\phi)}{l_i} \quad \text{(EQ 7.4)}$$

Toppling analysis procedure. The analysis procedure for a slope susceptible to toppling is as follows (refer to Figure 7.7A):

1. Draw the model cross section.
2. Calculate the dimensions and weights of the columns.
3. Determine the model step height, b.
4. Determine the friction angle, ϕ.
5. Check the stability of the last column (column 11 in Figure 7.7A). If $(t/h)_{11} > \tan\beta$, then $P_{11t} = 0$ for the column.
6. Continue analyzing t/h for columns downslope from the last column (columns 10, 9, 8, etc., in Figure 7.7A) until $(t/h)_i < \tan\beta$.
7. For the first column such that $t/h < \tan\beta$, solve for P_{it} by using the equation given earlier and setting P_{i+1} to 0.
8. Use the P_{it} value from step 7 as P_{i+1} for the next column, $(i-1)$, and solve for P_{it} and P_{is}. Set the greater of P_{is} or P_{it} as P_i for the column.
9. Continue for the remaining columns. Use P_{is} or P_{it}—whichever is greater—as P_i for the column being analyzed.
10. If $P_{1s} > 0$, then the first column will slide; if $P_{1t} > 0$, then the first column will topple.
11. Calculate the retaining force—and the angle of its application to the first column—required to stabilize the slope to an acceptable factor of safety.

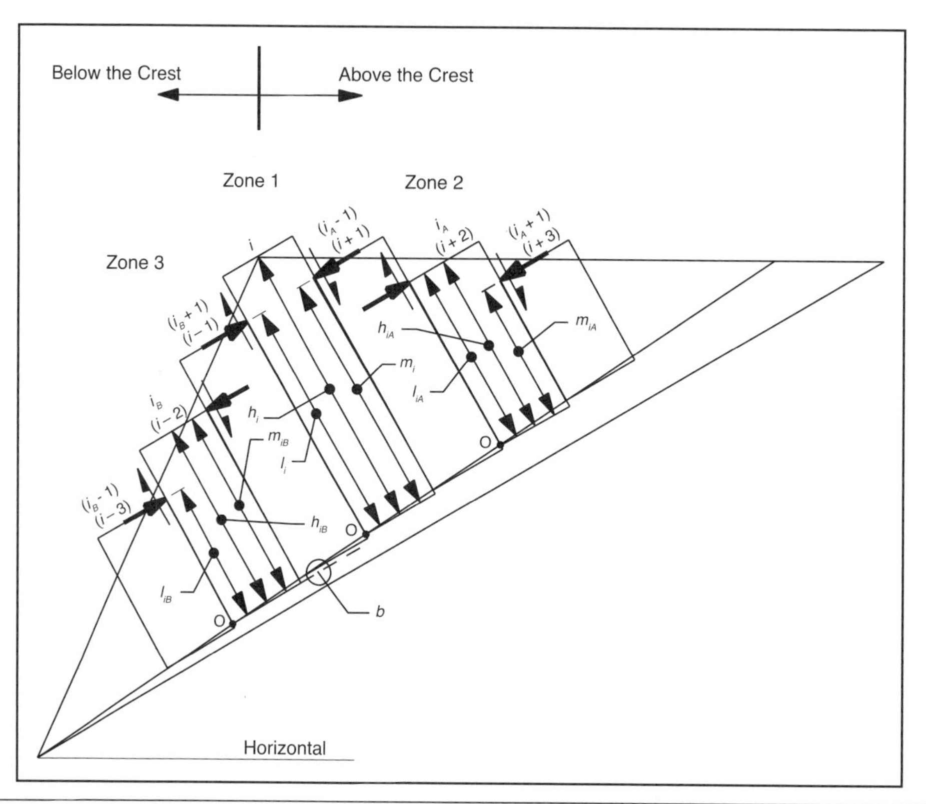

FIGURE 7.8 Moment arms l_i and m_i at the crest of the slope, above the crest, and below the crest

Determination of Moment Arms

The lengths of the moment arms, m_i and l_i (see Figure 7.5B), depend on the location of the block under consideration, i, with respect to the crest block. As Figure 7.8 shows, three zones can be observed:

- Zone 1: for the block at the crest, $n = i$ $\quad m_n = h_{n+1} + b$, $\; l_n = h_{n-1} - b$
- Zone 2: for the blocks above the crest, $n = i + 1, i + 2$, etc. $\quad m_n = h_{n+1} + b$, $\; l_n = h_n$
- Zone 3: for the blocks below the crest, $n = i - 1, i - 2$, etc. $\quad m_n = h_n$, $\; l_n = h_{n-1} - b$
- For block $n = 1$ $\quad m_1 = h_1$, $\; l_1 = h_1/2$

EXAMPLE: TOPPLING FAILURE

Consider the slope shown in Figure 7.7. Suppose the following geometrical and dimensional input parameters are known:

$$\beta = 30°$$

$$\delta = 33°$$

$$\delta - \beta = \theta = 3°$$

$$\alpha_1 = 64.31°$$

$$\alpha_2 = 0°$$

$$\alpha_1 - \delta = 31.31°$$

$$\delta - \alpha_2 = 33°$$

$$t = 1.5 \text{ m } (4.92 \text{ ft})$$

$$b = 1.5 \tan(3°) = 0.08 \text{ m } (0.262 \text{ ft})$$

$$\phi = 35°$$

$$\gamma_r = 2{,}600 \text{ kg/m}^3 \ (162.31 \text{ lb/ft}^3)$$

$$H = 9 \text{ m } (29.53 \text{ ft})$$

Column heights for the 11 columns of Figure 7.7A are given in Table 7.1. The individual column heights are calculated by using trigonometry and some of the angular relationships of Figure 7.7B. For example, the heights of blocks 1 to 6 (h_1 to h_6) are calculated at their respective midpoints along line C-B. That is, a perpendicular is drawn from C-B at the midpoint of each block and projected up to line C-A, forming a right triangle. By calculating the length of the perpendicular and subtracting off the cumulative step height (b), each block height is determined. Hence,

$$\begin{aligned} h_1 &= (t_1/2)\tan\left[(\alpha_1 - \delta) + \theta\right] \\ &= (1.5/2)\tan(31.31° + 3°) \\ &= 0.51 \text{ m } (1.68 \text{ ft}) \end{aligned}$$

$$\begin{aligned} h_2 &= \left[t_1 + (t_2/2)\right]\tan\left[(\alpha_1 - \delta) + \theta\right] - 1b \\ &= \left[1.5 + (1.5/2)\right]\tan(31.31° + 3°) - 0.08 \\ &= 1.46 \text{ m } (4.78 \text{ ft}) \end{aligned}$$

$$\begin{aligned} h_6 &= \left[t_1 + t_2 + t_3 + t_4 + t_5 + (t_6/2)\right]\tan\left[(\alpha_1 - \delta) + \theta\right] - 5b \\ &= \left[(5)(1.5) + (1.5/2)\right]\tan(31.31° + 3°) - (5)(0.08) \\ &= 5.24 \text{ m } (17.18 \text{ ft}) \end{aligned}$$

TABLE 7.1 Column heights for the 11 columns of Figure 7.7A, given the previous geometrical and dimensional input parameters

Column	Calculation	Height Value (h_i)
1	h_1 = 0.75 tan(34.31°)	0.51 m (1.69 ft)
2	h_2 = 2.25 tan(34.31°) – b	1.46 m (4.78 ft)
3	h_3 = 3.75 tan(34.31°) – 2b	2.40 m (7.88 ft)
4	h_4 = 5.25 tan(34.31°) – 3b	3.35 m (10.98 ft)
5	h_5 = 6.75 tan (34.31°) – 4b	4.29 m (14.08 ft)
6	h_6 = 8.25 tan(34.31°) – 5b	5.24 m (17.18 ft)
7	h_7 = 9.75 tan(34.31°) – 1.5[tan(34.31°) + tan(30°)] – 6b	4.29 m (14.08 ft)
8	h_8 = 11.25 tan(34.31°) – 3.0[tan(34.31°) + tan(30°)] – 7b	3.35 m (10.98 ft)
9	h_9 = 12.75 tan(34.31°) – 4.5[tan(34.31°) + tan(30°)] – 8b	2.40 m (7.88 ft)
10	h_{10} = 14.25 tan(34.31°) – 6.0[tan(34.31°) + tan(30°)] – 9b	1.46 m (4.78 ft)
11	h_{11} = 15.75 tan(34.31°) – 7.5[tan(34.31°) + tan(30°)] – 10b	0.51 m (1.69 ft)

After the peak of the crest—that is, for blocks 7 to 11 (h_7 to h_{11})—the portion of the block height extending above line A-B must be subtracted off to determine the correct block height. Other than that, the method is the same. So we have

$$
\begin{aligned}
h_7 &= \left[t_1+t_2+t_3+t_4+t_5+t_6+\left(t_7/2\right)\right]\tan\left[\left(\alpha_1-\delta\right)+\theta\right] \\
&\quad -\left[\left(t_6/2\right)+\left(t_7/2\right)\right]\left\{\tan\left[\left(\alpha_1-\delta\right)+\theta\right]+\tan\left(\beta\right)\right\}-6b \\
&= \left[(6)(1.5)+\left(1.5/2\right)\right]\tan\left(31.31°+3°\right) \\
&\quad -1.5\left\{\tan\left(31.31°+3°\right)+\tan\left(30°\right)\right\}-6(0.08) \\
&= 4.29 \text{ m } (14.08 \text{ ft})
\end{aligned}
$$

and

$$
\begin{aligned}
h_{11} &= \left[t_1+\cdots t_{10}+\left(t_{11}/2\right)\right]\tan\left[\left(\alpha_1-\delta\right)+\theta\right] \\
&\quad -\left[\left(t_6/2\right)+t_7+t_8+t_9+t_{10}+\left(t_{11}/2\right)\right]\left\{\tan\left[\left(\alpha_1-\delta\right)+\theta\right]+\tan\left(\beta\right)\right\}-10b \\
&= \left[(10)(1.5)+\left(1.5/2\right)\right]\tan\left(31.31°+3°\right) \\
&\quad -(5)(1.5)\left\{\tan\left(31.31°+3°\right)+\tan\left(30°\right)\right\}-10(0.08) \\
&= 0.51 \text{ m } (1.69 \text{ ft})
\end{aligned}
$$

A column is unstable versus toppling when the thickness-to-height ratio is less than the tangent of the dip of the base plane, β. The procedure is to start the analysis at the top block, block 11, determine (t/h_{11}), and compare this value to tanβ (tanβ = 0.577). When

t/h_i is less than tanβ, then block i is analyzed for P_{is} and P_{it}. For this example, column 8 is the first column that is unstable versus toppling.

Column	t/h value	Is t/h < tanβ?	Magnitude of P_i
11	2.941	No	$P_{11} = 0$
10	1.027	No	$P_{10} = 0$
9	0.624	No	$P_9 = 0$
8	0.448	Yes	$P_8 \neq 0$

Hence, the values of P_{is} and P_{it} must be calculated for column 8 down to column 1, as shown in Table 7.2. Also shown in Table 7.2 are the normal and shear forces, R_n and S_n, respectively, for each block along with a comment on the stability of each block. Shown in Table 7.3 are the l and m values used in the calculation of P_{it} in Table 7.2.

The force P_{1s} acts in the direction parallel to the base plane. The amount of force, T (with an appropriate factor of safety included), to stabilize the slope—as well as the angle of application of T with respect to the base plane, ψ—may be determined by simple trigonometry. The large negative value of P_{1t} indicates that block 1 is stable against toppling. However, the positive value of P_{1s} means that the cumulative forces from the slope acting on the block will result in the block sliding and thereby causing the slope to fail, like dominoes. Block 1 is known as the "key block." The force required to prevent block 1 from sliding, at the designated factor of safety, is also the force required to stabilize the slope. Hence, assume that T is to be applied at an angle ψ = 30° (note that the same positive and negative angle convention will be used as illustrated in Figure 6.9) and that the slope is to be stabilized to a factor of safety of 1.3. The required force, in kilograms per meter of slope width, to achieve the limiting equilibrium state is

$$F = 5{,}830/\cos 30° = 6{,}732 \text{ kg/m}$$

The magnitude of the required force, T, is then $1.3F = 8{,}751$ kg/m of slope width.

A computer-generated model of the slope in the preceding example is illustrated in Figure 7.9. This model was generated using Rocscience's RocTopple toppling analysis computer program (Rocscience 2015). A slight difference results between in the P_i values from the computer-generated and the preceding computed values due to the method used to generate the step heights and the block volumes for the computer program. However, the computer solution is a quick, easy way to determine the stability of a slope versus toppling.

REFERENCES

Goodman, R.E., and J.W. Bray. 1976. Toppling of rock slopes. In *Proceedings of Specialty Conference Rock Engineering for Foundations and Slopes*, Vol. II. New York: American Society of Civil Engineers.

TABLE 7.2 Dimension of magnitudes of forces P_{is} and P_{it}

Column	Calculation of P_{is} (using EQ 7.3b)	Calculation of P_{it} (using EQ 7.4)
8	$P_{8s}=0-\frac{13{,}065(\cos 30°\cdot\tan 35°-\sin 30°)}{1-\tan^2 35°}$ $=-2{,}727$ kg/m	$P_{8t}=\frac{0+(13{,}065/2)(3.35\sin 30°-1.5\cos 30°)}{3.35}$ $=733$ kg/m
7	$P_{7s}=733-\frac{16{,}731(\cos 30°\cdot\tan 35°-\sin 30°)}{1-\tan^2 35°}$ $=-2{,}759$ kg/m	$P_{7t}=\frac{733(3.43-1.5\tan 35°)+(16{,}731/2)(4.29\sin 30°-1.5\cos 30°)}{4.29}$ $=2{,}056$ kg/m
6	$P_{6s}=2{,}056-\frac{20{,}436(\cos 30°\cdot\tan 35°-\sin 30°)}{1-\tan^2 35°}$ $=-2{,}210$ kg/m	$P_{6t}=\frac{2{,}056(4.37-1.5\tan 35°)+(20{,}436/2)(5.24\sin 30°-1.5\cos 30°)}{4.21}$ $=4{,}827$ kg/m
5	$P_{5s}=4{,}827-\frac{16{,}731(\cos 30°\cdot\tan 35°-\sin 30°)}{1-\tan^2 35°}$ $=1{,}335$ kg/m	$P_{5t}=\frac{4{,}827(4.29-1.5\tan 35°)+(16{,}731/2)(4.29\sin 30°-1.5\cos 30°)}{3.27}$ $=6{,}946$ kg/m
4	$P_{4s}=6{,}946-\frac{13{,}065(\cos 30°\cdot\tan 35°-\sin 30°)}{1-\tan^2 35°}$ $=4{,}219$ kg/m	$P_{4t}=\frac{6{,}946(3.35-1.5\tan 35°)+(13{,}065/2)(3.35\sin 30°-1.5\cos 30°)}{2.32}$ $=7{,}944$ kg/m
3	$P_{3s}=7{,}944-\frac{9{,}360(\cos 30°\cdot\tan 35°-\sin 30°)}{1-\tan^2 35°}$ $=5{,}990$ kg/m	$P_{3t}=\frac{7{,}944(2.40-1.5\tan 35°)+(9{,}360/2)(2.40\sin 30°-1.5\cos 30°)}{1.38}$ $=7{,}434$ kg/m
2	$P_{2s}=7{,}434-\frac{5{,}694(\cos 30°\cdot\tan 35°-\sin 30°)}{1-\tan^2 35°}$ $=6{,}245$ kg/m	$P_{2t}=\frac{7{,}434(1.46-1.5\tan 35°)+(5{,}694/2)(1.46\sin 30°-1.5\cos 30°)}{0.43}$ $=3{,}315$ kg/m
1	$P_{1s}=6{,}245-\frac{1{,}989(\cos 30°\cdot\tan 35°-\sin 30°)}{1-\tan^2 35°}$ $=5{,}830$ kg/m	$P_{1t}=\frac{6{,}245(0.51-1.5\tan 35°)+(1{,}989/2)(0.51\sin 30°-1.5\cos 30°)}{0.255}$ $=-15{,}345$ kg/m

TABLE 7.2 Dimension of magnitudes of forces P_{is} and P_{it} *(continued)*

Column	P_i	$R_i = W_i(\cos\beta) + (P_{i+1} - P_i)\cdot\tan\phi$ (using EQ 7.2 and referring to Figure 7.5B)	$S_i = W_i \sin\beta - P_i + P_{i+1}$ $S_i = R_i\cdot\tan\phi$, for sliding block (using EQ 7.1 and referring to Figure 7.5B)	S_i/R_i	Mode
11	0	1,722.52	994.50	0.5774	Stable
10	0	4,931.15	2,847.00	0.5774	Stable
9	0	8,106.00*	4,680.00*	0.5774	Stable
8	733	10,801.37	5,799.50	0.5369	Toppling
7	2,056	13,563.10†	7,042.50†	0.5192	Toppling
6	4,827	15,757.82	7,447.00	0.4726	Toppling
5	6,946	13,005.73	6,246.50	0.4803	Toppling
4	7,944	10,615.81	5,534.50	0.5213	Toppling
3	7,434	7,748.89	4,170.00	0.5381	Toppling
2	6,245	4,098.60	2,869.87	0.7002	Sliding
1	5,830	1,307.52	915.54	0.7002	Sliding

* $R_n = [(2.4) \times (1.5) \times (1) \times (2{,}600)] \times \cos(30)$

$S_n = [(2.4) \times (1.5) \times (1) \times (2{,}600)] \times \sin(30)$

† $R_n = \{[(4.29) \times (1.5) \times (1) \times (2{,}600)] \times \cos(30)\} + [(733 - 2{,}056) \times \tan(35)]$

$S_n = \{[(4.29) \times (1.5) \times (1) \times (2{,}600)] \times \sin(30)\} + (733 - 2{,}056)$

TABLE 7.3 Column heights, m_i and l_i values

Column	Column Height, h	Moment Arm, m_i	Moment Arm, l_i
8	3.35 (10.98)	2.40 + 0.08 = 2.48	3.35
7	4.29 (14.08)	3.35 + 0.08 = 3.43	4.29
6	5.24 (17.18)	4.29 + 0.08 = 4.37	4.29 − 0.08 = 4.21
5	4.29 (14.08)	4.29	3.35 − 0.08 = 3.27
4	3.35 (10.98)	3.35	2.40 − 0.08 = 2.32
3	2.40 (7.88)	2.40	1.46 − 0.08 = 1.38
2	1.46 (4.78)	1.46	0.51 − 0.08 = 0.43
1	0.51 (1.69)	0.51	0.51/2 = 0.255

Hoek, E., and J.W. Bray. 1977. *Rock Slope Engineering*. London: Institution of Mining and Metallurgy.

Piteau, D.R., and D.C. Martin. 1982. Mechanics of rock slope failure. In *Stability in Surface Mining*, Vol. 3. Edited by C.O. Brawner. New York: SME-AIME.

Rocscience. 2015. RocTopple—Toppling analysis for rock slopes. Version 1.0. Toronto: Rocscience.

Wyllie, D.C., and C.W. Mah. 2004. *Rock Slope Engineering Civil and Mining*. New York: Spon Press.

Zanbak, C. 1983. Design charts for rock slopes susceptible to toppling. *Journal of Geotechnical Engineering*, 109(8):1039–1062.

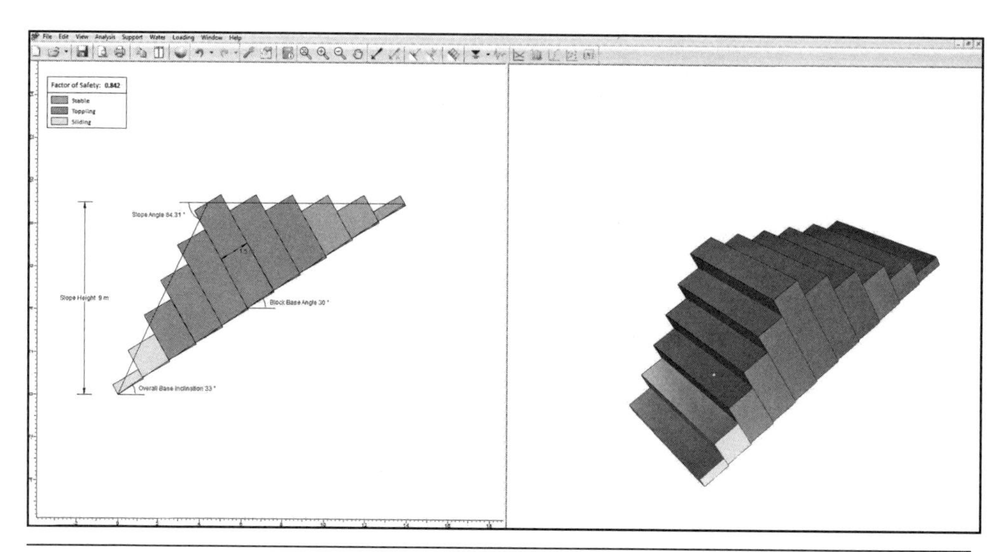

FIGURE 7.9 Computer-generated toppling model of the example of Figure 7.7

CHAPTER 8

Kinetic Slope Stability: Analysis of Wedge Failure

In the simplest case, a wedge failure can occur when a mass of rock slides on two intersecting planes, as shown in Figure 8.1. In this case, the failure block is assumed to be a tetrahedron, the sides of which are defined by the two intersecting planes, the slope face and the upper slope surface.

To analyze a wedge for possible instability, the two-dimensional limiting equilibrium model used for plane failure and toppling failure conditions must be replaced by a three-dimensional model. Again, the limiting equilibrium concept—that is, analysis of the factor of safety (FS) against sliding—can be used.

FIGURE 8.1 Typical wedge failure involving sliding on two discontinuities

Tetrahedron wedge failure can occur in one of the following ways (Piteau and Martin 1982):

- By sliding on both planes in a direction along the line of intersection
- By sliding along only one plane with separation across the other plane
- By rotational sliding on one plane and separation across the other plane
- By progressive raveling of rock along planes formed by the wedges in highly jointed rock

As was discussed in Chapter 5, in order for the wedge-type failure to be kinematically possible, the following three general conditions must be met:

1. The plunge of the line of intersection of the two intersecting discontinuities (β_i) must exceed the angle of friction for the rock surface, ϕ.
2. The plunge of the line of intersection of the two planes must be less than the dip of the slope face (ψ).
3. The trend of the line of intersection of the two planes must be approximately parallel to the dip direction of the slope face (within ±20°) and must daylight in the slope face.

Figure 8.2 depicts wedges for which failure based on the preceding criteria is kinematically possible, as well as wedges for which such failure is not possible. Figure 8.3 shows two stereographic representations: one of a wedge where failure is kinematically possible and one where failure is not kinematically possible. The top stereonet of Figure 8.3 shows the great circles representing two planes where the trend of the line of intersection is within ±20° of the dip directions of both the upper slope surface and the slope face. Also, the plunge of the line of intersection is less than the dip of the slope face and greater than the friction angle (see text regarding the Markland test in Chapter 5). The lower stereonet shows the great circles representing two intersecting planes that form a line of intersection that trends in a direction approximately opposite the dip direction of the slope face. That is, the line of intersection of the two planes in the lower stereonet plunges into the slope face, instead of toward the slope face as is the case with the upper stereonet.

Not all intersecting planes that meet the criteria for possible kinematic failure will actually fail. Discontinuities with a low degree of continuity (i.e., those with a high percentage of attachments across the planes) will exhibit higher strength in the form of cohesion and, possibly, a higher friction angle than discontinuities with high continuity. Intersecting major discontinuities such as faults, shear zones, and/or open joints, therefore, pose the greatest risk of wedge-type failure. Intersecting minor discontinuities, on the other hand, generally pose a low risk of failure.

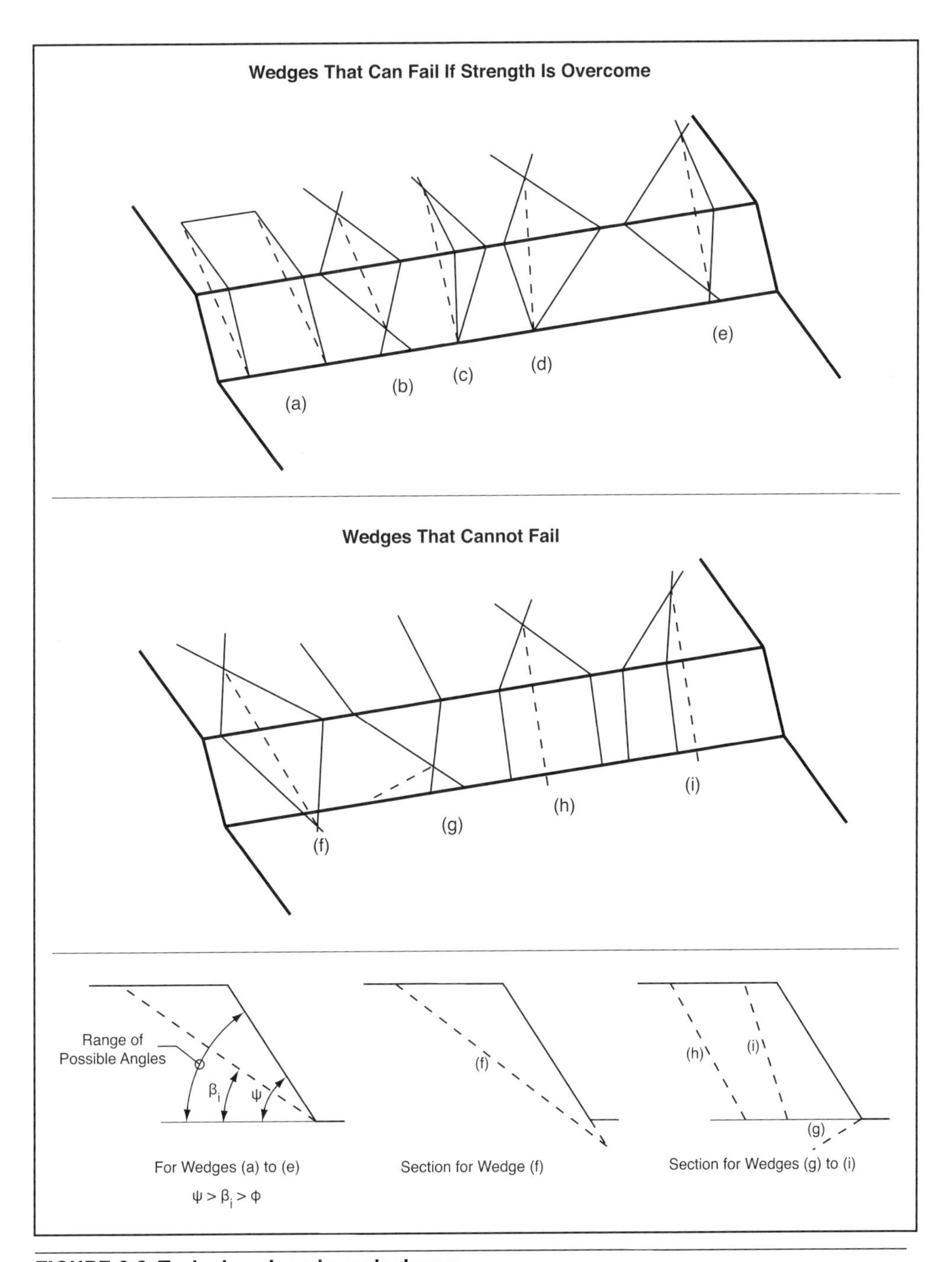

FIGURE 8.2 Typical wedges in rock slopes

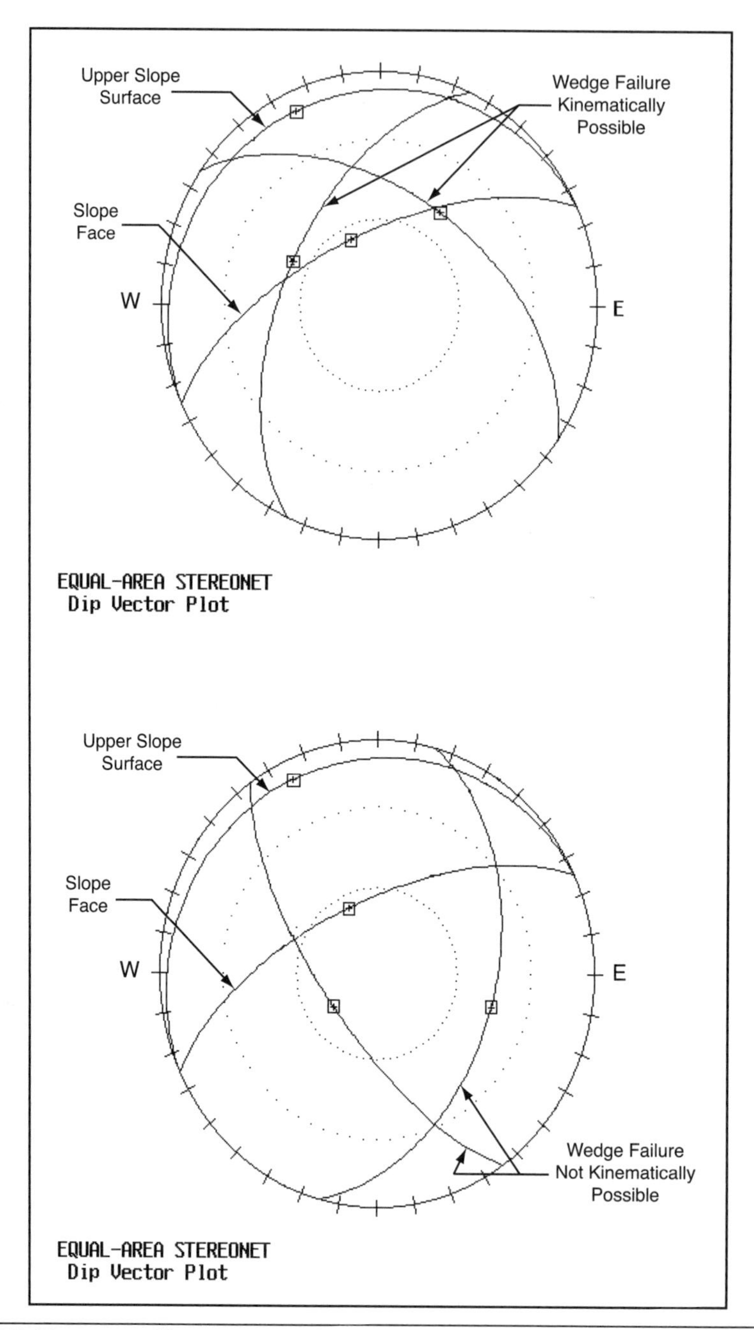

FIGURE 8.3 Stereographic representation of one wedge (top) for which failure is kinematically possible and one (bottom) for which failure is not kinematically possible

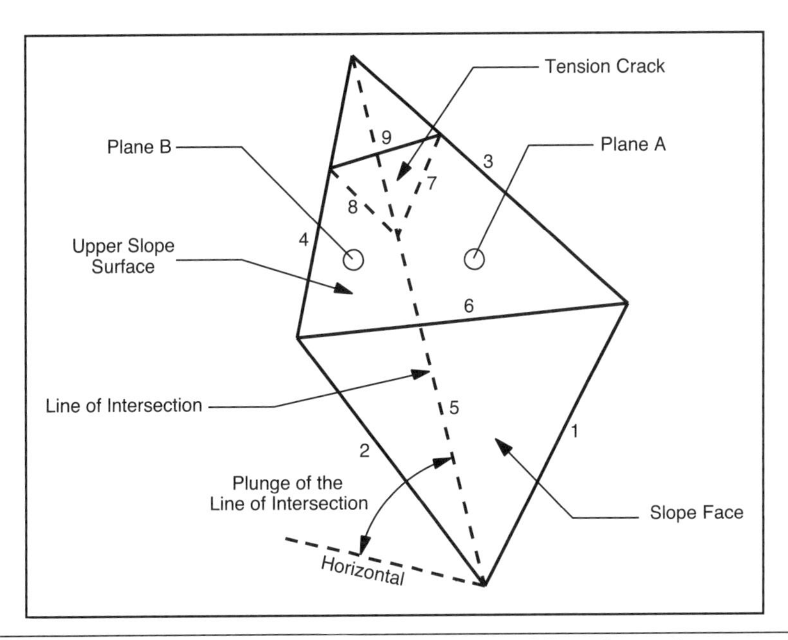

FIGURE 8.4 Notation used for designating the planes and lines for wedge geometry

WEDGE GEOMETRY

During manual calculations for the stability of a wedge, the stereographic projection is an invaluable tool. However, care must be taken to ensure that the correct angles—not conjugate angles—are measured. An incorrectly measured angle on a stereonet can lead to a gross error in wedge volume and wedge weight.

Stereographic Measurement of Wedge Angles

For the purpose of this discussion, the notation proposed by Hoek and Bray (1977) is used (Figure 8.4); that is,

- Plane A: The flatter of the two planes
- Plane B: The steeper of the two planes
- Line 1: The intersection of plane A with the slope face (SF)
- Line 2: The intersection of plane B with the SF
- Line 3: The intersection of plane A with the upper slope surface (USS)
- Line 4: The intersection of plane B with the USS
- Line 5: The intersection of planes A and B

If a tension crack exists in the USS, then additional numbering will be as follows.

- Line 6: The intersection of the USS with the SF
- Line 7: The intersection of the tension crack with plane A

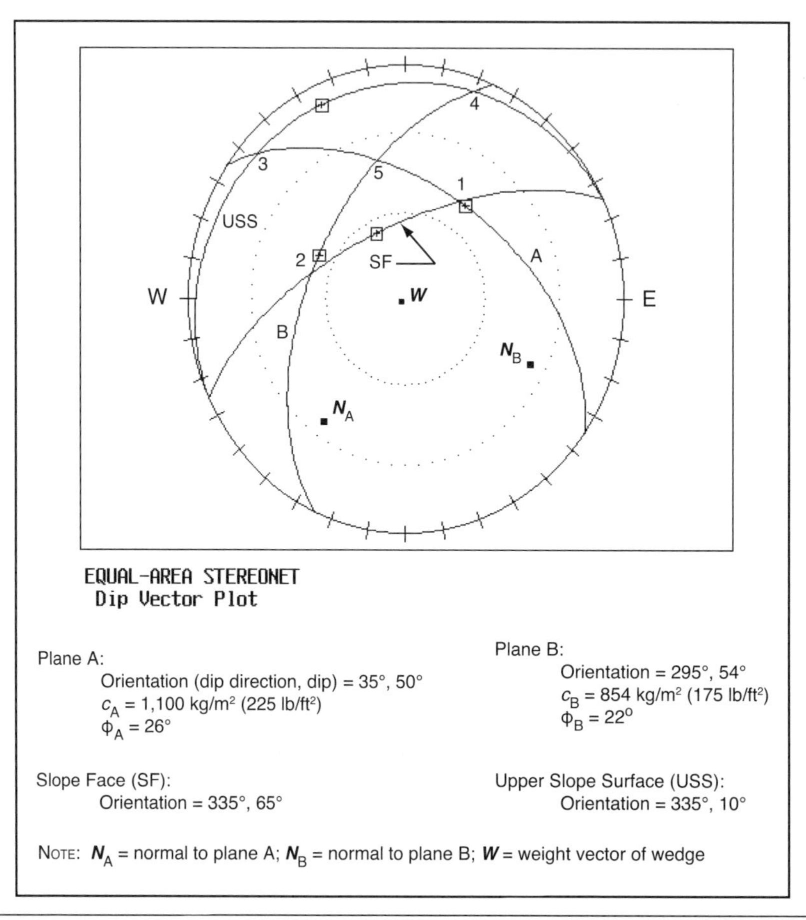

FIGURE 8.5 Notation of planes forming the wedge and the numbering of wedge end lines

- Line 8: The intersection of the tension crack with plane B
- Line 9: The intersection of the tension crack with the USS

This notation, without a tension crack, is shown on the stereonet of Figure 8.5 for the wedge geometry and strength attributes listed on the figure.

To start the analysis, the internal angles of the triangle forming plane A of the wedge must be measured with the stereonet. These internal angles include the angles between side 1 and side 5 (θ_{1-5}), between side 3 and side 5 (θ_{3-5}), and between side 1 and side 3 (θ_{1-3}). Likewise, the internal angles of the triangle forming plane B are measured (θ_{4-5}, θ_{2-5}, and θ_{2-4}).

To correctly interpret the angles measured with the stereonet (θ_{1-5}, θ_{3-5}, θ_{1-3}, θ_{4-5}, θ_{2-5}, and θ_{2-4}), we must be aware that we are measuring the angles between lines with trends in the same direction on the stereonet (i.e., lower-hemisphere projections). This means that, in most instances, the angle we measure between the SF and USS is the conjugate angle, not the internal angle. Except in the rare instances where the SF is dipping in the reverse direction (an overhanging bench condition exists), the angle between the intersection of plane A with the USS and the intersection of plane A with the SF (θ_{1-3})—as well as the angle between plane B and the same two surfaces (θ_{2-4})—should be greater than or equal to 90°. The angles measured off the stereonet of Figure 8.5 are as follows:

θ_{1-5}	32.5°
θ_{3-5}	44°
θ_{1-3}	76.5° (correct internal angle is 180° – 76.5° = 103.5°)
θ_{4-5}	42°
θ_{2-5}	45.5°
θ_{2-4}	87.5° (correct internal angle is 180° – 87.5° = 92.5°)
α_i, β_i	349°, 39° (orientation of the line of intersection of planes A and B—subscript i denotes the line of intersection)
angle($\boldsymbol{N}_A$, $\boldsymbol{N}_B$)	74°

Calculation of Rock Wedge Volume

To calculate the volume of the wedge, the length of at least one of the edges of the wedge (lines 1 to 5) must be known. This length can be obtained via a field measurement.

Another critical measurement needed is the internal angle between planes A and B (θ_w). The wedge angle θ_w is *not* equal to the angle between the poles of planes A and B—angle($\boldsymbol{N}_A$, $\boldsymbol{N}_B$). Instead, it is equal to 180° – angle($\boldsymbol{N}_A$, $\boldsymbol{N}_B$). This fact, however, is immaterial in the calculation, as we will see later.

For the example of Figure 8.5 presented earlier, assume that the vertical height, *H*, from the point at which the line of intersection of planes A and B daylights in the SF to the crest at the top of the SF is 25 m (82 ft). Also assume the rock mass has an in situ density of 2.6 g/cm^3 (162.24 lb/ft^3), which equates to an in situ tonnage factor of 2.6 t/m^3 (12.32 ft^3/st [short ton]).

Figure 8.6A shows a cross-sectional view of the wedge taken along the line of intersection of A-B. It is important to note that the dip angles of the USS and the SF as shown on the sectional view (10° and 65°, respectively) are the true dips of the respective planes and not the dip angles that would be actually measured in the field along the cross section (apparent dips). This is because the plunge direction of the line of intersection differs from the dip direction of the USS and the SF by 14° (349° vs. 335°). Nevertheless, the true dips of the respective planes were used on the cross section and in the following calculations. Figure 8.6A shows a view of plane A looking perpendicular to the plane, with the edge lines and internal angles noted. Figure 8.6B shows a cross-sectional view of the wedge

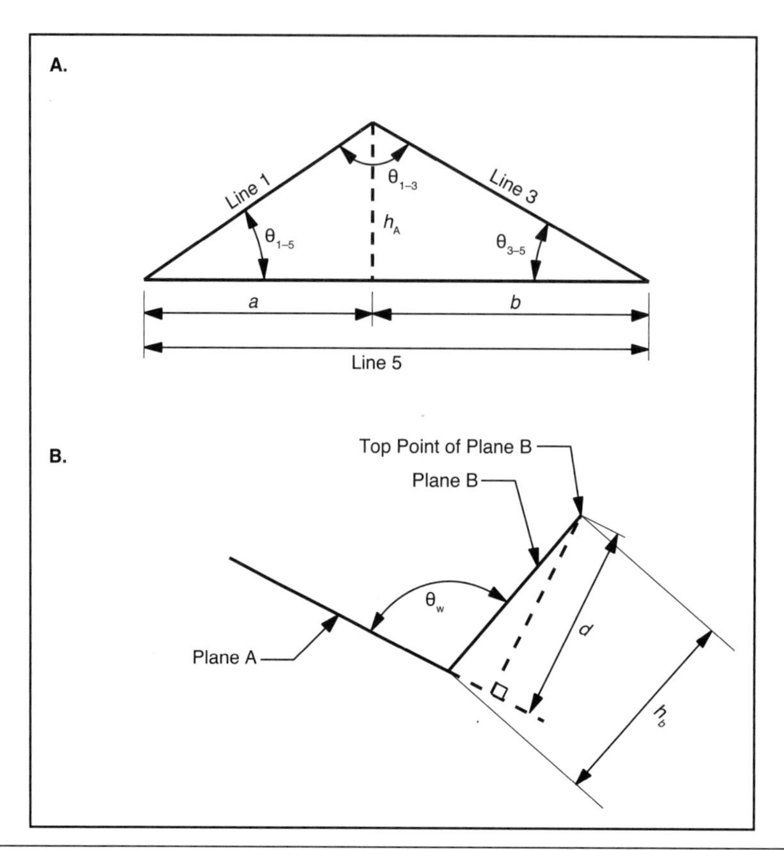

FIGURE 8.6 Notation for calculation of rock wedge volume: (A) view of plane A; (B) looking along the line of intersection

looking along the line of intersection. The equations that follow refer to the notation on these figures.

Given that we have measured H, the vertical height of the line of intersection, we can determine that the distance from the point where the line of intersection (L_5) daylights in the SF to the crest of the slope (SL) is

$$\mathrm{SL} = \frac{25}{\cos 25^\circ} = 27.58 \text{ m } (90.48 \text{ ft})$$

and by the law of sines

$$\frac{27.58}{\sin 29^\circ} = \frac{L_5}{\sin 125^\circ} = \frac{\mathrm{UL}}{\sin 26^\circ}$$

$$L_5 = 45.60 \text{ m } (152.87 \text{ ft})$$

$$\mathrm{UL} = 24.94 \text{ m } (81.81 \text{ ft})$$

where L_5 is the length of line 5, and UL is the length from where the line of intersection (L_5) daylights in the USS to the crest of the slope.

On plane A, we can determine the following:

$$\frac{h_A}{a} = \tan\theta_{1-5}$$

$$\frac{h_A}{b} = \tan\theta_{3-5}$$

$$a + b = L_5 = h_A\left(\frac{1}{\tan\theta_{1-5}} + \frac{1}{\tan\theta_{3-5}}\right)$$

$$\therefore h_A = \frac{L_5}{\left(\frac{1}{\tan\theta_{1-5}} + \frac{1}{\tan\theta_{3-5}}\right)}$$

It follows that, on plane B,

$$h_B = \frac{L_5}{\left(\frac{1}{\tan\theta_{2-5}} + \frac{1}{\tan\theta_{4-5}}\right)}$$

The area of plane A is then

$$A_A = \frac{h_A L_5}{2}$$

The height, d, from the apex of plane B (the intersection point of lines 2 and 4) normal to plane A is

$$d = h_B \cdot \sin\left(180° - \theta_w\right)$$

where θ_w is the internal angle between planes A and B. However, since θ_w is equal to 180° – angle($\boldsymbol{N}_A$, $\boldsymbol{N}_B$), we can use the measured angle between the poles of planes A and B to get

$$d = h_B \cdot \sin\left(\text{angle}\left(\boldsymbol{N}_A, \boldsymbol{N}_B\right)\right)$$

The volume of the wedge (volume of a prism) is

$$V_w = \frac{A_A d}{3}$$

and the weight of the wedge is

$$W_w = V_w \gamma_r$$

The same procedure can be used to calculate the volume of rock behind a tension crack. For the wedge attributes given previously, the volume and weight of the wedge are

$$V_{\mathrm{w}} = 2{,}972.02 \text{ m}^3 \ (104{,}956 \text{ ft}^3)$$
$$W_{\mathrm{w}} = 7{,}723{,}799 \text{ kg} \ (17{,}028{,}061 \text{ lb})$$

FACTOR-OF-SAFETY DETERMINATION

Numerous methods have been applied for the FS analysis of wedge failures. Most of these methods employ the use of stereographic projections and analytical techniques of static equilibrium. If friction alone is considered—that is, the effects of cohesion along the planes, water forces, vibratory forces, and stabilizing forces are ignored—the problem is much simplified. With this simplification, an assessment of the wedge stability by graphical means can rapidly be obtained.

For a more detailed solution that includes wedge geometry, wedge weight, friction, cohesion, water pressure, and other forces on the wedge, various techniques have been developed. These approaches include the following (Piteau and Martin 1982; Hoek and Bray 1977):

- **An algebraic solution.** This technique requires resolving the forces that act on the various failure planes by using simple algebraic and trigonometric methods.
- **An engineering graphics solution.** The algebraic solution just mentioned may also be solved by engineering graphics methods. This approach requires views of all planes and lines on the wedge, resolution of forces, and scaled force diagrams.
- **A stereographic projection solution.** This method uses a stereographic projection to determine angular relationships between the planes and forces on the wedge.
- **A vector solution.** In this method, all forces acting on the wedge are described in terms of vectors. Vector equations are then used to calculate the FS. This method is especially useful for computer applications.

This chapter shows how to determine the FS of a wedge with friction and cohesion by using the algebraic solution and stereographic projection solution techniques. With both methods, we must use force polygons to resolve components of force normal to and parallel to the line of intersection.

Algebraic Solution

The resisting forces on plane A, according to the Mohr–Coulomb criterion discussed earlier, can be expressed as

$$c_{\mathrm{A}} + N_{\mathrm{A}} \tan \phi_{\mathrm{A}}$$

Likewise, the resisting forces on plane B can be expressed as

$$c_B + N_B \tan\phi_B$$

The driving force parallel to the line of intersection of planes A and B can be expressed as $W\sin\beta_i$. Therefore, if the wedge forces are resolved into components normal to and parallel to the line of intersection of the two planes, then the factor of safety may be expressed as follows:

$$FS = \frac{c_A A_A + c_B A_B + N_A \tan\phi_A + N_B \tan\phi_B}{W\sin\beta_i}$$

where

N_A, N_B = magnitudes of the component forces normal to the failure planes
A_A, A_B = areas of the two failure planes
c_A, c_B = cohesions along planes A and B
ϕ_A, ϕ_B = friction angles along planes A and B
W = weight of the wedge
β_i = plunge of the line of intersection

The forces N_A and N_B are determined by (1) resolving the weight vector of the wedge, W, into a component along the line of intersection and on the same plane as N_A and N_B, then (2) resolving this component (N_i) into the forces N_A and N_B. To do so, the angle from W to N_i (β_i), as well as the angles from N_i to N_A and N_B (δ_A and δ_B), must be measured with the stereonet or calculated. Figure 8.7 is a stereonet showing the locations of W, N_A, N_B, and N_i, as well as the measured angles β_i, δ_A, and δ_B. Figure 8.8 shows the force polygons, along with the measured angles, used to determine the magnitudes of N_i, N_A, and N_B.

Resolving the forces via the law of sines, we get

$$N_i = 6{,}002{,}519 \text{ kg } (13{,}233{,}289 \text{ lb})$$
$$N_A = 4{,}096{,}704 \text{ kg } (9{,}031{,}691 \text{ lb})$$
$$N_B = 3{,}400{,}953 \text{ kg } (7{,}497{,}818 \text{ lb})$$

and the factor of safety with cohesion and friction is determined to be 0.879.

Stereographic Projection Solution

The general theory for the stereographic projection solution is given in Hoek and Bray (1977) and, to some extent, in Major et al. (1977). The explanation given here draws from those sources.

Consider again the wedge described in detail earlier in this chapter. As Figure 8.5 specifies, the friction angle for the wedge is 26° on plane A and 22° on plane B. The cohesion on

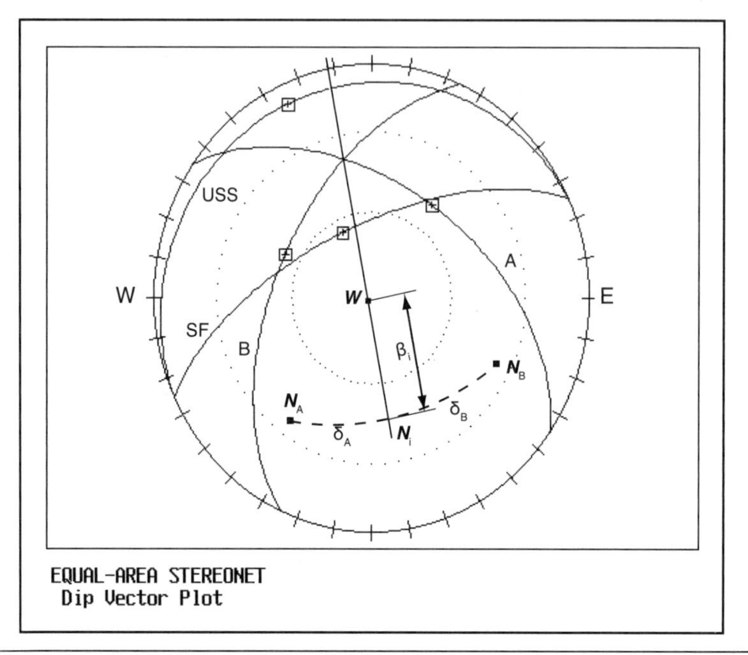

FIGURE 8.7 Location on the stereonet of wedge weight, *W*; the normals, N_i, N_A, and N_B; and angles to the normals, β_i, δ_A, and δ_B

plane A is 1,100 kg/m² (225 lb/ft²) and on plane B is 854 kg/m² (175 lb/ft²). The effective friction angles on planes A and B, incorporating cohesion, can be determined as

$$\tan\phi_{aA} = \tan\phi_A + \frac{c_A A_A}{N_A}$$

$$\tan\phi_{aB} = \tan\phi_B + \frac{c_B A_B}{N_B}$$

where

ϕ_{aA}, ϕ_{aB} = apparent friction as a result of cohesion and friction on plane A or plane B

ϕ_A, ϕ_B = friction angle of plane A or plane B

c_A, c_B = cohesion on plane A or plane B

A_A, A_B = area of plane A or plane B

N_A, N_B = magnitude of the normal force on plane A or plane B

The apparent friction angles on planes A and B are calculated from the previous equations as

$$\phi_{aA} = 30.94$$

$$\phi_{aB} = 28.12°$$

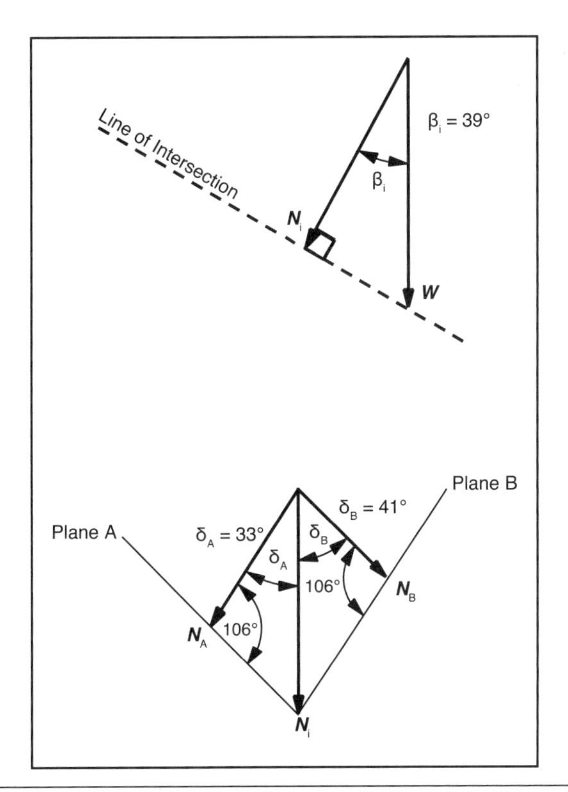

FIGURE 8.8 Force polygons to determine the magnitudes of N_i, N_A, and N_B

The normals to the shear planes, N_A and N_B, are plotted in Figure 8.9, and friction circles of radius ϕ and ϕ_a are drawn around the normals. These friction circles are drawn so that we can determine the factor of safety with and without cohesion.

The orientation of the line of intersection of the wedge (α_i, β_i) is 349°, 39° and is noted as $\mathbf{I}_{A\text{-}B}$ on Figure 8.9. Because sliding is in the direction of the line of intersection, $\mathbf{I}_{A\text{-}B}$ is also the orientation of the shear forces on each plane, S_A and S_B.

The resultant of the normal force N_A and the shear force S_A at the limiting equilibrium condition lies on the plane between N_A and $\mathbf{I}_{A\text{-}B}$. Similarly, the resultant of N_B and S_B lies on the plane between N_B and $\mathbf{I}_{A\text{-}B}$. The resultant on each plane is at an angle ϕ from the normal toward $\mathbf{I}_{A\text{-}B}$. The resultants—R_A and R_B for zero cohesion, R_{aA} and R_{aB} for nonzero cohesion—are shown on Figure 8.9.

The resultants of the forces—R_A, R_B, R_{aA}, R_{aB} —lie in the planes of angle(R_A, R_B) and angle(R_{aA}, R_{aB}). The plane of angle(R_A, R_B) is defined by the great circle that passes through R_A and R_B; the plane of angle(R_{aA}, R_{aB}) is defined by the great circle that passes through R_{aA} and R_{aB}. These two great circles define the limiting equilibrium conditions for sliding down and up the line of intersection for the two cases of $c = 0$ and $c \neq 0$.

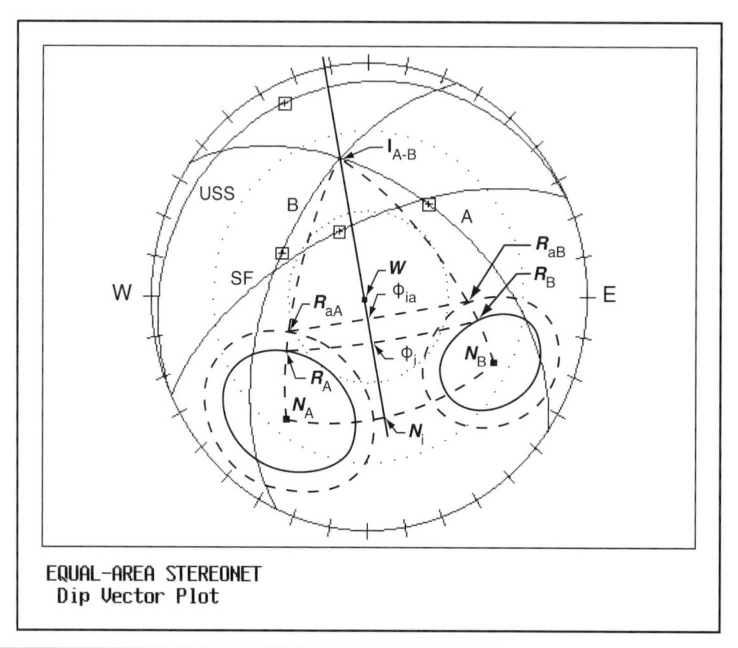

FIGURE 8.9 Stereonet for wedge factor-of-safety determination, with and without cohesion

The direction of the resultant weight of the wedge can be plotted directly on the net. If gravity alone is acting on the wedge, then the direction is vertically downward and the resultant weight force ($\boldsymbol{W}$) plots as the center pivot point on the stereonet. If other forces besides gravity are acting on the wedge (e.g., rock-bolt force and/or water force), then the use of force polygons and calculations is required to determine the magnitude and orientation of the resultant effective weight ($\boldsymbol{W}_e$).

To determine the factor of safety against sliding, the great circle containing the weight, $\boldsymbol{W}$, and the resultant shear force, which is defined by $\mathbf{I}_{A\text{-}B}$, is drawn. The intersection of this great circle with the great circle passing through the normals ($\boldsymbol{N}_A$ and $\boldsymbol{N}_B$) defines the position of the resultant normal force $\boldsymbol{N}_i$. The intersection of the great circle containing W and $\mathbf{I}_{A\text{-}B}$ with the great circle through the resultants ($\boldsymbol{R}_A$ and $\boldsymbol{R}_B$ for $c = 0$; $\boldsymbol{R}_{aA}$ and $\boldsymbol{R}_{aB}$ for $c \neq 0$) defines the position of the resultant friction angle (ϕ_i for $c = 0$; ϕ_{ia} for $c \neq 0$). For the values given in this chapter (see Figure 8.9), measuring along the great circle containing $\boldsymbol{W}$ and $\mathbf{I}_{A\text{-}B}$ gives an angle of 29.5° between $\boldsymbol{N}_i$ and ϕ_i (zero cohesion), and 37.5° between $\boldsymbol{N}_i$ and ϕ_{ia} (nonzero cohesion). It gives an angle of 40° between $\boldsymbol{N}_i$ and $\boldsymbol{W}$. (This angle is designated as δ_i; it should be equal to β_i, but there may be some error of measurement on the stereonet.)

The factor of safety can be defined as the ratio of the resultant resisting force, $\boldsymbol{N}_i\tan\phi_i$ or $\boldsymbol{N}_i\tan\phi_{ia}$, to the resultant driving force, $\boldsymbol{N}_i\tan\delta_i$:

TABLE 8.1 Wedge failure analysis for: (a) no water and $c \neq 0$; (b) no water and $c = 0$

	(a) Kroeger Wedge Failure Analysis, Dry and $c \neq 0$	(b) Kroeger Wedge Failure Analysis, Dry and $c = 0$
Rock unit weight	162.24 lb/ft^3 (2.6 gm/cm^3)	162.24 lb/ft^3 (2.6 gm/cm^3)
Unit weight of water	62.4 lb/ft^3 (1.0 gm/cm^3)	62.4 lb/ft^3 (1.0 gm/cm^3)
Slope crest height		
Slope face does not overhang toe		
Discontinuity 1:		
Dip direction	35°	35°
Dip angle	50°	50°
Cohesion	0	225 lb/ft^2 (1,100 kg/m^2)
Friction angle	26°	26°
Discontinuity 2:		
Dip direction	295°	295°
Dip angle	54°	54°
Cohesion	0	175 lb/ft^2 (854 kg/m^2)
Friction angle	22°	22°
Slope face:		
Dip direction	335°	335°
Dip angle	65°	65°
Upper slope surface:		
Dip direction	335°	335°
Dip angle	10°	10°
No tension crack		
Dry discontinuities		
Weight of wedge	15,881,755.6 lb (7,203,843.2 kg)	15,881,755.6 lb (7,203,843.2 kg)
Intersection of discontinuities:		
Trend	348.5°	348.5°
Plunge	39.3°	39.3°
There is contact on both discontinuities		
Factor of safety =	0.875	0.686

$$\text{FS}\ (\text{for}\ c = 0) = \frac{\tan(29.5°)}{\tan(40°)} = 0.67$$

$$\text{FS}\ (\text{for}\ c \neq 0) = \frac{\tan(37.5°)}{\tan(40°)} = 0.91$$

These results check closely with a computer solution (Kroeger 2011), which gives the value of the factor of safety with cohesion as 0.875 (Table 8.1a) and without cohesion as 0.686 (Table 8.1b).

FIGURE 8.10 Tension cracks at the head of a recent slope failure. Arrow indicates area where marker paint was painted at the crest of the slope a few hours before tension crack appeared.

OTHER CONSIDERATIONS

This section discusses other issues that may be important in assessing the potential for wedge failure: a tension crack, water pressure, and rock-bolt force.

Tension Crack

The presence of a tension crack in the USS behind the crest is a serious situation. It indicates that movement of the wedge has begun. The movement may have been only minor, thus opening up the tension crack, but with movement there is a loss of cohesion along the failure surface and a change from peak strength conditions toward residual strength conditions.

Figure 8.10 shows tension cracks in the crest of a slope that has recently failed. Notice the colored marker paint at the crest of the slope in the forefront. The geologist in charge had marked this tension crack with colored marker paint a few hours earlier. Naturally, the geologist did not mark the crack at the crest of the slope; rather, he had marked a crack some distance back from the slope. The slope continued to ravel (fail) for a significant period of time after the initial large-scale failure occurred. This picture was taken at the time the raveling had reached the tension crack marked by the geologist.

The geometry of the portion of the wedge behind the tension crack may be calculated in the same manner as the geometry of the entire wedge. Measurement (or calculation) of a length of one side of the portion behind the tension crack is essential to determine the remaining sides. Also, it is important to measure (or assume) the dip direction and dip of the tension crack. One important assumption is that the tension crack is open and continuous from the surface through the line of intersection of the wedge planes. The numbering scheme of the lines that define the edges of the tension crack is shown in Figure 8.4 and was discussed earlier.

The mass behind the tension crack is assumed to remain in place if failure of the wedge occurs. This event will then decrease the mass of the wedge by the mass of the portion behind the tension crack; it will truncate planes A and B and thus decrease the strength over these planes.

Water Pressure

The presence of groundwater in the slope, and thus possibly water pressure on the wedge failure planes, should also be a matter of concern. Water pressure will tend to decrease the normal forces on the failure planes and increase the driving force, thus decreasing the stability.

Five basic cases involving water pressure and the presence (or absence) of a tension crack are possible, namely (Major et al. 1977):

1. No tension crack or water is present.
2. There is no tension crack, but water is present.
3. A tension crack is present, but there is no water.
4. A tension crack is present with water within the tension crack.
5. A tension crack is present with the water table passing below the bottom of the tension crack.

For ease of computation, the water pressure distribution on the failure plane is normally assumed to be straight-line and tetrahedron-shaped. That is, the head reduces to zero where the tension crack daylights at the slope face, where the phreatic surface meets plane A or plane B or both, and/or where the phreatic surface intersects the tension crack. However, some practitioners assume a parabolic shape for the water pressure distribution. Table 8.2 and Figure 8.11 show the computer outputs from two wedge analysis programs. The first one (Table 8.2) assumes that the slope is completely submerged and that the water distribution has a parabolic shape. The calculated FS for this case is 0.457. The second one (Figure 8.11) assumes a straight-line, tetrahedron-shaped water distribution. The calculated FS for this case is 0.438, or a difference of about 4%.

The total force due to water is, therefore, equal to the maximum normal water pressure times the area over which the water pressure acts—or, in other words, it is equal to the volume of the tetrahedron or parabolic shape. The uplift forces, $\boldsymbol{U}_A$ and $\boldsymbol{U}_B$, are assumed to act in the opposite directions to the wedge normals, $\boldsymbol{N}_A$ and $\boldsymbol{N}_B$. The resultant force due to water pressure on the tension crack, $\boldsymbol{V}$, if present, is assumed to act horizontally. The water distribution and direction of $\boldsymbol{U}_A$, $\boldsymbol{U}_B$, and $\boldsymbol{V}$, with and without the presence of a tension crack (TC), are shown in Figure 8.12.

TABLE 8.2 Computer output of fully saturated wedge with parabolic water distribution (FS = 0.457)

	Kroeger Wedge Failure Analysis, Saturated and $c \neq 0$
Rock unit weight	162.24 lb/ft^3 (2.6 gm/cm^3)
Unit weight of water	62.4 lb/ft^3 (1.0 gm/cm^3)
Slope crest height	
Slope face does not overhang toe	
Discontinuity 1:	
Dip direction	35°
Dip angle	50°
Cohesion	0
Friction angle	26°
Discontinuity 2:	
Dip direction	295°
Dip angle	54°
Cohesion	0
Friction angle	22°
Slope face:	
Dip direction	335°
Dip angle	65°
Upper slope surface:	
Dip direction	335°
Dip angle	10°
No tension crack	
Water pressure:	
Discontinuity 1	989.8 lb/ft^2 (4,832.6 kg/m^2)
Discontinuity 2	989.8 lb/ft^2 (4,832.6 kg/m^2)
Weight of wedge	15,881,755.6 lb (7,203,843.2 kg)
Intersection of discontinuities:	
Trend	348.5°
Plunge	39.3°
There is contact on both discontinuities	
Factor of safety =	0.457

For the case of no tension crack, and assuming a tetrahedron-shaped water distribution, the maximum water pressure, $\mathbf{P}_{max}$, is

$$\mathbf{P}_{max} = \frac{h_w \gamma_w}{2}$$

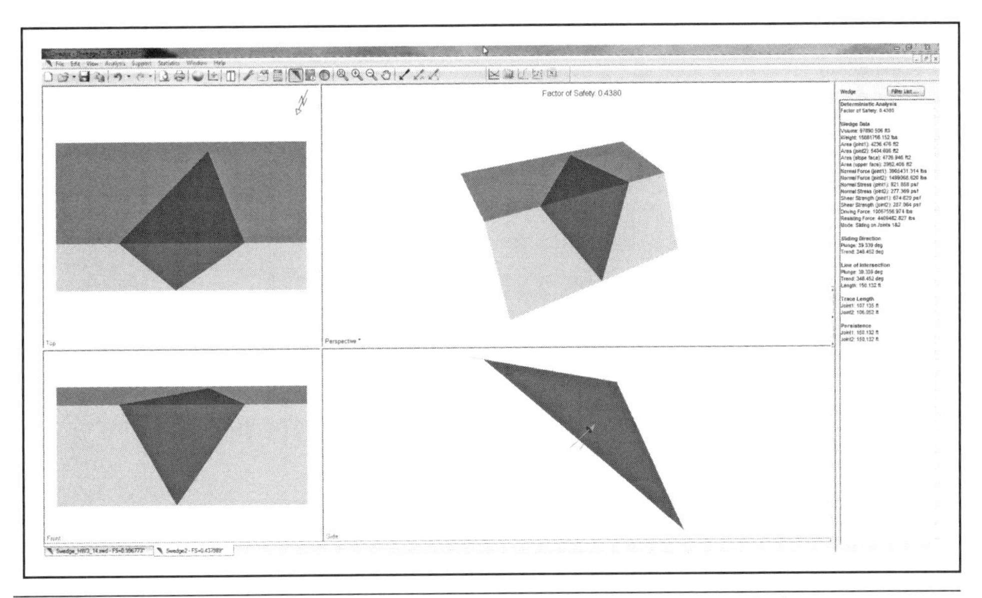

FIGURE 8.11 Computer output of fully saturated wedge with tetrahedral water distribution (FS = 0.438)

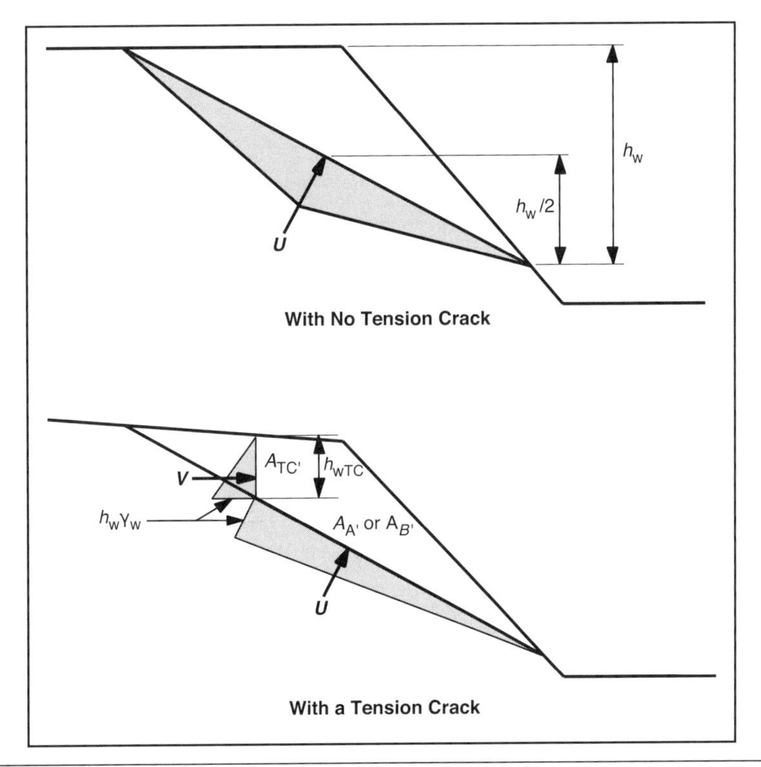

FIGURE 8.12 Water pressure distribution on a rock wedge

where

h_w = vertical height of water within the wedge from the toe of the wedge to the point where the phreatic surface intersects the line of intersection of the two planes

γ_w = unit weight of water, 1 t/m^3 (62.4 lb/ft^3)

For the case of a tension crack, the maximum water pressure, $\mathbf{P}_{max}$, is

$$\mathbf{P}_{max} = h_{wTC}\gamma_w$$

where h_{wTC} is the height of water on the tension crack.

The magnitudes of the uplift forces on planes A and B and of the horizontal force on the tension crack ($\boldsymbol{U}_A$, $\boldsymbol{U}_B$, and $\boldsymbol{V}$) are

$$\boldsymbol{U}_A = \frac{\boldsymbol{P}_{max} A_{A'}}{3}$$

$$\boldsymbol{U}_B = \frac{\boldsymbol{P}_{max} A_{B'}}{3}$$

$$\boldsymbol{V} = \frac{\boldsymbol{P}_{max} A_{TC'}}{3}$$

where

$A_{A'}$ = area of plane A over which the water pressure acts

$A_{B'}$ = area of plane B over which the water pressure acts

$A_{TC'}$ = area of the tension crack over which the water pressure acts

Rock-Bolt Force

The classical small- to medium-sized wedge failure mode involving one or two bench slope instabilities is quite amenable to stabilization by the use of rock bolts, rock anchors, or cable bolts. Larger wedges or multiple wedges, however, may be too large or too complex for mechanical support mechanisms (Seegmiller 1982).

The general stereographic method of analysis is to apply rock-bolt force as a line oriented in a direction approximately opposite to the trend of the line of intersection and plunging at an angle $\pm\theta$ to cross the line of intersection. The rock-bolt orientation is plotted as a point representing a line on the stereonet at the desired orientation. The effect of rock-bolt force is to increase the effective weight of the wedge, $\boldsymbol{W}_e$, and to shift the effective weight toward the rock-bolt force on the stereonet along the great circle between the weight and rock-bolt force (Figure 8.13). The magnitude of the effective weight, $\boldsymbol{W}_e$, can be obtained by measuring the angle between the wedge weight, $\boldsymbol{W}$, and the applied rock-bolt force, $\boldsymbol{T}$, on the stereonet and solving a force polygon for $\boldsymbol{W}_e$. Rock-bolt force tends to shift the effective

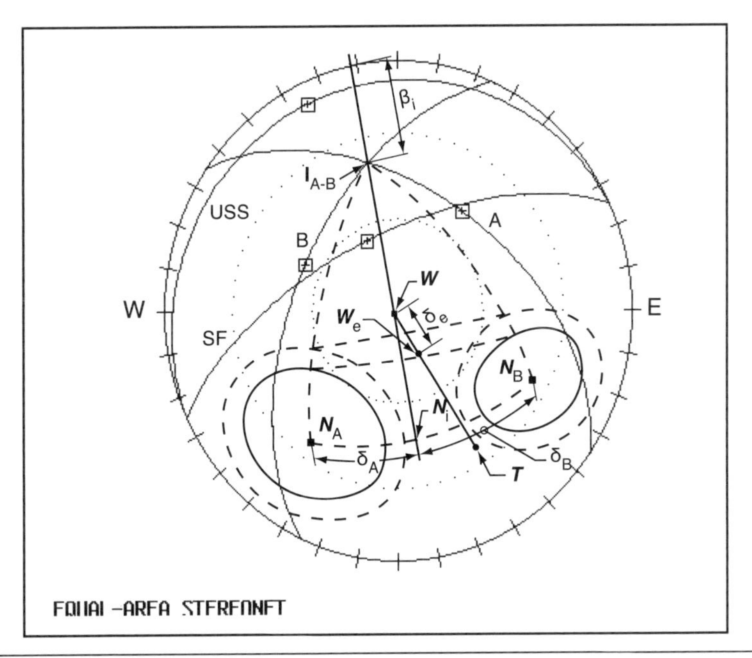

FIGURE 8.13 Rock-bolt force, *T*, and effective weight W_e

wedge weight toward stability, whereas water pressure and the resultant water force tend to shift it toward instability.

There are some drawbacks associated with this technique. For example, calculations of the minimum rock-bolt force and optimum applied angle for a desired factor of safety are quite tedious. In addition, rock-bolt force in the field is often applied in directions that best conform to the geologic conditions; that is, the bolts are "field fit."

REFERENCES

Brawner, C.O., and V. Milligan, eds. 1971. *Stability in Open Pit Mining: Proceedings of the First International Conference on Stability in Open Pit Mining.* New York: SME-AIME.

Hoek, E., and J.W. Bray. 1977. *Rock Slope Engineering.* London: Institution of Mining and Metallurgy.

Kroeger, E.B. 2011. *Wedge Failure Analysis Module v 2.0.* http://wedge-failure-analysis.software.informer.com/ (accessed July 2015).

Major, G., H-S. Kim, and D. Ross-Brown. 1977. *Pit Slope Manual Supplement 5-1: Plane Shear Analysis.* Report 77-16. Edited by D.F. Coates. Ottawa, ON: Canada Centre for Mineral and Energy Technology.

Piteau, D.R., and D.C. Martin. 1982. Mechanics of rock slope failure. In *Stability in Surface Mining,* Vol. 3. Edited by C.O. Brawner. New York: SME-AIME.

Rocscience. 2015. SWedge 6.0—3D surface wedge stability for rock slopes. Toronto: Rocscience.

Seegmiller, B.L. 1982. Artificial support of rock slopes. In *Stability in Surface Mining,* Vol. 3. Edited by C.O. Brawner. New York: SME-AIME.

CHAPTER 9

Rock Slope Stabilization Techniques

Some degree of slope instability can be expected with virtually any slope cut in rock, whether the slope is a mine highwall, is for a road cut, or is a part of some other construction project. When slope stability investigations indicate that the possibility of slope failure exists, there are a number of response options available (Call and Savely 1990):

- Leave the unstable area alone.
- Continue mining without changing the mine plan.
- Unload the slide through additional stripping.
- Leave a step-out (i.e., a bench or berm of unexcavated rock at the toe of the slope to increase the resisting forces of the slope).
- Conduct a partial cleanup.
- Mine out the failure.
- Support the unstable ground.
- Dewater the unstable area.

Three general principles of slope mechanics should be kept in mind in cases of slope instability (Call and Savely 1990):

1. **Slope failures do not occur spontaneously.** One or more of the forces acting on a potentially unstable rock mass must change for the mass to become unstable.
2. **Most slope failures tend toward equilibrium.** A slope fails because it is unstable under the existing conditions. Failure tends to bring the slope to some sort of equilibrium. It normally involves a reduction in the driving forces and/or an increase in the resisting forces of the failed zone.
3. **A slope failure does not occur without warning.** Prior to failure, measurable movement and/or the development of tension cracks will occur. These indications of failure can develop, indicating imminent slope failure, then subside for a long period of time, indicating apparent stability.

In the evaluation of the necessity and type of stabilization technique to be used, the first issue to be considered is the degree of urgency. If the slope has started to move, immediate remedial actions should be taken. These actions may include evacuating structures, closing roadways, or cordoning off a section of a mine.

Once the situation has become less urgent, determining the cause of the instability is necessary by the following means: visual observation based on experience, water level measurements, slope instrumentation, tests on the materials, and a survey of the discontinuity patterns (Golder 1971). Laboratory tests may be required on the rock and discontinuities to determine appropriate strength parameters.

Slope stabilization techniques can be divided into six general categories:

1. Grading
2. Controlled blasting
3. Mechanical stabilization
4. Structural stabilization
5. Vegetative stabilization
6. Water control

Frequently, many of the methods in these general categories are used together to achieve a stable, aesthetically appealing rock slope.

GRADING

Grading involves the shaping of the rock slope into a more stable configuration. It may include flattening the slope, leaving benches in the slope face, or sculpting the slope face to a more natural appearance.

Serrating

Serrating is the cutting of regular, low-height benches into a slope. Generally, the benches are 1–2 m (3.3–6.6 ft) in height and width and extend over the full height of the slope. This technique is normally used on soil slopes or slopes of loose, weathered rock that can be excavated by mechanical means. It is frequently used where vegetation is expected to become established.

Careful consideration must be given to drainage of water that may accumulate on the benches. An often advisable approach is to design the benches with a cross slope and back slope (Figure 9.1) so that the water drains off the slope or into preengineered channels or flumes.

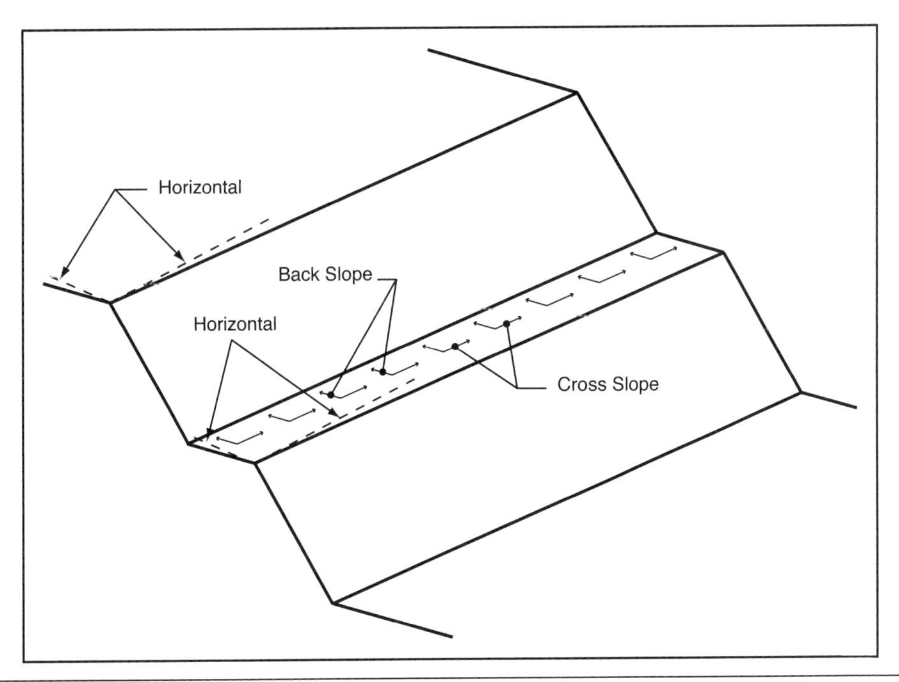

FIGURE 9.1 Bench designed with a cross slope and back slope to drain water away from the crest and across the bench to a disposal system

Benching

Historically, benching has been a popular technique used on cut slopes where an operating level is needed, as in mining, or where a break in slope is required to retain rockfall, as in road cuts. Benching is also useful in breaking up the expanse of a high rock cut to make the slope appear more natural.

Bench heights vary from about 3 to 10 m (10 to 33 ft) in highway corridor applications to more than 30 m (100 ft) for mine final highwalls. Bench widths also vary from a few meters to as much as 30 m (100 ft).

For years, benches were used along highway corridors as a means of intercepting rockfall before it reached the roadway. The original premise was that the benches would be cleaned periodically. However, experience has shown that the benches are rarely cleaned and that they act as launching ramps for rocks falling from the slope above the bench (Brawner 1994).

There are three main types of highwall benches in use:

1. **Level benches at regular intervals** (Figure 9.2). These benches should not actually be level; they should be designed and constructed with a cross slope and back slope to facilitate water drainage.

FIGURE 9.2 A level catch bench in the face of a rock slope along a highway

2. **Benches constructed parallel to the ground surface or to particular strata** (Figure 9.3). These benches may blend better with other topographic features and may have easier access for cleaning.
3. **Benches constructed at irregular intervals.** These benches may be designed and constructed at irregular vertical intervals, with variable bench widths, and to be discontinuous across the face of the slope. Their purpose is to blend in with the natural topography and to appear natural rather than constructed.

Catch bench (berm) design. Catch benches are used extensively in highway corridor construction through mountainous terrain and on mine highwalls. A catch bench is simply a cut bench on the rock slope, its main purpose being to "catch" rocks so that they do not continue unhindered over the crest of the catch bench and travel to the toe of the slope and beyond into the work area, as in the case of a mine, or into the roadway area, as in the case of a highway corridor. Figure 9.4 shows a wide catch bench constructed in the overburden highwall at a coal mine.

Open-pit bench geometry defines the steepest inter-ramp slope that can be mined while maintaining adequate catch bench widths. The two primary factors that control bench configuration are the type of mining equipment that is used and the bench face angles that can be achieved. The type of mining equipment determines the safe operating height of the bench. The achievable bench face angles are controlled by rock strength, geologic structure characteristics, and the mining techniques used to construct the slope (e.g., the blasting and digging practices; Ryan and Pryor 2000).

The final, benched highwall slope configuration is related to the operating bench height, the bench face angles, and the required catch bench widths to contain the spill adequately.

FIGURE 9.3 A catch bench in the face of a rock slope along a highway, constructed parallel to the surface topography

Courtesy of Brian Wenig, Cloud Peak Energy

FIGURE 9.4 Wide catch bench constructed in the overburden highwall at a coal mine

- **Bench height.** The major mining equipment used at the operation (excavators and drilling/blasting equipment) determines the bench height. Often, the maximum bench height is determined by safety and is a function of the maximum digging height of the major bench-mining excavator employed. Currently, most large mining operations use working bench intervals of 12–18 m (40–60 ft), with 15-m (50-ft) heights being very common.

Courtesy of Mine Safety and Health Administration

FIGURE 9.5 Catch bench at a coal operation

- **Bench face angles.** In soft rock operations, bench face angles are often controlled by the digging (crowd and hoist) geometry of the excavating equipment being used. Whereas in hard rock, the stable bench face angles are controlled primarily by the strength and orientation of the local geologic structure, and the orientation of the mine highwall with respect to the strike and dip of the various structures.
- **Berm or catch bench widths.** The catch bench width varies in width from the design width from point to point along its length because of the orientation of local geologic structure, drilling and blasting practices, and possible over-excavation. Figure 9.5 shows a catch bench at a coal operation that was narrower than design width in one location because of blasting overbreak, rock discontinuity intersections, or other factors, and which resulted in rock spilling over the crest of the catch bench to the pit floor below causing an unsafe condition.

Some of the first research undertaken in the area of catch bench design was published by Arthur M. Ritchie, chief geologist of the Washington State Department of Highways, in 1963. Ritchie rolled rocks off a variety of walls, observed the rock's motion during the fall, and tested a range of ditch configurations at the toe. The results of the study include a design guide that can be used to select the appropriate ditch depth and distance between the pavement and the wall toe (Ritchie 1963).

The work performed by Ritchie and the Washington State Department of Highways became the design standard for highway slopes, and the guidelines were used extensively until a change in highway regulations occurred.

A modified Ritchie criteria evolved from the original Ritchie criteria since the original Ritchie criteria was limited to a small number of slope-angle/slope-height geometries and therefore required extrapolation for use in open-pit mining. The modified Ritchie criteria is an empirical relationship between slope height and the average, or preferred, catch bench width of the form (Call 1992; Ryan and Pryor 2000):

$$\text{bench width (m)} = 0.2 \times \text{bench height (m)} + 4.5 \text{ m}$$

Catch bench design is the process of conducting stability analyses to estimate the bench face angles, selecting the bench width and, to a limited extent, the bench height. The bench height is controlled by the height of the mining levels, but it is possible to increase the height by leaving catch benches on every other level (double benching) or every third level (triple benching). Figure 9.6 shows (A) mine operating benches, (B) the final highwall with catch benches left at every level, (C) catch benches at every other level (double benching), and (D) catch benches at every third level (triple benching). As can be observed in Figure 9.6, increasing the spacing of the catch benches from every level to every other level or every third level results in an increase of the inter-ramp slope angle as long as the catch bench width, step-out, and bench slope angle remains unchanged from configuration to configuration. The "step-out" shown in Figure 9.6 is the lip of the bench remaining after drilling and blasting as close as possible to the highwall with the bench drilling equipment being used.

The catch bench is designed with a specific width relative to its height so that rocks will come to rest before falling over the crest. In addition, a "backbreak" distance is often incorporated into design as the bench crest will often fail from its (usual) vertical position, thus narrowing the usable catchment width. The design catch bench width is therefore always wider than the bench width required for safety, as shown in Figure 9.7 (Mathis 2009). Design catch bench width, w, can be estimated from the following equation (Bertuzzi 1999; Martin and Piteau 1977; Piteau and Martin 1982) (see Figure 9.8):

$$w = h\sqrt{x \cdot z \cdot \gamma_{\text{bulk}}}$$

where

h = bench height

$$x = \frac{1}{\tan\beta} - \frac{1}{\tan\alpha}$$

$$z = \frac{1}{\tan r} - \frac{1}{\tan\beta}$$

$\gamma_{\text{bulk}} = 1 + [(\text{material swell, \%})/100]$

and

β = dip of the discontinuity

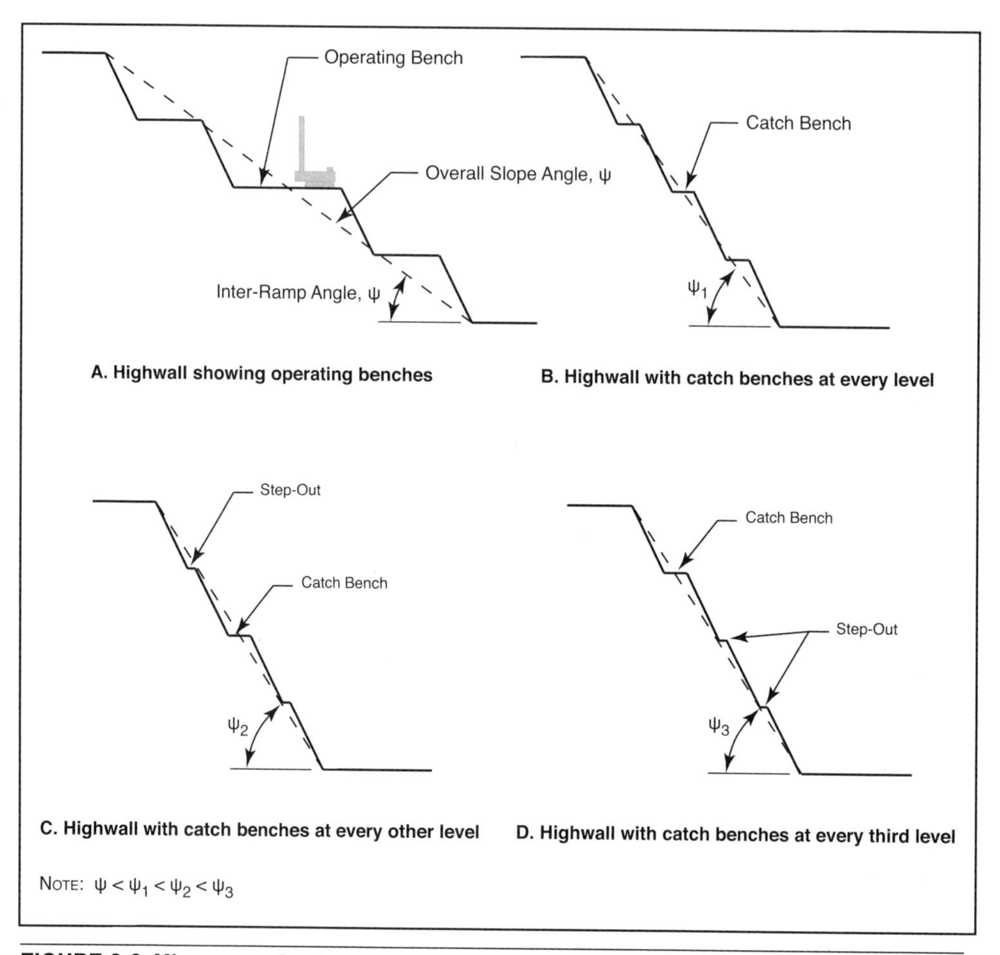

FIGURE 9.6 Mine operating benches and highwall catch bench configuration

$\alpha =$ bench slope angle

$r =$ angle of repose of broken rock; typically about 37°

Where the defect is short and affects only part of the slope (Figure 9.9), the catch bench width, w, can be estimated from

$$w = l \cdot \sin\beta \sqrt{x \cdot y \cdot \gamma_{\text{bulk}}}$$

where

$l =$ through-going discontinuity length

$x =$ defined previously

$y = \dfrac{1}{\tan r} - \dfrac{1}{\tan \alpha}$, where r and α are defined previously

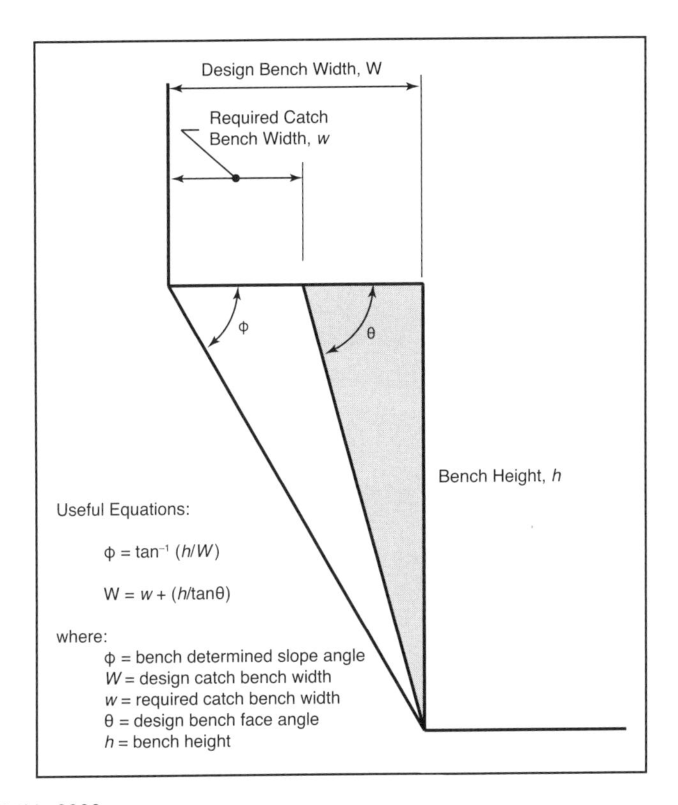

Adapted from Mathis 2009

FIGURE 9.7 Required catch bench width versus design bench width

Several factors contribute to the integrity and effectiveness of catch benches, including structure (discontinuity) orientation, continuity, length, spacing, and number of joint sets as well as blasting practices adjacent to the planned catch bench. Typical explosives used in mine production blasting generate detonation pressures exceeding 6,895 MPa (1,000,000 psi or 68.95 kb). Besides the initial shock pressure from the detonating explosives, the explosives generate large quantities of expanding pressurized gases, which can be forced into preexisting tight discontinuities, thereby fracturing attachments across the discontinuity and significantly weakening the shear strength of the discontinuity. Calculations using Equation 9.4 later in this chapter, and an explosive detonation pressure of 6,894 MPa with full borehole coupling and typical discontinuity undisturbed strength, have demonstrated that the discontinuity can be weakened for a distance exceeding 30.5 m (100 ft). Figure 9.10 shows an adversely oriented discontinuity in the highwall of a mine with loaded blastholes detonating adjacent to the discontinuity and pressurized gases venting into the discontinuity causing it to open. Figure 9.11 shows failures in the highwall at a coal mine caused by backbreak from the last row of a cast blast. Bedding planes dipping less than the dip of the highwall and striking approximately parallel to the strike of the highwall can plainly be observed. A weak layer (rider coal seam) at the bottom of the line of failures can also be observed. Plainly, gases from detonation of the last row of holes vented

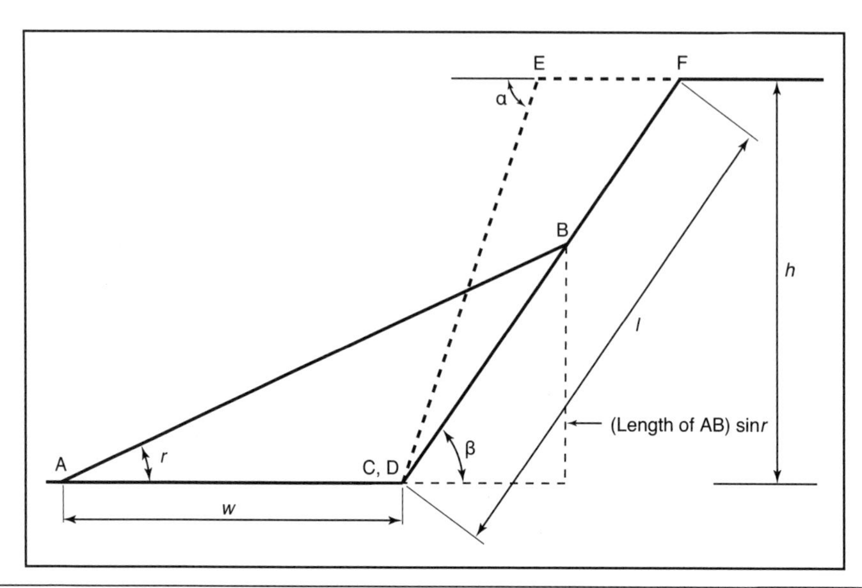

FIGURE 9.8 Geometry for catch bench design when the defect is long and affects all of the slope

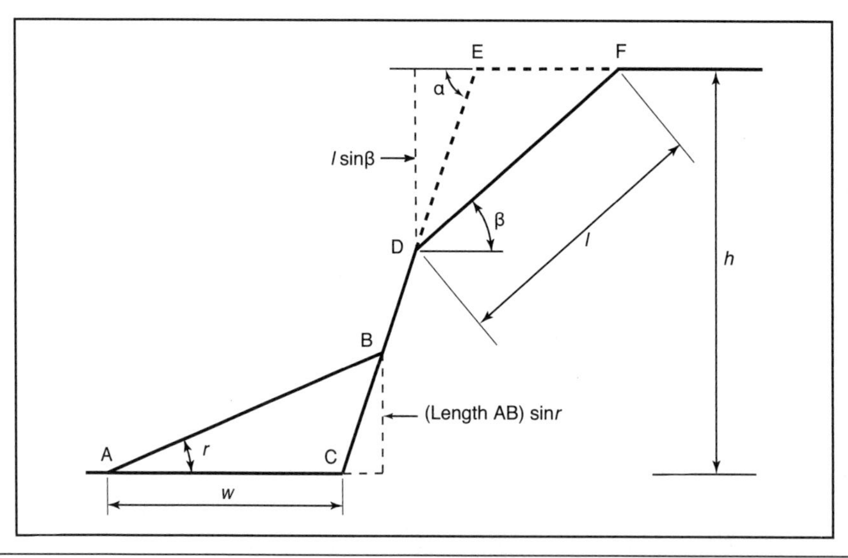

FIGURE 9.9 Geometry for catch bench design when the defect is short and affects only part of the slope

through the weak layer and traveled back for a distance into the intact overburden along the bedding planes. Over time, the weight of the hanging rock exceeded the strength of the shale-like material, and small planar-type failures occurred, with the base of the plane being the bedding planes.

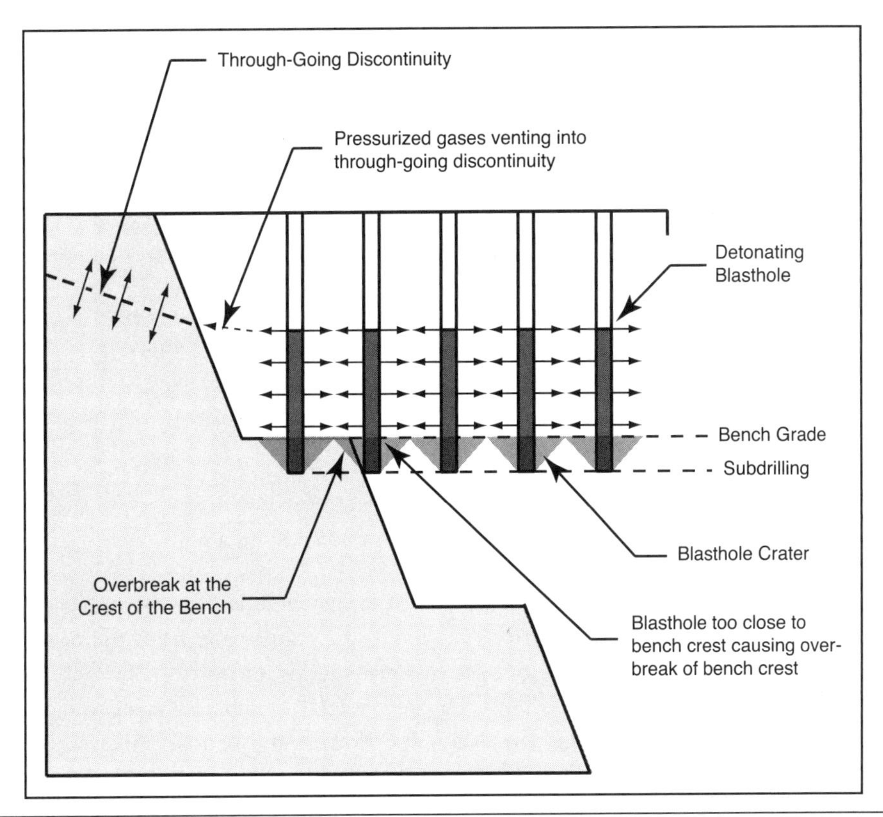

FIGURE 9.10 Bad positioning of blastholes with respect to bench crest and blast gases venting into through-going discontinuity leading to backbreak

Courtesy of Peabody Energy

FIGURE 9.11 Failures in the highwall at a coal mine caused by backbreak from the last row of a cast blast

FIGURE 9.12 Rock slope cut to match the dip of the beds along a highway

Also shown in Figure 9.10 is the situation where a blasthole is placed too close to the designed crest of a bench. In the situation shown, the crater cone from the detonating explosives within the subdrilling zone has caused breakage over a portion of the crest of the bench adjacent to the hole. This is a likely cause of the catch bench narrowing in Figure 9.5, which resulted in material spilling over the crest of the catch bench. This is an all-too-common problem when using some sort of modified production blasting adjacent to the highwall to establish catch benches instead of the controlled blasting techniques discussed in the following sections.

Matching the Dip of Bedding

In certain rock strata, cutting the slope to match the dip of the beds may be an effective method of achieving stability (Figure 9.12). This approach may be especially effective in thinly bedded shales, slates, limestones, and so on. For high slopes, benches may be included to break up the expanse of a high, smooth face (Figure 9.13). In situations where the strata are steeply dipping, other stabilization methods, such as rock anchors or cable bolts, may need to be employed in conjunction with this technique to prevent failure by flexural bending of the long, thin strata columns (see Figure 1.11).

Rock Sculpting

Rock sculpting is a technique employed at some mining operations for final reclamation and along some highway corridors to achieve a stable as well as aesthetically pleasing highwall. In essence, drilling and blasting, mechanical excavation, vegetative stabilization, and other techniques are employed to shape the highwall so that it appears natural. This technique may result in more potential for rockfall, however, and is often more expensive than other similar techniques.

Courtesy of Homestake Mining Company

FIGURE 9.13 A mine highwall with overall slope angle equal to the dip of the beds

Instead of a straight, steep highwall, the slope is designed and cut to include undulations in the strike and dip of the face, midwall benches where appropriate, and talus accumulations on the benches and at the toe. Sculpting and the establishment of vegetative cover on the slope produces a more natural appearance.

CONTROLLED BLASTING (OVERBREAK CONTROL)

The purpose of overbreak control is to achieve a stable highwall by limiting the damage from production blasting beyond the cut limit. Often, a secondary purpose is to achieve an aesthetically appealing wall. Several drilling and blasting techniques have been developed for overbreak control to accomplish this purpose, including (McKown 1984; Floyd 1998)

- Modified production blasts,
- Presplit blasting,
- Trim (cushion) blasting, and
- Line drilling.

Two aspects of final wall design should be considered in selecting the appropriate controlled blasting technique, or combination of techniques, for a particular job: (1) defining rock damage criteria and (2) developing a procedure to design blasts that will minimize rock damage without seriously affecting production.

Modified Production Blasts

In modified production blasting, the energy level is decreased adjacent to the wall to reduce overbreak. This decrease in energy level is often achieved for competent rock simply by

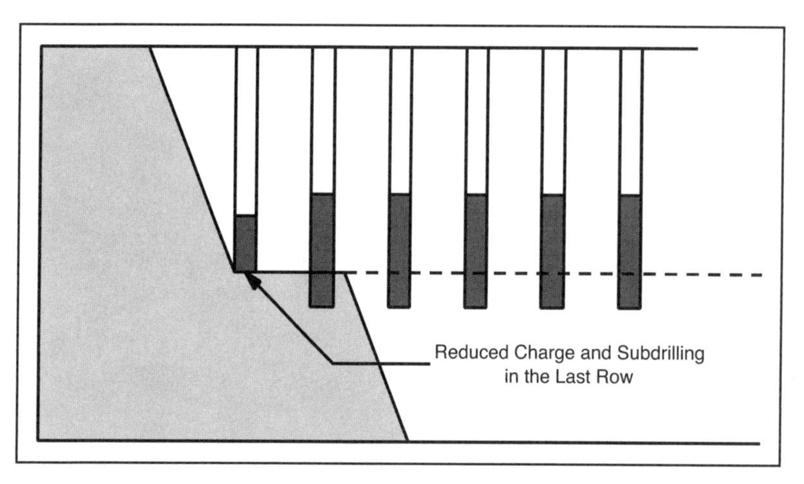

Adapted from Floyd 1998

FIGURE 9.14 Modified production blast design in favorable conditions

reducing the charge weight in the row nearest the slope by about 30%–60% (Floyd 1998), as shown in Figure 9.14.

For less competent rock masses, additional modifications to the blast design may be required to minimize overbreak damage. These modifications can include using decked charges (as explained later), reducing the burden (i.e., the distance to the nearest free face) and spacing of the last row, minimizing subdrilling, and increasing the delay interval between the last two rows of blastholes.

The primary advantage of the modified production blasting technique is that it requires few design changes. The primary disadvantage is that the wall rock is not protected from crack dilation, gas penetration, and block heaving (Floyd 1998).

Presplit Blasting

Presplitting uses lightly loaded, closely spaced drill holes that are fired before the production blast to form a fracture plane across which the radial cracking from the production blast cannot travel (Konya 1995). As a secondary benefit, the fracture plane formed may be aesthetically appealing. A typical presplit blast is shown in Figure 9.15. The presplit blast may be fired a considerable amount of time before the production blast (hours, days, weeks, etc.) or shortly before as on a prior delay. Delayed blasting techniques are frequently used to separate the detonation times of explosive charges (i.e., individual holes, or a series of holes as in a row of holes). A charge on a prior delay means that the charge is detonated earlier than the subsequent charge (normally, anywhere from a few milliseconds to 1 to 2 seconds). Figure 9.16 shows a mine highwall where presplit blasting has been used to control overbreak and produce a stable final wall.

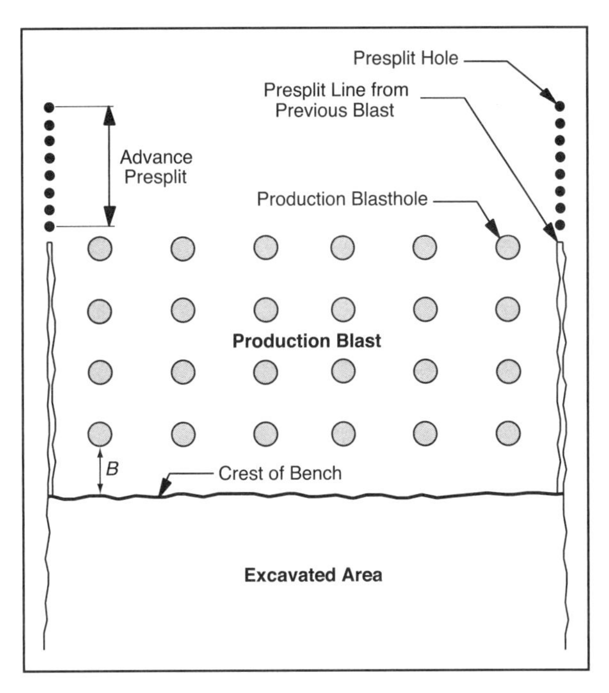

Courtesy of International Society of Explosives Engineers

FIGURE 9.15 Typical presplit blast

During presplitting, the production blast has not yet been fired, and for all practical purposes the burden is infinite. However, as a rule, the presplit row is commonly positioned approximately $0.5B$ to $0.8B$ behind the last row of the production blast, where B is the production blast burden.

The approximate explosive load per meter of presplit borehole length that will not damage the wall but that will produce sufficient pressure to cause the splitting action to occur can be determined from the following equation (Konya 1995):

$$d_{ec} = \frac{D_h^2}{12.14} \text{ in metric units}$$
$$= \frac{D_h^2}{28} \text{ in U.S. customary units} \qquad \text{(EQ 9.1)}$$

where

d_{ec} = explosive load, in g/m (lb/ft)

D_h = diameter of the empty hole, in mm (in.)

NOTE: Notice the traces of the presplit holes (commonly called half-casts).

FIGURE 9.16 Presplit blasting used to provide a stable final main highwall

If this approximate explosive load is used, then the spacing between presplit holes, S (in millimeters), can be approximated by

$$10D_b < S < 14D_b \quad \text{(EQ 9.2)}$$

Another method for determining the spacing between the presplit holes is based on the theory of a pressurized thick-walled cylinder (Figure 9.17). For a pressurized cylinder with an infinite outer radius, the radial and tangential stresses may be determined as follows (Jaeger and Cook 1979):

$$\sigma_r = \sigma_0 + (P_i - \sigma_0)\frac{a^2}{r^2} \quad \text{(EQ 9.3)}$$

$$\sigma_\theta = \sigma_0 - (P_i - \sigma_0)\frac{a^2}{r^2} \quad \text{(EQ 9.4)}$$

where

σ_r = radial stress, in MPa (psi)
σ_0 = in situ stress field, in MPa (psi)
P_i = decoupled borehole pressure, in MPa (psi)
a = radius of the hole, in m (in.)
r = distance from the center of the hole to the point of interest, in m (in.)
σ_θ = tangential stress, in MPa (psi)

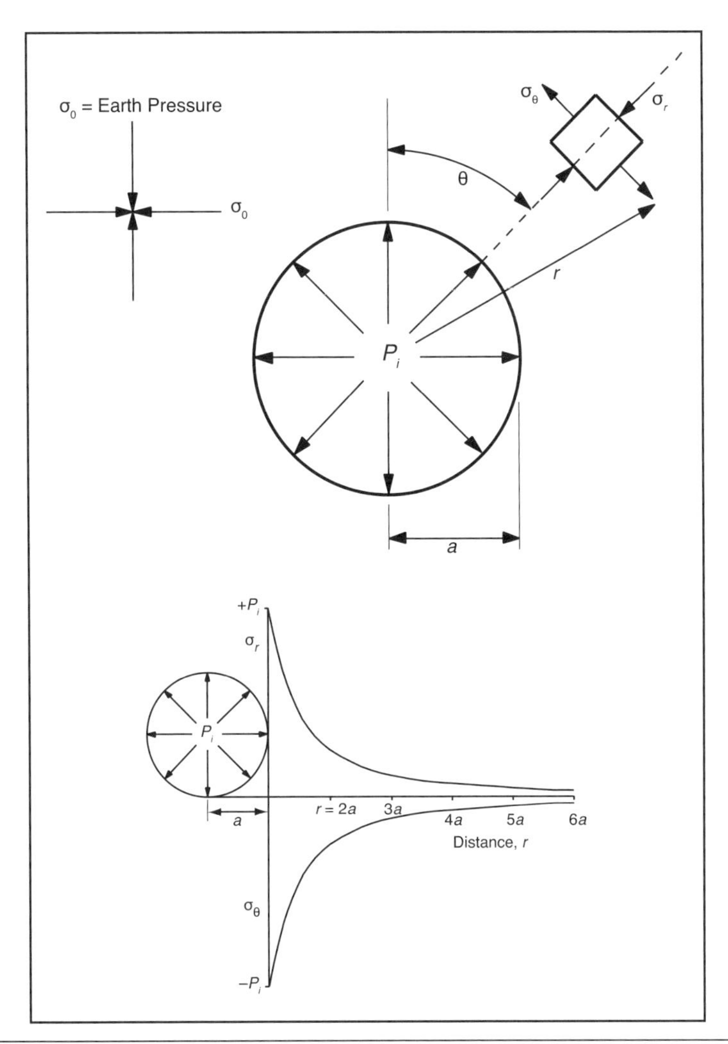

FIGURE 9.17 Radial and tangential stresses at distance r from the center of a pressurized thick-walled cylinder

If the radial stress from the expanding gases is greater than the compressive strength of the rock, then crushing around the borehole will occur. If such crushing should not be allowed, then the charge load per unit of borehole length can be reduced, the charge diameter can be reduced, or the explosive type can be changed.

If the tangential stress from the expanding gases is greater than the tensile strength of the rock, then the rock will crack as a result of tensile failure. The maximum length of the crack—and thus one-half the maximum spacing between adjacent holes—is the distance r for which σ_θ just equals T_0, the tensile strength of the rock.

Decoupling and/or decking the explosive column can reduce borehole pressure (Workman and Calder 1991). Decoupling occurs when the diameter of the explosive is less than the diameter of the borehole. The decoupling ratio is the ratio of the diameter of the explosive to the diameter of the borehole. The pressure generated by a decoupled explosive is a function of the volume of explosive and the volume of the hole. The ideal gas equation for adiabatic expansion can be applied to explosive gases (Coates 1981):

$$P_1 = P_0 \left(V_0 / V_1\right)^y \quad \text{(EQ 9.5)}$$

where

P_1 = new pressure
P_0 = original pressure
V_0 = original volume
V_1 = new volume
y = the specific heat ratio, which has been found for many explosives to be close to 1.2 (Cook 1958)

For a unit height of explosive in a unit height of borehole, the new pressure is

$$P_1 = P_0 \left(V_0 / V_1\right)^{1.2} = P_0 \left(r_0^2 / r_1^2\right)^{1.2} \quad \text{(EQ 9.6)}$$

where

r_0 = radius of the explosive
r_1 = radius of the hole

Decking is another means of reducing borehole pressure. *Decking* is the separating of individual explosive charges by a gap (air deck or stemming) within the borehole. The decked percentage is the ratio of the total length of explosive column to the total length of the borehole.

Example. Consider a situation where presplit blasting is to be used. We want to determine the proper explosive loading density and presplit spacing with Konya's equations (Equations 9.1 and 9.2) and then verify the results by using the equation for a pressurized thick-walled cylinder (Equation 9.4). The following details are applicable to this situation:

- Explosive has a detonation pressure of 50 kbar = 4,971 MPa (721,000 psi).
- Explosive specific gravity, G_e, is 0.85.
- Explosive is loaded for 100% of column length.
- Hole diameter, D_h, is 76.2 mm (3.0 in.). (Hole radius is therefore 38.1 mm, or 1.5 in.)
- Tensile strength of the rock, $T_0 = \sigma_\theta$, is 6.9 MPa (1,000 psi).

The explosive load, according to Konya's formula given earlier (Equation 9.1), is calculated as

$$d_{ec} = \frac{(76.2)^2}{12.14} = 478.29 \text{ g/m} = 4.78 \text{ g/cm}$$

Given that we are intending to load the hole for 100% of the column length, it is necessary to determine an equivalent explosive diameter that will give us the loading density calculated previously. This equivalent diameter will be the diameter of explosive that, when continuously line-loaded for the entire length of the hole, will weigh 4.78 g/cm of hole length. The equivalent explosive diameter is determined by finding the radius of a unit-high (1-cm-high) cylindrical charge of the required explosive of the given explosive density ($G_e \cdot \gamma_w$) that results in 4.78 g per the 1.0-cm height.

The volume of a unit-high (1.0-cm-high) cylinder of the explosive is given by

$$V = \pi (r_e)^2 \cdot (1.0)$$

where

V = cylinder volume

r_e = unknown equivalent explosive radius

The loading density of the unit-high cylinder of explosive can be determined by

$$d_{ec} = \pi (r_e) 2 \cdot G_e \gamma_w$$

where γ_w is the unit weight of water (1.0 g/cm^3 or 62.4 lb/ft^3). Therefore, based on the loading density calculated previously with Konya's equation,

$$d_{ec} = \pi (r_e) 2 \cdot G_e \gamma_w = 4.78 \text{ g/cm}$$

If we insert known values into the preceding equation, we get

$$(3.14159)(r_e)^2 \cdot (0.85)(1.0 \text{ g/cm}^3) = 4.78 \text{ g/cm}$$

Solving for r_e, we get

$$r_e = 1.38 \text{ cm} = 13.8 \text{ mm}$$

which results in the equivalent explosive diameter of

$$d_e = 2r_e = 2.67 \text{ cm (1.05 in.)}$$

The spacing, then, may be determined either by Equation 9.2 or by Equation 9.4. In this example, the spacing will be determined by both methods as a check. First, by using Equation 9.2 and the low-end value to estimate the spacing, we get

$$S = 10(76.2) = 762 \text{ mm} = 0.76 \text{ m (29.92 in.)}$$

Alternatively, according to the theory of a pressurized thick-walled cylinder (Equation 9.4), the spacing is calculated as follows:

$$\sigma_\theta = \sigma_0 - (P_i - \sigma_0)\frac{a^2}{r^2}$$

where

σ_θ = tangential stress = tensile strength of the rock, T_0 = 6.9 MPa (1,000 psi)
σ_0 = in situ stress field = 0 MPa (0 psi)
P_i = decoupled borehole pressure, in MPa (psi)
a = radius of the hole = 38.1 mm (1.5 in.)
r = unknown distance from the center of the hole to the point of interest = $S/2$, in mm (in.)

To use Equation 9.4, we must first determine the decoupled borehole pressure, P_i. The explosive pressure for a decoupled charge is given by Equation 9.5 and is

$$P_1 = 4{,}971(13.8^2/38.1^2)^{1.2}$$
$$= 434.44 \text{ MPa } (63{,}011 \text{ psi})$$

The value of 434.44 MPa is the expected pressure at the hole radius (38.1 mm) generated by the explosive charge of diameter 13.8 mm and detonating pressure of 4,971 MPa. This is the equivalent borehole pressure assuming full coupling of an explosive charge with the 76.2-mm-diameter hole. That is, a 76.2-mm-diameter charge with a detonation pressure of 434.44 MPa would generate the same pressure at a distance of 38.1 mm from the center of the charge as a charge of 27.6-mm diameter with a detonation pressure of 4,971 MPa.

If we let P_1 of 434.44 MPa from the preceding solution equal P_i in Equation 9.4 and solve for r, we get

$$6.9 = 0 - (-434.44 - 0)\frac{38.1^2}{r^2}$$

$$r = 302.32 \text{ mm } (11.90 \text{ in.})$$

The unknown spacing, S, is twice r, giving

$$S = 2r = 0.60 \text{ m } (23.80 \text{ in.})$$

We can also calculate the spacing based on the equivalent explosive radius, r_e. Referring again to Equation 9.4, we have

$$\sigma_\theta = \sigma_0 - (P_i - \sigma_0)\frac{a^2}{r^2}$$

where

σ_θ = tangential stress = tensile strength of the rock, T_0 = 6.9 MPa (1,000 psi)
σ_0 = in situ stress field = 0 MPa (0 psi)
P_i = decoupled borehole pressure = explosive detonation pressure = 4,971 MPa (721,000 psi)
a = radius of the hole = equivalent explosive radius, r_e = 13.8 mm (0.54 in.)
r = unknown distance from the center of the hole to the point of interest = $S/2$, in mm (in.)

Inserting the variables and solving for r, we get

$$6.9 = 0 - (-4{,}971 - 0)\frac{13.8^2}{r^2}$$

$$r = 370.40 \text{ mm } (14.58 \text{ in.})$$

$$S = 2r = 0.74 \text{ mm } (29.13 \text{ in.})$$

Trim (Cushion) Blasting

Trim blasting is a control blasting technique that is used to clean up a final wall after the production blast has taken place (Konya 1995). The trim blast may be on a later delay of the production blast—tied to the timing of the adjacent production blast so that the trim blast follows the detonation of the last hole of the production blast—or at a much later date, possibly years after blasting. The purpose of the trim blast is twofold: (1) to create an aesthetically appealing final wall and (2) to enhance the stability of the final wall by removing overbreak from the production blasting. Because the trim row is shot *after* the final production row, the trim blast does little to protect the stability of the final wall from production blasting. It does, however, provide enhanced stability by removing the loose material caused by overbreak from the production blasting.

In trim blasting, cost is incurred from extra drilling, longer blasthole-loading time, and some reduced mine production, but these short-term costs are often offset by decreased future costs in terms of stripping and slope failure. Both of these potential future costs—stripping and slope failure—can be incurred as a result of overbreak into the final highwall from the production blasting.

For a single-row trim blast fired after the production blast, the following general design equations can be used, with the same approximate explosive load per unit of borehole length as presented earlier for presplitting (Konya 1995):

$$S = 16D_h$$

$$B \geq 1.3S$$

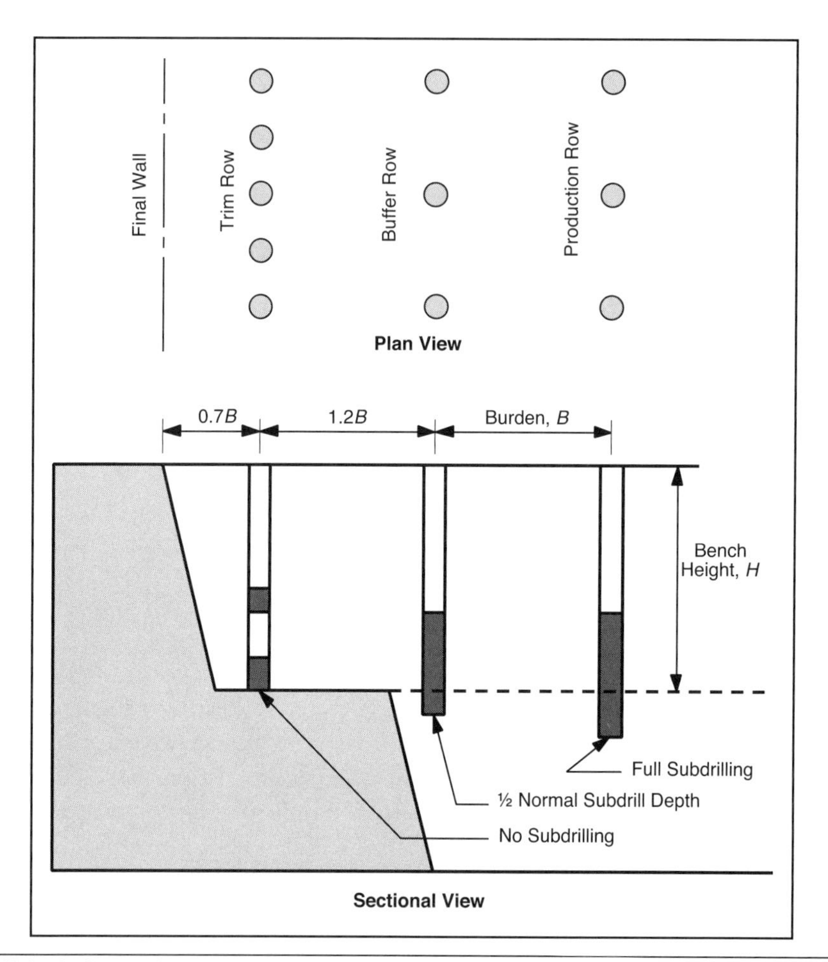

FIGURE 9.18 Generalized trim blast design using production drills

where

S = spacing, in mm
D_h = diameter of the hole, in mm
B = burden (distance to the production blast), in mm

A blasthole's burden, in short, is the distance from the hole to the nearest free face. In presplit blasting, since the presplit blast is shot before the adjacent production blast, in essence the burden is infinite. In the case of trim blasting, since the trim blast is shot after the adjacent production blast, a free face exists. This free face is the rock face at some distance B in front of the trim blast.

Savely (1986) and Floyd (1998) discussed a technique for trim blasting to improve the stability of the final pit wall. The general designs presented by Floyd and Savely incorporate the trim row as the final row of the production blast. In this technique, production

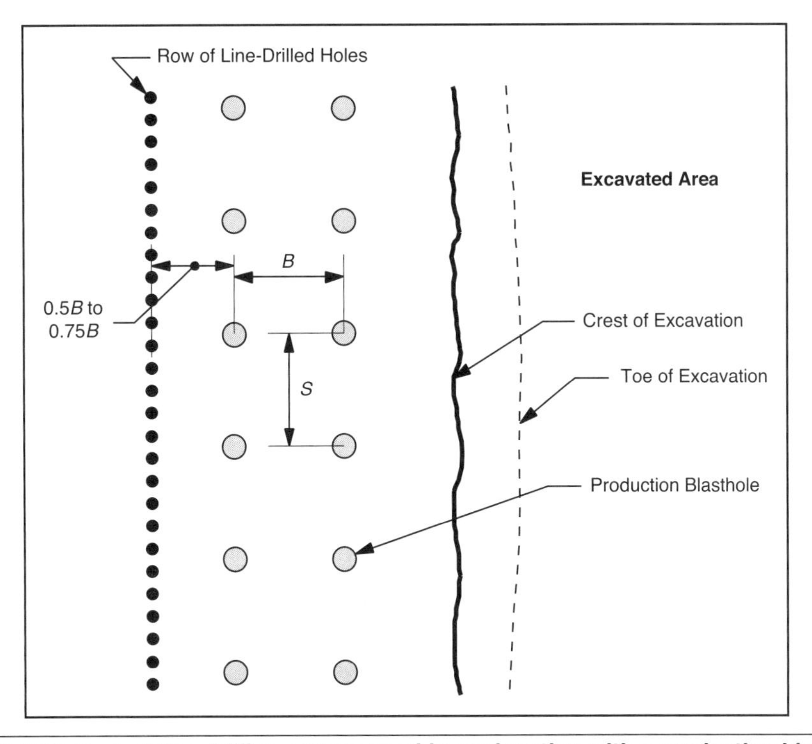

FIGURE 9.19 Typical line-drilling pattern used in conjunction with a production blast

blastholes, buffer holes, and the trim holes are all drilled the same diameter (Figure 9.18). However, the buffer holes and the trim holes are drilled with little or no subdrilling to minimize damage to the lower bench. The explosive load is decreased from the production row to the buffer row and from the buffer row to the trim row. Decking of the explosive is advisable in at least the trim row. Also, the burden and spacing for the trim row should be reduced to compensate for the reduced explosive load in the row.

Line Drilling

Line drilling is often included as one of the controlled blasting techniques. However, it is not a blasting technique per se. Line drilling uses a single row of unloaded, closely spaced drill holes at the perimeter of the excavation, as shown in Figure 9.19. When a production blast adjacent to a series of line-drilled holes is detonated, the shock wave from the detonating holes will cause a stress concentration around the unloaded, line-drilled holes. If the stress exceeds the rock strength, failure will occur in the form of a crack extending from one line-drilled hole to the adjacent line-drilled hole.

Line drilling is an expensive perimeter control technique because of the number of holes required for the technique to work properly. Holes are normally drilled from two to four drill-hole diameters apart and up to 10 m (33 ft) in depth. The distance between the line-drilled holes and the last row of production blastholes is normally $0.5B$ to $0.75B$, where B

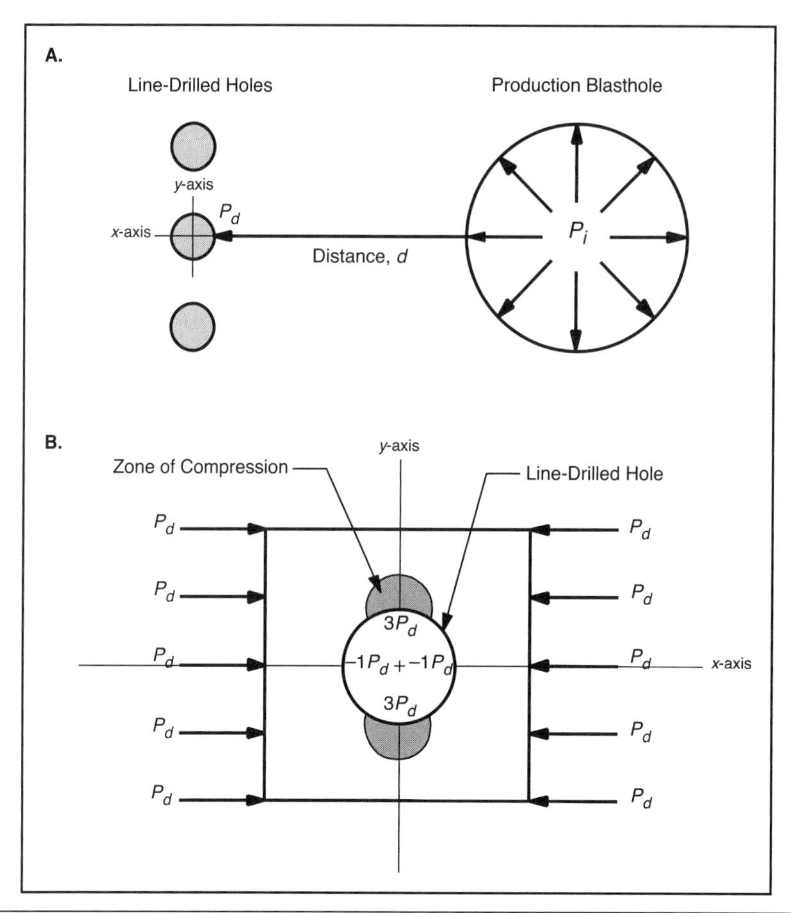

FIGURE 9.20 Overview of line-drilled holes: (A) production blasthole detonating adjacent to a line-drilled hole; (B) stresses on the line-drilled hole from the detonation of the blasthole

is the production blast burden. Because of the close spacing of the line-drilled holes, careful alignment of holes in the same vertical plane is essential; otherwise holes may cross.

The general theory of rock breakage across the web between line-drilled holes is based on the theory for the two-dimensional (2-D) distribution of stresses around a circular opening in an elastic body. When a production blasthole adjacent to a line-drilled hole detonates (Figure 9.20A), the shock and expanding gases from the detonating explosive subject the rock mass to high stress, P_i. This stress decreases with distance away from the blasthole, as shown in the bottom portion of Figure 9.17.

The stress, P_d, from the detonation acts uniaxially on the adjacent line-drilled hole, as shown in Figure 9.20B. Along the x-axis, which is the principal stress trajectory, a zone of tension will develop at the edge of the line-drilled hole's opening; along the y-axis, a zone of compression will develop at the edge of the opening. The magnitudes of the major and minor principal stresses at these edge points will be $3.0P_d$ and $-1.0P_d$, respectively (see Figure 9.20B). If the magnitude of the compressive stress at the edge of the line-drilled hole is sufficiently high, compressive and shear failure will occur outward from the edge of the hole.

The stress-concentrating effect of the hole dies away fairly rapidly, and at $r = 3a$, where a is the hole radius, the ratio of induced to applied stress is very close to unity (Hoek and Brown 1980). This implies that compressive and shear failure will occur for a distance of less than $3a$ and that the maximum spacing between line-drilled holes should be less than $6a$.

The following are some other noteworthy considerations in the context of line-drilled holes and production blastholes:

- P_d will act on the line-drilled hole dynamically, impulsively, and almost instantaneously.
- Stresses from other detonating production blastholes will also influence the line-drilled hole.
- The strength of the rock mass will be the dynamic strength (as opposed to the static strength).
- The magnitude of P_d will be a function of the explosive's detonation pressure, decoupling, and the distance between the production hole and the line-drilled hole.

MECHANICAL STABILIZATION

Mechanical methods of slope treatment are those that alter or protect the slope face to reduce erosion, prevent rockfall, or reduce raveling. Common treatment methods include (1) protective blankets and geotextiles and (2) wire net or mesh.

Protective Blankets and Geotextiles

Protective blankets made from jute, excelsior, burlap, cotton, or other natural or manufactured materials have been used for many years for erosion control and to prevent or reduce raveling on cut slopes. The blankets are usually pinned to the slopes and combined with seed and fertilizer. The purpose of pinning the blanket is to hold it in place until the vegetation takes root. The blankets are often expected to deteriorate and thus biodegrade over time as the vegetation takes hold.

A *geotextile* is defined by ASTM International as "any permeable textile material used with foundation, soil, rock, earth, or any other geotechnical engineering-related material, as an integral part of a man-made project, structure, or system" (Christopher and Holtz 1985).

Strictly speaking, the term *geotextile* refers to knitted, woven, and nonwoven fabrics. However, the term is also currently being applied generically to many materials that do not technically conform with the definition of a textile but that are used in combination with, or in place of, geotextiles. These other materials should instead be referred to as "geotextile-related materials"; they include such materials as webs, mats, nets, grids, and formed plastic sheets.

Geotextiles and geotextile-related materials are generally classified on the basis of their manufacturing process. The following is a list of 10 types of geotextiles and related products; the first four types are geotextiles as defined by ASTM International, and the last six are geotextile-related products (Rankilor 1981):

1. **Knitted geotextiles:** used mostly as filters in pipe wrap applications.
2. **Woven geotextiles:** used in a wide variety of applications.
3. **Nonwoven geotextiles:** used in a wide variety of applications.
4. **Composite geotextiles:** usually designed to perform a specific function.
5. **Webs or webbing:** typically used for erosion control, bank protection, and soil reinforcement.
6. **Mats:** typically used for erosion control or combined with woven or nonwoven geotextiles acting as filters to form a drainage structure.
7. **Nets:** typically used for soil reinforcement and in fabricating gabions.
8. **Grids:** typically used for soil reinforcement.
9. **Formed plastic sheets:** typically used in combination with woven or nonwoven geotextiles acting as filters to form a drainage structure.
10. **Prefabricated composite structures:** typical examples are mats, nets, or formed plastic sheets combined with one or two geotextiles acting as filters to form prefabricated drainage structures.

Geotextile applications can be divided into four primary functions: separation, drainage, reinforcement, and filtration (Christopher and Holtz 1985). In separation, layers of different sizes of solid particles are separated from one another by the geotextile (e.g., landfill covers). In drainage, the geotextile allows water to pass; in the special case of "drainage transmission," the geotextile itself acts as a drain to transmit water through soils of low permeability (e.g., horizontal drains below heap leach pads). In the case of reinforcement, the geotextile acts as a reinforcing element in the earth through either stress distribution or an increase in soil modulus (e.g., a net against rockfalls). For filtration, the fabric acts similar to a 2-D sand filter, allowing water to move from the soil while retaining the soil (e.g., silt screens).

Wire Net or Mesh

Another method of slope stabilization involves draping or pinning wire netting over the slope face to prevent rockfalls from bouncing outward from the toe region. Two types of wire mesh are commonly used for this purpose: (1) welded wire fabric, such as that used in concrete reinforcement, or (2) chain-link mesh, as is commonly used for fencing. A typical welded wire mesh application would be to use mesh with a 100-mm × 100-mm

FIGURE 9.21 Chain-link wire mesh anchored at the crest and draped over a slope to prevent rockfalls from passing the toe region

or 150-mm × 150-mm (4-in. × 4-in. or 6-in. × 6-in.) opening and a wire size from 9 to 4 gauge (Seegmiller 1982). Chain-link fence is often coated with a galvanizing agent and will therefore better withstand adverse environmental conditions. Also, because of the nature of its construction, chain-link fence tends to be more flexible and stronger.

Pinning the wire mesh to the face holds the rock in place and reduces rock removal at the toe. The pins (typically rock bolts, rock dowels, or a reinforcing bar) must be strong enough and spaced close enough to hold large, loose rocks and prevent them from dislodging and tearing the mesh.

Draping the wire mesh (Figure 9.21) involves anchoring the mesh to the crest of the slope and hanging it down over the face of the slope. Adjacent sections of the mesh are often either overlapped or tied together with wire. Weights, such as old tires, concrete blocks, or timbers, may be attached to the bottom of the mesh to contain rockfall to the toe region. Accumulated rockfall should be periodically cleaned up.

Rockfall nets and barriers have been in use in mining and highway construction for many years to provide safety and protection from falling rocks for highways, buildings, conveyors, crusher facilities, maintenance facilities, mine haul roads, mine access roads, and parking areas.

Rockfall barriers consist of I-beams or steel posts spaced at certain intervals with a wire rope support system running from post to post with special friction devices placed in the support ropes to allow for give when the barrier is struck by falling or rolling rocks. Wire rope nets are attached to the support system and backed with a high-tensile steel wire mesh. The barriers (Figure 9.22) are designed catch and deflect rockfalls, and to progressively yield under loads, absorbing the impact from debris slides or rockfalls. Systems offer

Courtesy of Geobrugg

FIGURE 9.22 Barrier designed to catch and deflect rockfalls

protection against dynamic loads up to 8,000 kJ (5.9M ft-lb) and come in heights up to 10.5 m (34.4 ft; Geobrugg 2017).

Flexible ring nets and cable nets for rockfall are designed to withstand high static and dynamic loads. They can be used to stabilize individual boulders or in conjunction with other systems, such as rockbolts, wire mesh, and cable lashing to stabilize large areas. Figure 9.23 shows a cable rockfall net made from 9.5-mm (⅜-in.)-diameter wire rope used to stabilize a large boulder overhanging a critical facility. (The powerline in the figure has been de-energized to allow safe work in the area.) Figure 9.24 shows a close-up of the rockfall net of Figure 9.23 after installation of additional stabilizing devices, including cable lashing and epoxy-coated, fully grouted threadbars. The rockfall net in these figures was constructed by splicing 2 m × 2 m (6 ft × 6 ft) panels of the cable nets together using 9.5-mm (⅜-in.) wire rope threaded through the end squares of the adjoining panels.

STRUCTURAL STABILIZATION

Structural stabilization includes those methods that reinforce the structure of the rock at the slope face or provide a structure that supports the slope. Methods available include the use of gunite or shotcrete, rock bolting, and construction of rock buttresses or retaining walls.

Gunite or Shotcrete

One common method utilizes pneumatically applied mortar and concrete (generally known as gunite or shotcrete) sprayed or pumped onto the slope face to seal the face and

FIGURE 9.23 Cable rockfall net

FIGURE 9.24 Rockfall net after installation of additional stabilizing devices

bind together small fragments on the face (Figure 9.25). This approach is used primarily to prevent weathering and spalling of a rock surface, as well as to knit together the surface of a slope. Generally, for rock slope stabilization, the material is applied in one 50- to 75-mm (2- to 3-in.) layer (Brawner 1994).

Shotcrete is applied by either wet or dry application. For dry mix shotcrete, additives and mortar are mixed on-site and pumped via compressed air to the nozzle, where the water is added. The wet mix is premixed at a central plant to specifications and then transported to the site in bulk.

Courtesy of Doug Hoy

FIGURE 9.25 Application of shotcrete over wire mesh and rock bolts

One disadvantage of shotcrete is its low tensile strength. For this reason, welded wire mesh, anchored to the rock, is often used to reinforce the shotcrete. A problem with using wire mesh as reinforcement for shotcrete is the difficulty of molding the mesh to a rough surface. Where the surface is irregular, large gaps may develop between the mesh and the rock, making bonding of the shotcrete to the rock difficult.

Additives can be added to either the wet or dry mix to provide additional strength and durability. Steel fibers, when added to the mix, increase the tensile strength of the shotcrete by providing numerous bonding surfaces within a small area. The fiber reinforcement also reduces the risk that shrinkage cracks will develop during curing. In many cases, the addition of fibers can replace wire mesh as reinforcement, thus reducing the overall cost.

Coloring can be added to the mix to provide a surface that more closely blends in with the surrounding rock. Shotcrete, however, naturally weathers over a period of years.

The addition of drain holes through the shotcrete is essential to eliminate water pressure behind the shotcrete. Small lengths of steel or PVC (polyvinyl chloride) pipe, inserted prior to shotcrete application into joints in the rock face where seeps have been noted or where seeps may occur, will provide partial drainage. Other drain holes should be created at regular intervals along the slope face.

Rock Bolts, Rock Anchors, and Rock Dowels

Steel reinforcement members in the form of rock bolts, cable bolts, resin-grouted threadbars, or rock dowels are used to tie together the rock mass so that the stability of a rock cut or slope is maintained. Rock bolts are commonly used to reinforce the surface or near-surface rock of the excavation, and rock anchors are used for supporting deep-seated instability modes in which sliding or separation on a discontinuity is possible.

A rock anchor generally consists of a bar or cable of high-strength steel tensioned inside a borehole to about 60%–70% of its yield strength. Tension in the member is transmitted to the surrounding rock mass by anchorage points at the ends. The length of the rock anchor can be from 3 m (10 ft) to more than 100 m (330 ft; Sage 1977).

Steel reinforcement members are either active or passive at the time of installation. An active member is one that loads the strata in which it is placed immediately upon installation. A passive member is one that is placed in the rock but does not play a role in ground support until the mass moves and subsequently loads the fixture. The main advantage of the active system over the passive system is that no movement has to occur before the active system develops its full capacity; thus, deformation and possible tension cracking of the slope are minimized.

Both active rock-bolt force and passive rock-bolt force tend to increase the frictional resistance along the potential failure plane. However, in the case of active rock-bolt force, the assumption is that the resulting force parallel to the failure plane tends to increase the resisting forces along the potential failure plane. In the case of passive rockbolt force, the assumption is that the resulting force parallel to the failure plane tends to decrease the driving forces of the potential failure mass.

Another classification of bolts or anchors is whether they are point-anchored or are in full contact with the rock along their length. An example of a point-anchor bolt is an expansion shell rock bolt; an example of a full-contact bolt is a friction anchor or "split set" rock bolt.

Rock bolts. Rock bolts have been used for years in construction and mining applications, most notably for reinforcement of rock in the wall and back (in underground mining) and for tunnel reinforcement. In rock slope applications, rock bolts are primarily used to reinforce the surface and near-surface rock of an excavated or natural slope. They come in variable lengths, normally from 1.2 m (4 ft) to more than 12 m (40 ft). Rod diameters vary from 10 mm (⅜ in.) to 51 mm (2 in.), with steel yield strengths ranging from 4,445 kg (9,800 lb) to 140,600 kg (310,000 lb). Drill-hole diameters range from 32 mm (1¼ in.) to 89 mm (3½ in.).

The use of wire mesh or straps complements the reinforcement achieved with a rock-bolt pattern. In areas of highly jointed or fractured rock, wire mesh can be used to hold in place the small blocks of rock between the faceplates. Where the rock mass is very slabby (i.e., where most of the weakness planes run in only one direction), straps may be a more effective means of face support than mesh.

A general rule for rock-bolt spacing is that the distance between faceplates should be approximately equal to three times the average spacing of the planes of weakness in the rock mass; the bolt length should be twice the bolt spacing (Hoek and Wood 1988).

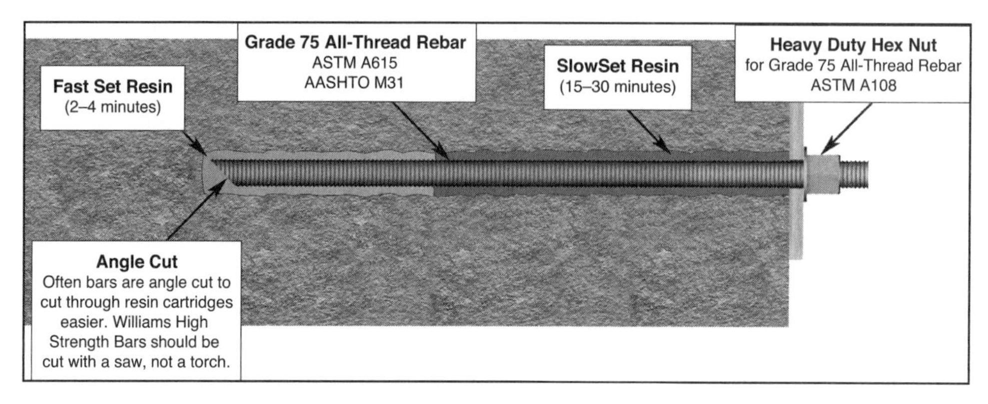

Courtesy of Williams Form Engineering Corporation

FIGURE 9.26 Resin-anchored threadbar

Rock anchors. Resin-grouted, tensioned threadbar and grouted cables provide a means to control large failure blocks. Lengths of the units may be as short as 3 m (10 ft) or as long as 100 m (330 ft), depending on the specific application. Holes for installation of the anchors are normally drilled well past the potential failure plane; then the anchors are inserted and grouted in place with or without tensioning.

Grade 60 (414 MPa, or 60,000 psi) and Grade 75 (517 MPa, or 75,000 psi) resin-grouted, tensioned threadbar (Figure 9.26) comes in diameters ranging from 22 mm to 64 mm (0.86 in. to 2.5 in.) and in nominal lengths of 12 m (40 ft). Such threadbar can be cut to lengths shorter than 12 m (40 ft) or made longer by coupling two or more units together. The resin plus catalyst for anchoring the bolts comes in sausage-like packages for ease of installation in the hole. Normally, the bottom one-third of the hole is line-loaded with quick-set resin cartridges, and the top two-thirds of the hole is loaded with slow-setting resin cartridges. The threadbar is then inserted into the hole and rotated, usually via a rock drill, to break the plastic containers and mix the resin and catalyst. After the resin has set, the anchor plate, bevel or flat washers, and the end nut are added. Wedge washers may also be used where the end plate is not perpendicular to the threadbar. The threadbar can be tensioned by tightening the end nut to load the anchor. Figure 9.27 shows a rock wedge anchored with 6.1-m (20-ft) lengths of no. 14 size resin-grouted threadbar.

Grouted cables were introduced to mining for reinforcement of the backs of cut-and-fill stopes. Cable bolting, whether tensioned or untensioned, is widely used in mining applications. The cable bolts should be made from high-strength steel (about 1,380-MPa or 200,000-psi yield strength, typically) because the steel will creep in tension; as a result, a gradual decrease in anchor load will occur over time. This loss in strength is approximately the same for all types of steel (Sage 1977). Mild steel will, therefore, lose almost all the load that can initially be applied to it, making it useless for prestressing applications. For high-strength steel, the loss represents only a small portion of the total load.

FIGURE 9.27 Resin-grouted threadbars used to anchor a rock wedge in an unstable cut slope

Courtesy of Homestake Mining Company

FIGURE 9.28 Grout and air-bleeding tubes inserted with a cable bolt

Cable anchors are available in several sizes. They can be constructed with just a single strand or with multiple strands tied together. Generally, the cable anchor with the fewest strands giving the same strength will be the cheapest and easiest to install.

Two plastic tubes should be taped to the bolt and inserted with the bolt into the hole, as shown in Figure 9.28. One tube is used for placement of the grout, and the other is used to bleed out the air. Because the anchor is usually inserted manually, a nose cone may be fitted to the end to prevent snagging.

A special prestressing jack is required for tensioning the bolt. Also, for the tensioned cable bolt, a special bearing plate, anchor block, and cable-gripping cones are required. Figure 1.11 shows cable bolts used to stabilize the highwall of an open-pit mine.

Rock dowels. Grouted rock dowels essentially consist of steel reinforcing bars (rebars) that are cemented into boreholes; these bars may or may not be subjected to post-tensioning. Untensioned dowels, therefore, do not provide any additional normal force across the failure plane; they do, however, provide additional shearing resistance across the potential failure surface.

Rebar dowels can be fashioned to virtually any length by coupling two or more together. Bar diameter ranges from 13 mm to 51 mm (0.5 in. to 2 in.) of grade 60 steel.

Rock dowels are commonly used to provide support for steeply dipping, slabby rock formations. They are also used to provide anchor keys and tiebacks for shearing resistance at the toe and flanks of retaining walls. Dowels can also be used to anchor draped wire mesh, to pin wire mesh to the face of a highwall, to hold strapping in place, or to anchor restraining nets or cables.

The rock dowels can be anchored in place with pumped grout or by using cartridges of resin similar to the threadbar application discussed previously. When resin cartridges are being used, the rebar is rotated and driven through the cartridges with the drill, thus breaking the packages and mixing the resin.

Buttresses and Retaining Walls

The best application of rock buttresses and retaining walls is for stabilizing soil-like slopes. Both should be founded on bedrock or on good soil below the potential failure surface.

Cast-in-place reinforced concrete buttresses have been used successfully to provide support for overhanging rock slopes. To ensure that the rock and buttress act as a single unit, concrete buttresses must be reinforced and anchored to the rock wall and foundation with grouted dowels.

Buttresses constructed from rock or earth fill increase resistance to sliding by (Richards and Stimpson 1977):

- Providing additional weight at the toe of the slope,
- Increasing the shear strength in the toe area greater than that of the in situ material, and
- Improving drainage.

Retaining walls are structures constructed to provide stability for earth, soil banks, coal, ore or waste stockpiles, water, or other material where conditions do not allow the material to assume its natural slope. Retaining walls are classified into six principal types based on the method of achieving stability (Bowles 1982):

- **Gravity wall:** Depends on its own weight for stability.
- **Cantilever wall:** A reinforced-concrete wall that uses cantilever action to prevent the mass behind the wall from assuming a natural slope. Stability of this type of wall is partially achieved from the weight of soil on the heel portion of the base slab.
- **Counterfort wall:** Similar to a cantilever retaining wall, except that (1) it is used where the cantilever is long or for very high pressures behind the wall and (2) it has counterforts, which tie the wall and base together, built at intervals along the wall to reduce the bending moments and shears.
- **Buttressed wall:** Similar to a counterfort wall, except that the bracing is in front of the wall and is in compression instead of tension.
- **Crib wall:** Built-up members of lengths of precast concrete, metal, or timber. These types of walls are normally supported by anchors embedded in the rock or soil behind the wall.
- **Semigravity wall:** Intermediate between a true gravity wall and a cantilever wall.

Other methods of stabilization include the following:

- Gabion walls, which are rock-filled wire or synthetic mesh baskets.
- Reinforced earth, which is the use of strips of metal, concrete, or synthetic material placed between layers of fill for reinforcement.
- Tie-back walls, which include steel sheet piling, precast concrete panels, or sheet steel anchored to the rock face with rock bolts.

VEGETATIVE STABILIZATION

Vegetative techniques are most frequently used for aesthetic purposes, such as slope reclamation (Figure 9.29). However, there are many treatment methods that use vegetation to improve the stability of a slope. Generally, these methods are most successful when minor or shallow instability (such as raveling or erosion) is involved, as is usually the case for soil slopes or highly fractured rock slopes (Buss et al. 1995). The establishment of vegetation on steep soil slopes or loose-rock slopes is often enhanced by the construction of benches or stairstep terraces in the slope face. These arrangements act to hold the seed mix in place, encourage infiltration, and impede water flow to minimize erosion and sedimentation.

Mulched materials are often used for temporary erosion control of bare soils and to simultaneously improve the soil environment for establishing vegetation quickly by augmenting germination and seedling growth. Chemical binders and soil stabilizers may also be used to temporarily tack the mulch to the soil. Hydromulching is the most practical method for applying mulch, seed, fertilizer, erosion control compounds, and soil amendments on steep slopes. Commercial hydromulch fibers are often dyed with a fugitive green dye that lasts only a few hours or days. This visual aid assists in obtaining an even distribution on the slope. Rates of hydromulching vary from 0.6 to 3.4 t/ha (0.27 to 1.52 st/acre; Kay 1978).

Courtesy of Wharf Resources

FIGURE 9.29 A graded, shaped, and reclaimed mine highwall that appears similar to the surrounding natural topography

Natural or manufactured mats, webs, or fabrics can also be used for erosion control and to hold seeds in place, though their cost and effectiveness often limit their use. They require high labor inputs for installation and cost much more than tacked or hydromulched straw; in addition, some are not well adapted to fitting to rough surfaces. They must also be heavy enough or anchored in enough spots to prevent wind whipping.

The planting of trees or other large woody plants on rock slopes is beneficial in "softening" the appearance of the cut slope to make it appear more natural (Buss et al. 1995). To make access easier for planting trees or shrubs, benches, berms, or furrows must be constructed in the slope face. If access allows, the use of a tree spade enables the transplanting of large, mature trees. The plants are placed in holes that have been previously excavated on the reclamation site. (A single hole drilled and blasted in rock will provide an excellent site for a transported tree or shrub.) The plants are transplanted with a minimum of root disturbance. Tree spades, however, are expensive to use, and their use should be reserved to transporting hard-to-establish trees or to achieve an objective of immediate stocking with mature trees.

WATER CONTROL

Water *pressure* decreases the stability of a slope; therefore, drainage of water from a slope is an effective method of increasing the stability. Water in a slope may come from two primary sources: surface water and groundwater.

Control of Surface Water

Grading and shaping are major considerations in the control of surface water. Surface water can be controlled through a combination of topographic shaping and runoff control structures (Glover et al. 1978). Surface water allowed to flow down a slope, or to pond on benches of a slope, can infiltrate into the ground along discontinuities and thereby cause an increase in the driving forces on an unstable area through a buildup in pore pressure.

Topographic shaping is used to control the rate and direction of surface water flow by manipulating the gradient, length, and shape of the slope. One way this objective may be accomplished is by grading benches to divert water away from the slope face and off the bench. Another method is to flatten the gradient of the slope to encourage sheet runoff as opposed to channel flow.

Runoff control structures include dikes, waterways, diversion ditches, diversion swales, and chutes or flumes (Glover et al. 1978). The purpose of these structures is to intercept surface water flow before it reaches a critical area and to divert it to a disposal area. Waterways and ditches constructed in erosive materials, such as in loose soil, should be lined with concrete, rock or riprap, or vegetation. Chutes or flumes are similar to interceptor ditches constructed behind the crest of a slope to divert water away from the slope face. They are normally lined with rock or concrete and can be constructed at a steeper gradient than unlined ditches.

Water should be prevented from entering open tension cracks above the crest of a slope. This objective can be accomplished by diverting the water away or, where practical, by plugging the tension cracks. When the tension cracks are being plugged, the lower portion should be filled with a porous medium, and the upper portion should be sealed so that an impenetrable water barrier is not formed (such a barrier would cause a buildup in subsurface water pressure).

Control of Groundwater

The purpose of subsurface drainage (i.e., groundwater control) is to lower the water table and, therefore, the water pressure to a level below that of the potential failure surfaces. As discussed in Chapter 3, controlling groundwater is an effective means of increasing the stability of a slope. Methods of subsurface drainage include drain holes, dewatered (pumped) wells, and drainage galleries or adits.

Drain holes. Horizontal drain holes drilled into the face of the slope from the toe region offer an effective method of slope drainage. Normally, the holes are 50–150 mm (2–6 in.) in diameter and are drilled at an inclination of +3° to +5° from the horizontal. The length of the holes should extend beyond the critical failure surface. The direction of the drain holes depends on the orientation of the critical discontinuities; the optimum design is to intersect the maximum number of significant discontinuities for each unit length (meter, foot, etc.) of hole drilled.

The holes should be thoroughly cleaned after drilling; improperly cleaned holes may have their effectiveness reduced by as much as 75% (Piteau and Peckover 1978). If the holes tend to collapse, then perforated drain pipe should be inserted. The outer one-third of the pipe, up to 20 m (66 ft), should be left unperforated to ensure that water that enters the pipe remains in the pipe and is discharged at the pipe outlet.

Spacing of the drain holes can range from about 7–30 m (20–100 ft), and lengths into the slope should not exceed one-half the slope height, with a minimum length of 15 m (49 ft) and a maximum length of 100–125 m (330–410 ft) (Brawner 1982). For high rock cuts, installation of drain holes at different levels is recommended. Where rock is taken out in several lifts, drain holes should be drilled at the toe of every lift (Brawner 1994).

Dewatering wells. Dewatering wells are designed primarily to lower the groundwater level to a predetermined depth and to maintain that depth until all belowground activities have been completed. The main purposes for construction dewatering include the following (Driscoll 1986):

- Intercepting seepage that would enter an excavation site and interfere with construction activities
- Improving the stability of slopes, thus preventing sloughing or slope failures
- Preventing the bottoms of excavations from heaving because of excessive hydrostatic pressure
- Improving the compaction characteristics of soils in the bottoms of excavations
- Drying up borrow pits so that excavated materials can be properly compacted in embankments
- Reducing earth pressures on temporary supports and sheeting

Investigations necessary to determine the feasibility of using dewatering wells are generally extensive and costly. Primary basic data needed include the following (Bureau of Reclamation 1995):

- Aquifer transmissivity and storage coefficient
- Areal extent, thickness, and homogeneity of the aquifer
- Mode of occurrence of the groundwater—whether unconfined, confined, or leaky
- Estimated recharge caused by precipitation, irrigation, and other sources
- Boundary conditions
- Water quality data

If the transmissivity and storage coefficient of the aquifer can be determined, the principle of superposition can be applied to determine approximate well capacity, depth, and spacing required to accomplish the dewatering. This principle (as discussed in Chapter 3) states that "if the transmissivity and storativity [storage coefficient] of an ideal aquifer and the yield and duration of discharge or recharge of two or more wells are known, then the combined drawdown or buildup at any point within their interfering area of influence may be estimated by adding algebraically the component of drawdown of each well" (Bureau of Reclamation 1995). The principle of superposition can be used to determine desirable spacing of wells in well fields, effects of recharging wells, and in the evaluation of boundary conditions. Refer to Figure 3.18 for an illustration of this principle.

Drainage galleries or adits. Drainage adits or galleries driven under a pit—or into a slope or highwall to intercept the groundwater—can provide an effective method of drainage. Where employed, drain holes should be drilled from the adit upward in a fan pattern to increase drainage effectiveness.

An adit can be used not only for drainage but also as a means of obtaining detailed discontinuity information. This type of data is useful for future slope stability predictions. Additional information, such as water quality and the variations in permeability along the length of the adit, can also be gathered.

The cost of driving an adit is high. However, it need only be large enough to allow efficient excavation and to properly drain the problem area, generally 1.0–1.5 m × 2.0–2.2 m (3.3–4.9 ft × 6.5–7.2 ft).

REFERENCES

Bertuzzi, R. 1999. Technical note for estimating catch bench width. *Australian Geomechanics*, (Sept.):85–88.

Bowles, J.E. 1982. *Foundation Analysis and Design*. New York: McGraw-Hill.

Brawner, C.O. 1982. Stabilization of rock slopes. In *Stability in Surface Mining*, Vol. 3. Edited by C.O. Brawner. New York: SME-AIME.

Brawner, C.O. 1994. *Rockfall Hazard Mitigation Methods Participant Workbook*. NHI Course No. 13219. FHWA SA-93-085. McLean, Va.: U.S. Department of Transportation, Federal Highway Institute.

Bureau of Reclamation Ground Water and Drainage Group. 1995. *Ground Water Manual*, 2nd ed. Final Report no. PB96-207394. Denver, CO: U.S. Department of the Interior, Bureau of Reclamation, Technical Service Center.

Buss, K., R. Prellwitz, and M.A. Reinhart. 1995. *Highway Rock Slope Reclamation and Stabilization Black Hills Region, South Dakota Part II, Guidelines*. Report SD94-09-G. Pierre, SD: South Dakota Department of Transportation.

Call, R.D. 1992. Chapter 10.4, Slope stability. In *SME Mining Engineering Handbook*, 2nd ed., Vol. 2. Edited by H.L. Hartman. Littleton, CO: SME.

Call, R.D., and J.P. Savely. 1990. Open pit rock mechanics. In *Surface Mining*, 2nd ed. Littleton, CO: SME.

Christopher, B.R., and R.D. Holtz. 1985. *Geotextile Engineering Manual*. FHWA-TS-86-203. Washington, D.C.: Federal Highway Administration, National Highway Institute.

Coates, D.F. 1981. *Rock Mechanics Principles*. Monograph 874. Ottawa, ON: Canada Centre for Mineral and Energy Technology.

Cook, M.A. 1958. *The Science of High Explosives*. ACS Monograph No. 139. New York: Reinhold Publishing.

Driscoll, F.G. 1986. *Groundwater and Wells*. St. Paul, MN: Johnson Division.

Floyd, J.L. 1998. The development and implementation of efficient wall control blast designs. *Journal of Explosives Engineering*, 15(3):12–18.

Geobrugg AG. 2017. *RXE Rockfall Protection Barriers from 500 to 8000 kJ*. Romananshorn, Switzerland: Geobrugg AG. https://www.geobrugg.com/datei.php?src=portal/download center/dateien/downloadcenter/level1-brochures/RXE-barrier/ROCK_brochure_RXE_en_ch_72dpi_160824.pdf

Glover, F., M. Augustine, and M. Clar. 1978. Grading and shaping for erosion control and rapid vegetative establishment in humid regions. In *Reclamation of Drastically Disturbed Lands*. Edited by F.W. Schaller and P. Sutton. Madison, WI: American Society of Agronomy, Crop Science Society of America, and Soil Science Society of America.

Golder, H.Q. 1971. The stabilization of slopes in open-pit mining. In *Stability in Open Pit Mining, Proceedings of the First International Conference on Stability in Open Pit Mining*. Edited by C.O. Brawner and V. Milligan. New York: SME-AIME.

Hoek, E., and E.T. Brown. 1980. *Underground Excavations in Rock*. London: Institution of Mining and Metallurgy.

Hoek, E., and D.F. Wood. 1988. Rock Support. *Mining Magazine* (Oct.):282–287.

Jaeger, J.C., and N.G.W. Cook. 1979. *Fundamentals of Rock Mechanics*, 3rd ed. London: Chapman and Hall.

Kay, B.L. 1978. Mulch and chemical stabilizers for land reclamation in dry regions. In *Reclamation of Drastically Disturbed Lands*. Edited by F.W. Schaller and P. Sutton. Madison, WI: American Society of Agronomy, Crop Science Society of America, and Soil Science Society of America.

Konya, C.J. 1995. *Blast Design*. Montville, OH: Intercontinental Development Corporation.

Konya, C.J., and E.J. Walter. 1990. *Surface Blast Design*. Englewood Cliffs, NJ: Prentice-Hall.

Martin, D.C., and D.R. Piteau. 1977. Select berm width to control local failures. *Engineering and Mining Journal*, June.

Mathis, J.I. 2009. Bench—inter-ramp—overall: A guide to statistically designing a rock slope. Presented at Slope Stability 2009, Santiago, Chile.

McKown, A.F. 1984. Some Aspects of Design and Evaluation of Perimeter Control Blasting in Fractured and Weathered Rock. In *Proceedings of the 10th Conference on Explosives and Blasting Technique*. Cleveland, OH: Society of Explosives Engineers.

Piteau, D.R., and D.C. Martin. 1982. Mechanics of rock slope failure. In *Stability in Surface Mining*, Vol. 3. Edited by C.O. Brawner. New York: SME-AIME.

Piteau, D.R., and F.L. Peckover. 1978. Engineering of rock slopes. In *Landslides Analysis and Control*. Special Report 176. Edited by R.L. Schuster and R.J. Krizek. Washington, DC: Transportation Research Board, Commission on Sociotechnical Systems, National Research Council, National Academy of Sciences.

Rankilor, P.R. 1981. *Membranes in Ground Engineering*. London: John Wiley and Sons.

Richards, D., and B. Stimpson. 1977. *Pit Slope Manual Supplement 6-1: Buttresses and Retaining Walls*. Report 77-4. Ottawa, ON: Canada Centre for Mineral and Energy Technology.

Ritchie, A.M. 1963. Evaluation of rockfall and its control. In *Highway Research Record*, no. 17. Washington, DC: Highway Research Board, National Research Council. pp. 13–28.

Ryan, T.M., and P.R. Pryor. 2000. Designing catch benches and inter-ramp slopes. In *Slope Stability in Surface Mining*. Edited by W.A. Hustrulid, M.K. McCarter, and D.J.A. Van Zyl. Littleton, CO: SME.

Sage, R. 1977. *Pit Slope Manual Chapter 6: Mechanical Support*. Report 77-3. Ottawa, ON: Canada Centre for Mineral and Energy Technology.

Savely, J.P. 1986. *Designing a Final Wall Blast to Improve Stability*. Preprint No. 86-50. Littleton, CO: SME.

Seegmiller, B.L. 1982. Artificial support of rock slopes. In *Stability in Surface Mining*, Vol. 3. Edited by C.O. Brawner. New York: SME-AIME.

Workman, J.L., and P.N. Calder. 1991. A method for calculating the weight of charge to use in large hole presplitting for cast blasting operations. In *Proceedings of the 17th Conference on Explosives and Blasting Technique*, Vol. II. Cleveland, OH: Society of Explosives Engineers.

CHAPTER 10

Geotechnical Instrumentation and Monitoring

When a rock or soil mass is disturbed, either by the actions of people or by natural events, it undergoes a redistribution of stresses, resulting in a change in shape. This readjustment is reflected in displacements, deflections, pressures, loads, stresses, and strains, which can be detected and measured. Many of the same measurement methods and instrumentation techniques can also be used to investigate the mechanical properties of the mass, the interaction between the mass and any associated artificial structures, and the effectiveness of remedial measures proposed to correct defects in either the mass or the structures.

Monitoring is the surveillance of engineering structures either visually or with the aid of instruments (Brown 1993). The objectives of a rock slope monitoring program are as follows (Call 1982):

1. To maintain a safe operation for the protection of personnel and equipment.
2. To provide advance notice of instability, thus allowing for the modification of the excavation plan to minimize the impact of the instability.
3. To provide geotechnical information to analyze the slope failure mechanism, design appropriate remedial measures, and/or conduct a redesign of the rock slope.

Monitoring can be done from within the rock mass and/or on the excavation boundary. The techniques available to measure the various components of rock deformation may be placed in two general categories: observational and instrumentation (Windsor 1993). Observational techniques include simple visual observations, photographic recording, and electronic and optical surveying. Instrumentation techniques include the application of mechanical and electronic instruments, such as extensometers, inclinometers, strain gauges, and crack gauges.

An overview of both observational and instrumental techniques, as well as the types of instruments for monitoring deformation, is given in Table 10.1.

TABLE 10.1 Classification of instruments for measuring the various components of rock deformation

Deformation Measurement Technique	Measurement Access	Measurement Method	Measurement Sensitivity
Observational Techniques			
Global Positioning System	Exposure	Manual	Medium
Terrestrial surveying	Exposure	Manual	Medium
Electronic distance metering and automatic surveillance	Exposure	Automatic	Medium
Slope stability radar	Rock face	Automatic	High
Instrumentation Techniques			
Movement indicators:			
Axial	Borehole, rock face	Observational	Low
Shear	Borehole, rock face	Observational	Low
Convergence indicators:			
Wire/tape	Rock face	Manual	Medium
Rod	Rock face	Manual/automatic	Medium
Strain meters:			
Resistance strain gauges	Borehole	Automatic	High
Vibrating wire strain gauges	Borehole	Automatic	High
Joint meters:			
Glass plates	Rock face	Manual	Medium
Pin arrays	Rock face	Manual/automatic	Medium
Strain gauges	Borehole	Automatic	High
Proximity transducer	Rock face	Automatic	High
Fiber optic	Borehole	Automatic	High
Potentiometers	Borehole, rock face	Automatic	High
Extensometers:			
Fixed extensometer:			
Wire/rod	Borehole	Manual/automatic	High
Reference point sensing	Borehole	Manual/automatic	High
Strain sensing	Borehole	Automatic	High
Portable extensometer:			
Magnetic anchor	Borehole	Manual	High
Magneto-strictive	Borehole	Manual	High
Sliding micrometer	Borehole	Manual	High
Inclinometers:			
Fixed inclinometers	Borehole	Automatic	High
Portable inclinometers	Borehole	Manual	High
Deflectometers:	Borehole	Automatic	High
Extensometer-inclinometer	Borehole	Manual	High
Extensometer-deflectometer	Borehole	Manual	High

Adapted from T. Szwedzicki (ed.), Geotechnical Instrumentation and Monitoring in Open Pit and Underground Mining. In *Proceedings of the Australian Conference*, Kalgoorlie, June 21–23, 1993.

Photo by Don Berger, courtesy of Homestake Mining Company

FIGURE 10.1 Remote pit slope data acquisition system

INSTRUMENTATION TO MEASURE ROCK DEFORMATION

Many sites use sophisticated electronic instrumentation along with field data collection systems to transmit the data to a base station via phone lines, radio telemetry, or a hard-wired system. Figure 10.1 shows a remote pit slope monitoring system installed in the field to collect rock deformation data and transmit it to a computer in the mine office.

An important point to remember is that, since many of the sites where field instrumentation is installed are located in areas of abundant wildlife, care should be taken to protect the instrumentation cables from animals. Some species of wildlife can—and often do—chew off the sheathing of cables and wires. A case from the early to mid-1990s involving an open-pit gold mine in the Black Hills of South Dakota illustrates this point. Many marmots make their home in and around the open-pit mine. They began chewing the insulation on the wires that connect the remote pit slope monitoring system to the base computer system at the mine office. Very soon after a pit slope monitoring instrument was installed, the marmots would chew through the cables and short-circuit the instrument. The problem was finally solved, through trial and error, by suspending the cables over the pit benches. The wires were hung on rock bolts inserted in the slope face just below the bench crest. Installing the rock bolts and then suspending the cables from the bolts required more effort on the part of the pit geotechnical engineer, but this approach saved time and money in the long run by eliminating the constant splicing or replacement of cables.

Observational Techniques

In general, observational techniques involve observing the deformations that occur in the exposed rock mass. They usually employ some sort of direct observation of the location of recording points to determine relative movement. Observational techniques that are commonly used to measure rock deformation include

- Global Positioning System (GPS) surveillance,
- Optical and electronic surveying, and
- Slope stability radar (SSR).

Global Positioning System surveillance. The U.S. Department of Defense (DOD) began construction of the GPS in the 1970s to allow military ships, aircraft, and ground vehicles to determine a location anywhere in the world. Even though the GPS was meant primarily for classified operations, the designers of the system made provision for civilians to use it, though at a far reduced accuracy. This "selective availability" was achieved by slightly altering (i.e., dithering) the satellites' atomic clocks according to a specific code. This alteration has resulted in a slight error concerning the exact time when the satellites send their signals. The modified signals allow all civilians to locate themselves reasonably well; navigational fixes will be off by no more than 100 m (330 ft).

The GPS system utilizes the concept of ranging—the measurements of distances to several satellites—to determine position. The concept of ranging is quite simple in concept—if, for example, one is able to determine that a particular satellite is 15,000 km (9,320 mi) away, then one's position with respect to the satellite must be somewhere on a large sphere of radius 15,000 km (9,320 mi) that surrounds the satellite. If the satellite is traveling in a stable, predetermined orbit, then the location of the satellite at any given instant and the imaginary sphere around it is known exactly.

If the person taking the range to the satellite can also do so for a second satellite, then a second sphere of position can be determined. The intersection of these two spheres gives a locus of locations forming a circle along which, somewhere, is the GPS user. If three satellites are used, then the intersections of the spheres yield two points. Therefore, a minimum of four satellites is required for a unique solution (Herring 1996).

However, since the receiver's clock is not exactly running at the same time as the satellites' clocks, this dithering results in an error with respect to the calculated range, called the pseudo-range. The four pseudo-ranges from the four satellites should meet at a single point but will not meet exactly because the receiver's clock and the satellites' clocks are not exactly synchronized.

Nevertheless, scientists and engineers working outside of the armed services have devised ways to circumvent the purposeful degradation of the GPS signals, and means are now available to achieve much better accuracy than the DOD had anticipated. One technique employed is known as differential GPS and—with specialized GPS receivers—can achieve precision to within about 1 cm (0.4 in.) (Herring 1996). Another technique, known as carrier tracking, allows locations to be determined to within a few millimeters (Herring 1996).

With differential GPS, two (or more) GPS receivers simultaneously take readings from at least four satellites. One receiver is operated as a base station at a point of known position, and the second is operated at a point for which the position is required. At the base site, corrections to the measured pseudo-ranges are calculated to achieve the correct solution for absolute pseudo-range position. These corrections are then applied to the pseudo-ranges measured by the remote receiver to give a differential position with accuracy typically better than ±5 m (±16 ft) (Collier 1993). A data communication link between the base receiver and remote receiver can be used to transmit the corrections and thus facilitate positioning in real time.

Carrier tracking gets its name from the satellite broadcasts that convey GPS signals on a set of so-called radio carrier waves. It works by determining which part of the radio wave strikes the antenna at a given instant—the "phase" of the received emission (Herring 1996). Carrier tracking allows a resolution that is a tiny fraction of a wavelength, typically on the order of a few millionths of the baseline length.

Two variations of carrier tracking have been developed: static carrier phase positioning and kinematic phase positioning (Collier 1993). A major disadvantage of static carrier tracking is the length of time the receiver must be placed at the position of the point being located to achieve a solution. To overcome this disadvantage, the kinematic technique was developed. Kinematic GPS surveys use at least two receivers and commence with a period of initialization, where the receivers are placed at locations of known three-dimensional (3-D) position (Collier 1993). The purpose of the initialization stage is to determine the phase of the incoming GPS carrier signals and thus establish the number of integer wavelengths between the receiver and the satellite. This unknown quantity is known as the integer cycle ambiguity and must be determined computationally.

A study conducted by Collier (1993) ascertained that kinematic GPS monitoring provided a viable, efficient, and accurate alternative to deformation monitoring by conventional techniques. That author reported that kinematic GPS could be used in deformation monitoring to reliably detect movements of 5 mm (0.20 in.) or greater.

A disadvantage of both static and kinematic GPS positioning using carrier phase data is that position vectors can be determined only through post-processing. New systems, on the other hand, use a radio telemetry link to transmit data from a roving receiver to a base site for real-time processing.

Photo by Kjell Moe, courtesy of Newmont Mining Company

FIGURE 10.2 Wooden survey hubs for monitoring movement across a tension crack

Photo by Kjell Moe, courtesy of Newmont Mining Company

FIGURE 10.3 Survey prism for monitoring slope movement

Optical and electronic surveying. The most widely used method of monitoring for movement employs standard optical or electronic survey equipment in conjunction with a survey network of targets located on a rock slope. The survey network may be as simple as wooden hubs or steel rebar driven into the ground (Figure 10.2), or it may consist of permanently mounted survey prisms (Figure 10.3). Locations of the targets should be chosen

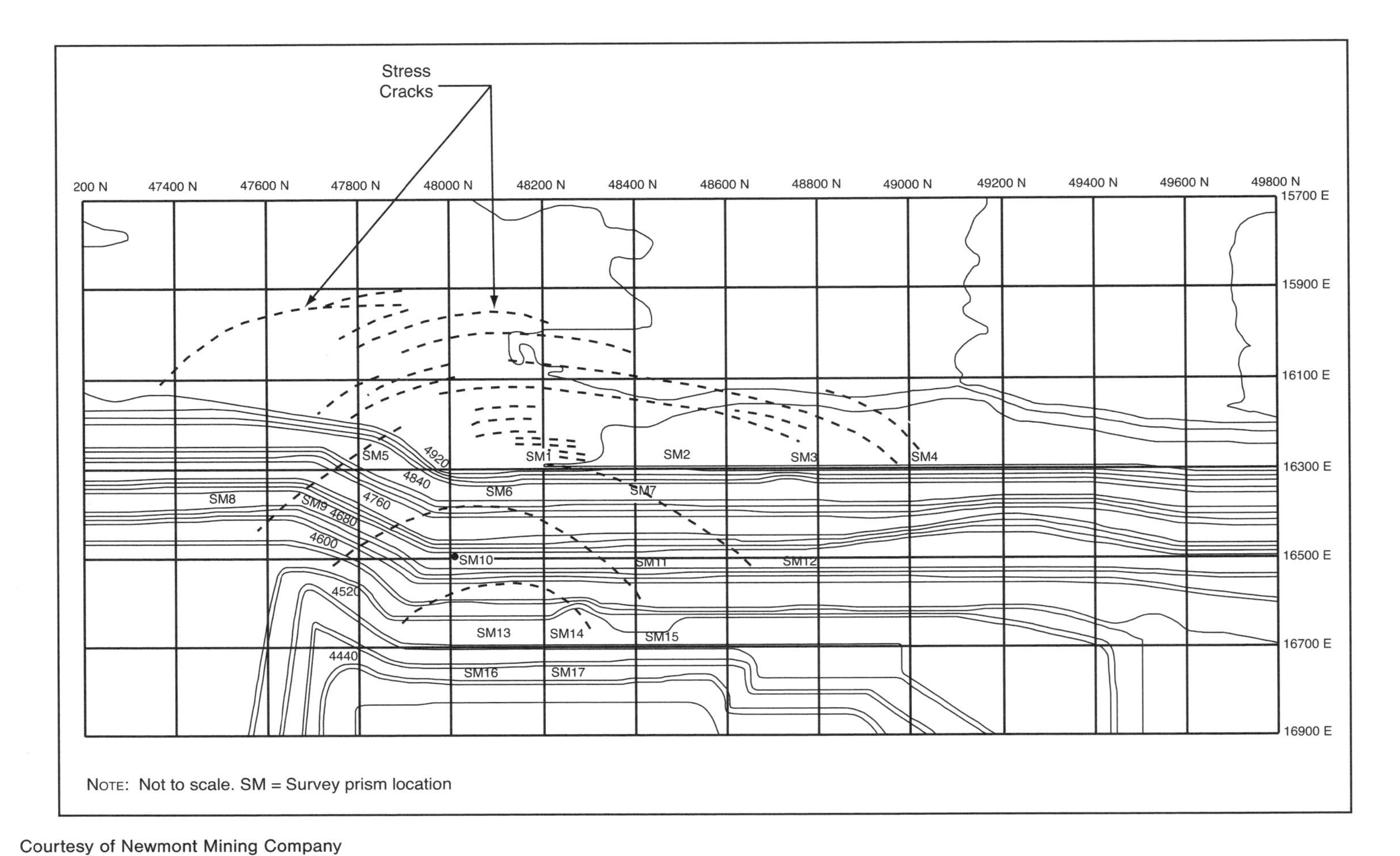

Courtesy of Newmont Mining Company

FIGURE 10.4 Survey net for the Twin Creeks mine in Nevada: Slope monitoring program west highwall (old dispatch site)

Photo by Travis Nice, courtesy of Peabody Energy

FIGURE 10.5 Real-beam slope stability radar unit

so that relative movement of the unstable area can be monitored. Additionally, the permanent control point(s), from which the targets are shot (i.e., observed), must be located on stable ground outside the slide area and within view of the targets. The network should be a set of well-conditioned triangles, with each vertex point being visible from two other points and the length of each line of sight within the measuring range of the equipment (Windsor 1993). A typical survey net for monitoring is shown in Figure 10.4.

Modern programmable electronic surveying theodolites fitted with an electronic distance meter make it possible to record displacements—at preprogrammed time intervals—of an array of reflecting prisms set on an unstable rock slope area. Once the initial location of each prism is recorded and stored in the computer system linked to the survey instrument, the survey instrument is capable of shooting each prism in the array on a continuous cycle and downloading each reading to the remote computer system.

Slope stability radar. SSR is a state-of-the-art technology used for slope stability monitoring of open-cut mine walls (Figures 10.5 and 10.6). It provides continuous, precise, and real-time online measurement of rock wall movements across the entire face of a wall. It remotely scans a rock slope to continuously monitor the spatial deformation of the face. Slope stability monitoring radars use a phase-based technique called *interferometry* to measure small changes in range (i.e., in the line-of-sight direction toward the radar) associated with precursor movements to mine slope failures. These small changes in phase (corresponding to the appropriate fraction of the radar wavelength in millimeters) are accumulated over time for each subsequent scan and converted to millimeters of displacement. Radar interferometry can operate at night and through smoke, haze, rain, fog, and dust without requiring reflectors on a highwall face. Consequently, this allows workers to avoid the hazards they might face while installing reflectors on a potentially unstable highwall.

Courtesy of Barrick Nevada–Goldstrike Mine

FIGURE 10.6 Synthetic aperture slope stability radar unit

In simple terms, radar measures the path length between the antennas and each resolved portion of the highwall face. Changes in the path length will be due to movement on the highwall face (McHugh et al. 2006). However, numerous factors can give rise to apparent path length changes—humidity fluctuations, liquid water, snow, freezing, and so on.

Two basic types of SSR are in common use today. The first type is real-beam radar equipped with a conventional dish antenna that mechanically scans a pencil beam in raster fashion over the region of interest (Figure 10.5); and the second type is synthetic aperture radar where a small antenna slides along a rail collecting data that are processed to form multiple fan-shaped beams (Figure 10.6).

In the first approach, a single pencil-beam antenna is scanned in two dimensions over the highwall face, as illustrated in Figure 10.7A. At each scan location, a radar signal is transmitted, and the radar echo received and processed. The radar signal phase recorded at each scan location is preserved. The face is repeatedly scanned in time. The phase measured at each location is compared to the phase of the previous scan (or a time average of previous scans). Differences in phase between scans are related to face movement with an estimated correction based on weather conditions. This approach requires a high-precision, two-dimensional (2-D) scanning system and exceptionally phase-stable radar, both conditions that add to the expense of the system. This approach forms the basis of the monitoring system developed in Australia (University of Queensland 2002).

The second approach uses a fan-beam transmitting antenna to illuminate the entire vertical face over a narrow horizontal distance. The face is scanned repeatedly in a horizontal (one-dimensional, or 1-D) sweep, as illustrated in Figure 10.7B. Two receiving antennas on the unit are separated by a short baseline. The radar's range resolution enables vertical

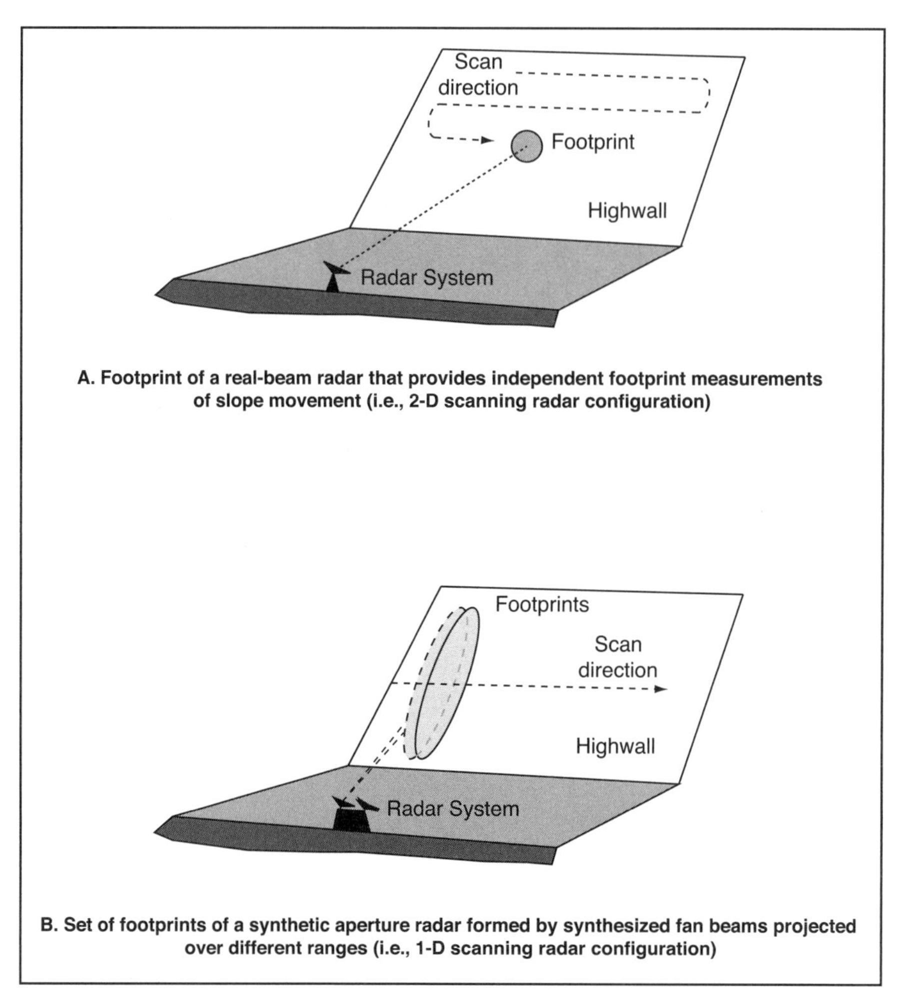

Source: McHugh et al. 2006

FIGURE 10.7 Two approaches to the application of radar inferometry

resolution of the face. The interferometric phase difference between the receiving antennas (which is due to the small path length difference between the target and each antenna) is recorded for each scan position. Given that the radar is stationary, the differential interferometric phase between scans can be easily computed. The differential phase is very sensitive to changes in the surface in the plane of the baseline orientation (e.g., either azimuth [horizontal] or range [vertical]). (Note that with three receiving antennas, azimuth and range interferometry can be done simultaneously.) However, because the interferometric path length for a given scan is nearly the same, the atmospheric effects are similar for both channels, and interference is minimized. Furthermore, this approach does not have the stringent long-term phase stability requirements of the pencil beam, 2-D scan design, although two receivers and antennas are required. In addition, since scanning is in a single dimension, the scanning system is simpler than a 2-D system.

TABLE 10.2 Summary of differences between real-beam and synthetic aperture radar units

Constraints	Real-Beam Radar	SAR
Mapping displacement	The most robust and most accurate in 3-D	The least robust and accurate in 3-D
Coverage of slope	100% of scanned area	Typically 70% of scanned area—some of the data has artifacts and is systemically unreliable
Maximum sector scanned	Typically ±120° in front, left side, and right side directions with potential for 360° coverage	Up to ±30° in front direction only
Monitoring steep slopes relative to line of sight	Suitable	Not suitable
Monitoring shallow slopes at long range	Suitable, but not for extremely long ranges (many kilometers)	Suitable for all ranges

Source: Longstaff n.d.

Using interferometry to make the fine measurements in range, if a reflector moves closer by half a wavelength toward the radar, the path length there and back is shortened by a whole wavelength. Path length changes can be measured from the change in phase of the reflected signal, with 360 degrees corresponding to 15 mm (0.6 in.) for the typical radar wavelength of 30 mm (1.2 in.). Phase changes of just a few degrees can be easily measured by modern signal processing technology, giving the potential for very accurate range measurements. In practice, external factors limit the accuracy to a millimeter or so. Such factors include propagation anomalies (analogous to the shimmer seen on a hot day), the movement of vegetation cover, and interfering returns from reflectors at other angles.

Another common feature of these radars is they all display the measured displacements on an image of the rock face, with color representing the degree of movement. It is the technology behind the image formation process that differentiates these radars and their suitability for particular applications. Table 10.2 summarizes the differences between the real-beam techniques and the synthetic aperture techniques for SSR units and their impact on performance for safety-critical decisions (Longstaff n.d.).

Some important and diverse applications of SSR include the following (Noon and Harries 2007):

- **Safety monitoring during production**—The radar is used during mining as a monitoring tool of a specific unstable slope area adjacent to excavation.
- **Background geotechnical monitoring**—It is used to rapidly scan large sections of a mine pit to identify developing geotechnical hazards to determine which specific areas should be targeted with other monitoring systems.
- **Slope failure risk management**—The radar is used to continuously monitor a moving failure mass from a stand-off position. This allows the operation to also identify any smaller sections within the larger mass that are detaching and moving separately.

- **Post-failure recovery**—The real-time monitoring with SSR has been successfully used to monitor the slope during cleanup after a failure during the "rescue" of mining equipment buried by the failure.
- **Safety monitoring of access routes**—It is often used to "safeguard" access routes, which is very important in pits with a single access ramp.
- **Subsidence monitoring**—The system is used in monitoring equipment working over mine voids.
- **Waste dump monitoring**—The radar can be used for monitoring the stability of waste repositories.

Instrumentation Techniques

Numerous instruments are available to measure and/or record rock deformation within the rock mass and on the excavation boundary. Geotechnical instrumentation serves many purposes, including site investigation, design verification, construction control, quality control, safety monitoring, legal protection, and performance monitoring (Slope Indicator Company 1994).

Instrumentation techniques can be grouped according to the following types of instrumentation (Windsor 1993):

- Movement indicators
- Convergence meters
- Strain meters (or gauges)
- Joint meters
- Extensometers
  - Fixed borehole extensometers
  - Portable wire line extensometers
- Inclinometers
  - Fixed borehole inclinometers
  - Portable borehole inclinometers
- Deflectometers
- Extensometer-inclinometers
- Extensometer-deflectometers

Extensometers. Extensometers are one of the principal instruments used to measure rock deformation. Two general types of extensometers are in use for rock slope stability monitoring: the fixed borehole extensometer and the portable wire line extensometer.

The fixed borehole extensometer measures only axial displacement between a fixed number of reference points on the same measurement axis. When more than two reference points are used, the instruments are referred to as multiple-position or multipoint extensometers.

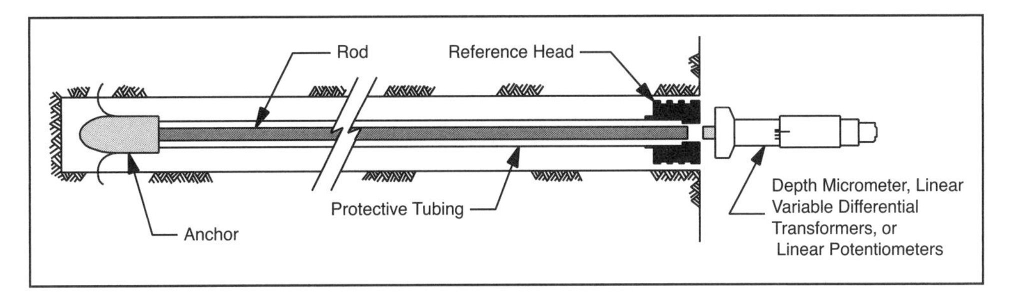

Courtesy of Slope Indicator Company

FIGURE 10.8 Single-point rod extensometer

Multipoint extensometer data can reveal the relative movement between anchor points and the distribution of displacement, in addition to the magnitude, rate, and acceleration of displacement (Slope Indicator Company 1994).

The basic components of a fixed borehole extensometer are an anchor, a linkage, and a reference head. The reference head is installed at the borehole collar. The linkage system may be composed of wires or of solid rods; it spans the distance between the reference head and the anchor. A change in this distance indicates that ground movement has occurred. Measurements are taken at the reference head with a depth micrometer or an electronic sensor and are used to determine the displacement. A typical rod extensometer is shown in Figure 10.8. Figure 10.9 shows a cable extensometer built at a mine site and installed in an exploration drill hole.

Holes for the extensometers should be drilled to intersect probable deformation zones at angles approximately normal to the strike of the potential failure surface(s) and at a dip angle that crosses an expanse of ground—including the potential failure surface(s)—rather than narrowly intersecting a single discontinuity. Figure 10.10 shows a schematic of the installation of a multipoint extensometer to monitor magnitude and rate of movement of two discontinuities.

Extensometer output data are in the form of magnitude of displacement, but they can also include direction of displacement (for multiple extensometer placements), rate of displacement, and acceleration. On a graph of displacement versus time, velocity is the slope of the curve, and acceleration is the rate of change of the slope. Acceleration is an important parameter because if it continues, it invariably leads to failure. The single most important precept of extensometer data interpretation is the accurate detection and recognition of acceleration. A typical displacement–time graph for a multi-position extensometer is shown in Figure 10.11. Also shown on the bottom of Figure 10.11 is a plot of acceleration versus time for the displacement–time data over the full length of the extensometer. Acceleration in this context is measured as the rate of change of strain (in micrometers per meter per hour) per unit of time (hour).

Photo by Don Berger, courtesy of Homestake Mining Company

FIGURE 10.9 Cable extensometer installed in exploration drill hole

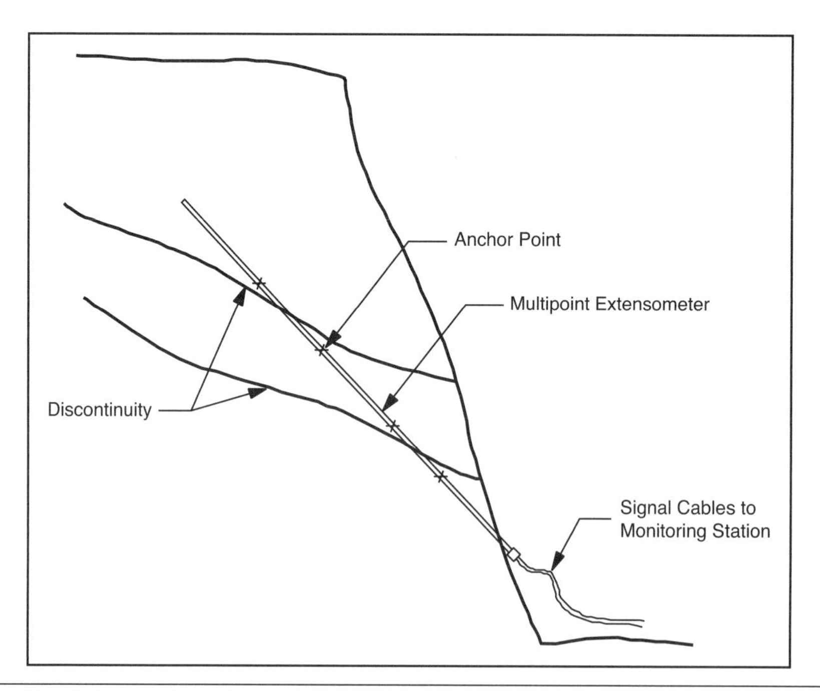

FIGURE 10.10 Multipoint extensometer for monitoring magnitude and rate of movement along two discontinuities

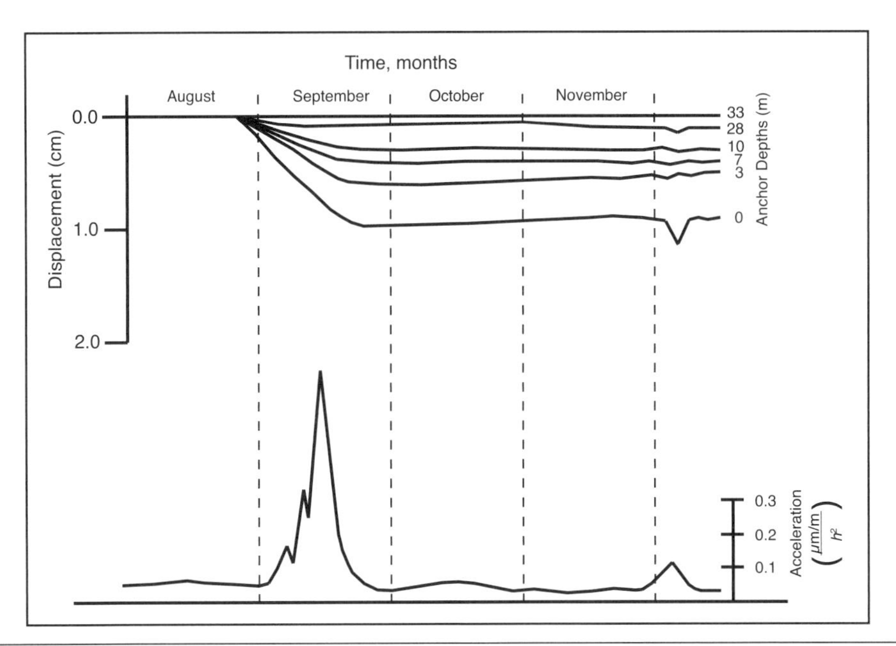

FIGURE 10.11 Displacement versus time for a multi-position extensometer, as well as rate of change of displacement (acceleration) over the entire length of the anchor

The portable wire line extensometer (Figure 10.12) has been used for decades to monitor dump or wall rock displacement, often across a tension crack. In this system, the monitor is set on stable ground behind the last visible tension crack and the wire should extend out to the unstable area (Call 1982). As the unstable ground moves, the wire is pulled over the head pulley on the extensometer and movement is recorded either manually or electronically. Long lengths of wire can lead to errors due to sag or to thermal expansion, so readjustments and corrections are often necessary.

A more recent development, the automated slope monitoring system, adds the ability to record data automatically and then transmit the data to a central computer system for analysis. The system is used for real-time displays of slope movement, as well as for long-term analysis of all recorded information (Martin 1992). The automated system shown in schematic form in Figure 10.13 is composed of two main components: the slope monitor unit, which is located in the field, and the central computer and radio, which are located in the mine office. The two most important features of the slope monitor unit are the electronics box and the wire spool. An attached solar panel can be used to recharge the battery power supply. The electronics box records and transmits to the central computer—at a user-defined time interval—such attributes as movement, velocity, date, time, location, and battery power (Martin 1992). Alarms may also be transmitted if certain conditions occur, such as broken wire, excessive movement or velocity, or communication failures. For instance, if the wire breaks or the slope anchor probe pulls loose, the wire spool falls to the base of the mounting tripod; the falling spool pulls a magnet off the electronics box, which immediately radios a

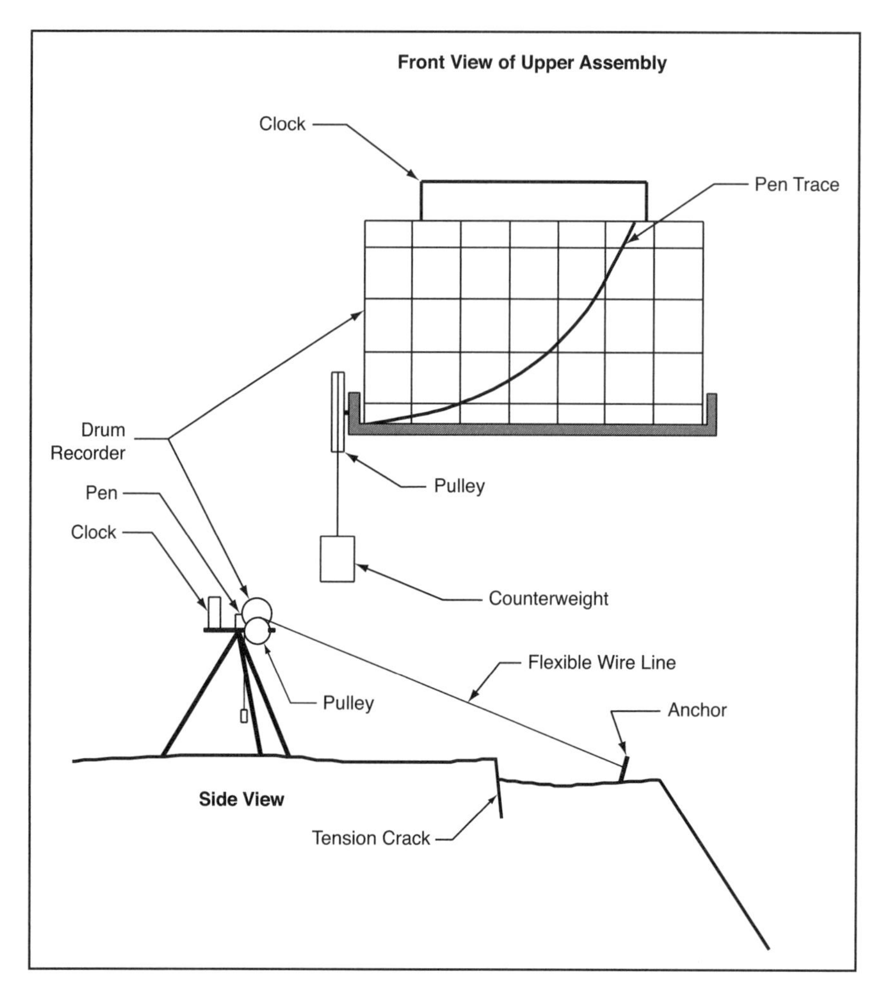

Adapted from Call 1982

FIGURE 10.12 Original design of the portable wire line extensometer with continuous recorder

warning to the central computer (Martin 1992). The central computer system collects and processes the data and generates screen displays and/or reports on slope movement status.

Figure 10.14 shows a solar-powered, automated slope extensometer that sends readings via a radio modem link to the mine dispatch terminals. Several factors are automatically corrected in this system (i.e., there is a temperature sensor at the station to correct for wire expansion and contraction). The wire length and gauge are input into the software, and corrections are transmitted back to the system. The user can set the movement velocity tolerance to transmit data, and an alarm will sound if the movement exceeds a particular specified velocity. The system is capable of transmitting a reading every 3 seconds if necessary. This system was used to monitor movement across a large tension crack (also shown on Figure 10.14) on a slope failure at a mine in eastern Nevada.

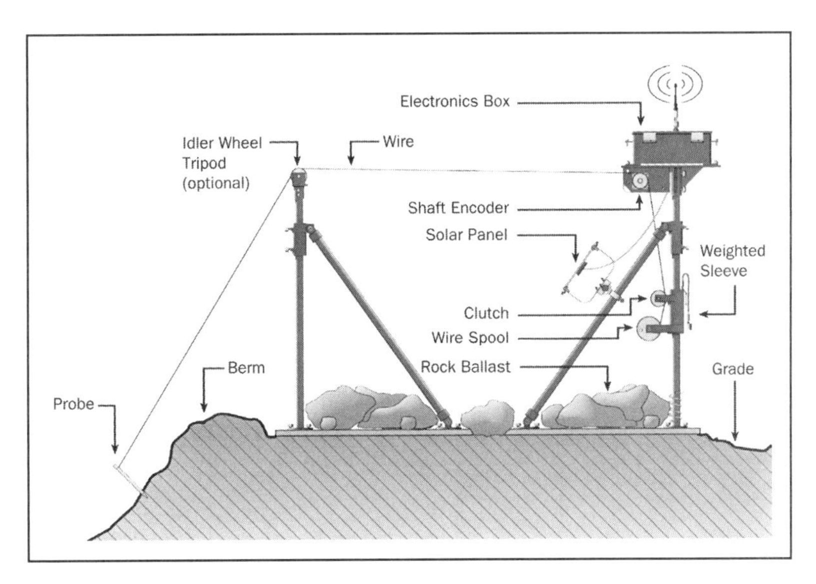

FIGURE 10.13 Schematic of an automated wire line extensometer

Courtesy of BHP

FIGURE 10.14 An automated wire line extensometer monitoring movement across a stress crack at a mine in eastern Nevada

Inclinometers. An inclinometer measures the change in inclination (or tilt) of a borehole and thus allows the distribution of lateral movements to be determined versus depth below the collar of the borehole as a function of time (Wilson and Mikkelsen 1978). Therefore, the application of inclinometers to slope stability studies is important for the following reasons:

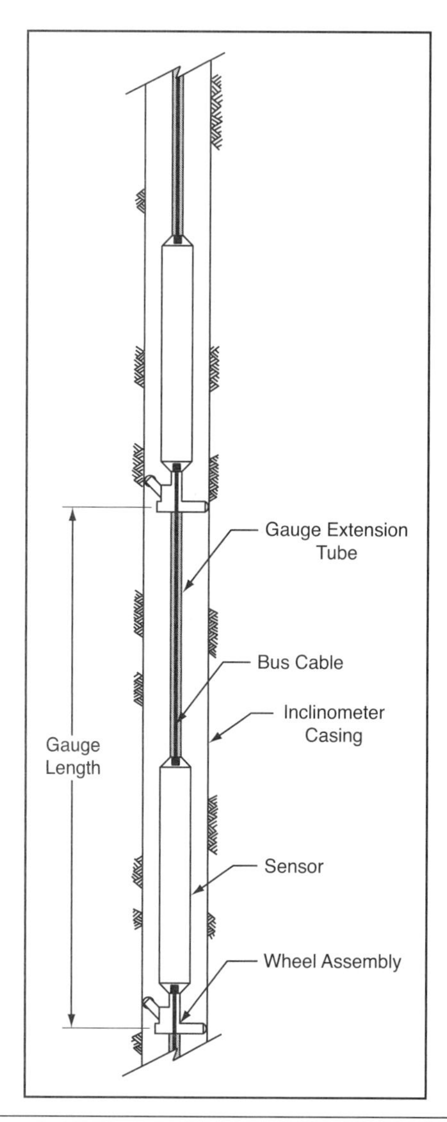

FIGURE 10.15 An in-place inclinometer system

- To locate shear zone(s)
- To determine whether the shear along the zone(s) is planar or rotational
- To measure the movement along the shear zone(s) and determine whether the movement is constant, accelerating, or decelerating

The in-place inclinometer was developed for automated monitoring. It is composed of a string of inclinometer sensors permanently mounted in the casing. The sensors are normally positioned within the casing to span the zone where movement is anticipated. The string of sensors is usually attached to a data acquisition system that is programmed to trigger an alarm if certain boundary conditions are exceeded. The in-place inclinometer is

Photo by Kjell Moe, courtesy of Newmont Mining Company

FIGURE 10.16 A traversing probe inclinometer system

expensive; hence, its use is generally limited to only the most critical applications (Boisen and Monroe 1993). A typical in-place inclinometer system is illustrated in Figure 10.15.

The traversing probe type of inclinometer (Figures 10.16 and 10.17) was developed to address this problem of expense. It employs a single sensor that can be used to monitor any number of inclinometer casings. Tilt readings are typically obtained at 2-m intervals (2-ft intervals are used with probes that utilize the U.S. customary system of measure) as the probe is drawn from the bottom to the top of the casing. The main drawback to the system is that it is slow and requires an on-site operator (Boisen and Monroe 1993). The traversing probe inclinometer system consists of four main components (Slope Indicator Company 1994):

1. **The inclinometer casing.** This is a special-purpose grooved pipe that deforms with the adjacent ground. It is normally installed in boreholes but can be placed in fill. Casing comes in outside diameters of 48 mm (1.9 in.), 70 mm (2.75 in.), and 85 mm (3.34 in.), and in lengths of 1.52 m (5 ft) and 3.05 m (10 ft). The internal grooves control the orientation of the wheeled sensor unit.
2. **The inclinometer probe.** The wheeled inclinometer probe tracks the longitudinal grooves in the casing, thus orienting the sensor in the desired plane or planes of measurement. Two force-balanced servo-accelerometers measure tilt in two perpendicular planes: the plane of the wheels and 90° from the wheel plane. The probe wheel base is 500 mm for the metric probe and 24 in. for the U.S. customary system probe.

Photo by Kjell Moe, courtesy of Newmont Mining Company

FIGURE 10.17 Close-up of inclinometer casing, cable, and data collector

3. **A control cable.** This cable is used to transmit power and signals from the probe to the readout unit and to control the depth of the probe. It is graduated in 0.5-m intervals for the metric unit and 1-ft intervals for U.S. customary units.
4. **A readout unit.** This device may be a special-purpose unit for inclinometer probes, a data acquisition unit, or a computer. It records and displays inclination measurements obtained from the inclinometer probe. Computer software is available for graphing, listing the data, generating reports, and performing other advanced features.

Figure 10.18 shows an inclinometer installed to monitor differential settling of a tunnel constructed to allow mine haul trucks to cross a busy highway. Figure 10.19 illustrates idealized plots of inclinometer deformation versus depth for planar and rotational shear failures. The inclinometer profile for the planar slope failure shows forward displacement of the inclinometer casing at the shear zone but no rotation between the shear zone and the upper slope surface. The inclinometer profile for rotational failure shows forward displacement at the shear zone as well; however, it also shows tilting backward from the shear zone to the upper slope face. The three profiles shown for each mode illustrate relative movement at three different measurement times.

Other instrumentation. Strain gauges, crack gauges, and tiltmeters have been used to monitor rock deformation at sites with potentially unstable rock slopes.

Photo by Don Berger, courtesy of Homestake Mining Company

FIGURE 10.18 Inclinometer installed to monitor settlement of a reinforced concrete tunnel built to allow mine haul trucks to cross a busy highway

Electrical resistance strain gauges, consisting of a grid of strain-sensitive metal foil bonded to a plastic backing material, are commonly used to measure rock deformation on laboratory- and even bench-scale specimens. The gauges are bonded to the rock surface with a special adhesive and connected electrically to a measuring instrument. As the rock deforms under stress, the gauge deforms with the rock. The resistance of the wires within the gauge also change proportionally with the deformation of the gauge. The change in resistance is proportional to the change in strain; therefore, the strain within the rock may be determined. Field-scale strain gauges are usually short base-length devices that are bonded to a rock surface or embedded in the rock within an encapsulating medium such as concrete or grout (Windsor 1993). They, like the laboratory-scale strain gauges, possess high sensitivity, resolution, linearity, and repeatability.

Designs for field-scale strain gauges may also be based on the vibrating steel strip principle. A typical gauge based on this principle consists of a steel strip and an electromagnetic coil sealed inside a steel tube body. The strip is held in tension between the two ends of the body. Strain is transferred from the rock to the steel strip. An increase in tensile strain increases tension in the strip, and a decrease in tensile strain decreases tension in the strip. The coil is used to vibrate the strip at a frequency relative to its tension. The vibration of the strip within the magnetic field of the coil induces a frequency signal that is transmitted to a readout device. The readout device processes the signal and gives a number representing the microstrain in the member.

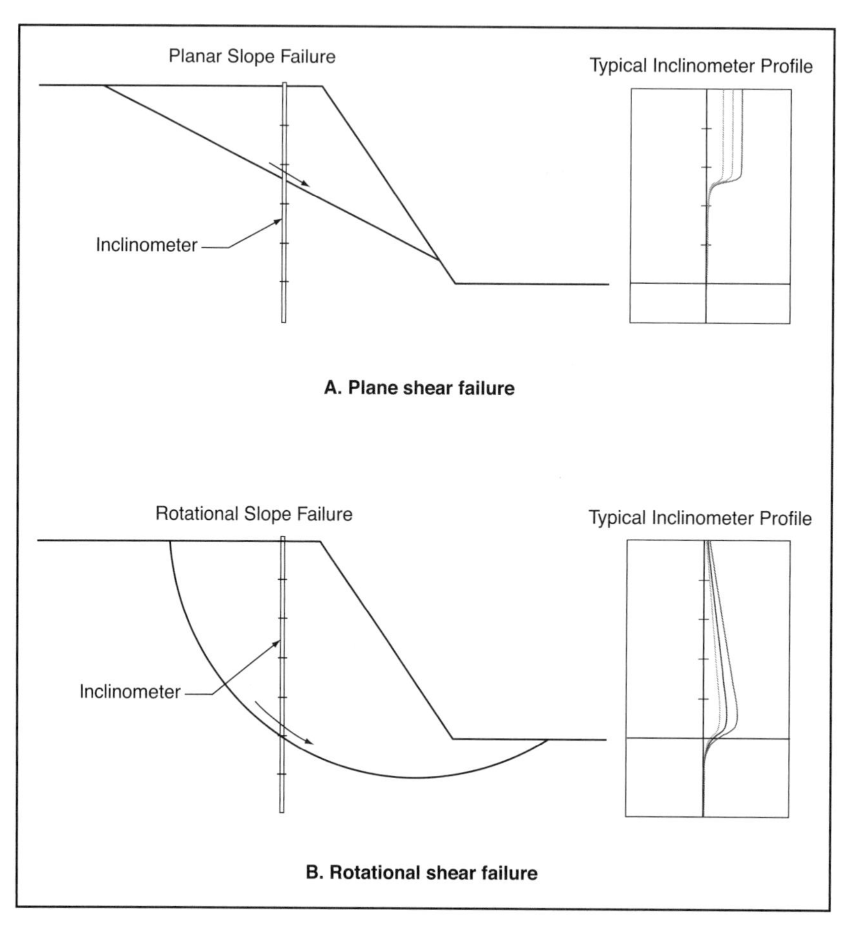

Courtesy of Slope Indicator Company

FIGURE 10.19 Typical plots of inclinometer depth versus deformation

Crack gauges are used to measure displacement (opening or closing) across cracks in the rock mass. One low-cost method commonly employed is the placement and surveying of wood or steel hubs on either side of the crack. Alternatively, a simple mechanical crack gauge can be constructed by mounting a dial gauge between steel plates attached to either side of a crack; this device would, however, require manual reading periodically.

A similar mechanical crack gauge employing a linear measuring scale, rather than a dial gauge, is shown in Figure 10.20.

A simple electrical crack gauge can be constructed by mounting the body of a linear variable differential transformer (LVDT) to a frame on one side of a crack and then attaching the core via a core extension rod to a frame on the other side of the crack. A crack gauge of this type is shown in Figure 10.21. The LVDT is an electromechanical device that produces an electrical output directly proportional to the displacement of a separate movable

Photo by Don Berger, courtesy of Homestake Mining Company

FIGURE 10.20 Mechanical crack gauge

Photo by Don Berger, courtesy of Homestake Mining Company

FIGURE 10.21 Electrical crack gauge using an LVDT to measure displacement

core. The linear displacement range for an LVDT is normally taken from a null position in the core with respect to the LVDT coil. That is, output voltage is zero midway in the linear range, becomes negative as the core moves into the coil, and becomes positive as the core moves out of the coil. The device requires calibration of output voltage versus displacement. LVDTs are available to measure displacement ranges of smaller than ±1 mm (0.039 in.) to greater than ±1 m (3.28 ft). Precision LVDTs (those that measure minute displacements) are often used to measure strains in laboratory-scale rock specimens, whereas LVDTs with large displacement ranges can be used to measure opening or closing across

Photo by Don Berger, courtesy of Homestake Mining Company

FIGURE 10.22 Electrical tiltmeter for monitoring tilt of a rock overhang

fractures in rock. Two common applications for the larger-displacement LVDTs in rock slope stability monitoring are as crack gauges and as displacement transducers in extensometers. Output from the LVDT can be tied directly to a data recorder for processing on a periodic basis, or it may be linked to a central data acquisition and monitoring system for real-time processing.

Tiltmeters can be used where the failure mode of a mass may be expected to contain a rotational component. They can be used to measure (1) the tilt of rock outcrops, benches of open-pit mines, or highway and railway cuts; (2) the surface tilt of bridge piers or retaining walls; (3) profiling of surface settlements; (4) the tilt of vertical walls and structures; and (5) subsidence in landslide areas, around earth or rockfill dams, and adjacent to building excavations (Slope Indicator Company 1994). A portable tiltmeter uses a force-balanced servo-accelerometer to measure inclination. The tiltmeter, along with several tilt plates mounted at different locations, can be used to detect differential movement, but the resulting data cannot provide absolute displacement and settlement profiles (Slope Indicator Company 1994). Changes in tilt are found by comparing the current reading with the initial reading. Figure 10.22 shows a tiltmeter mounted in the field to monitor rotation and tilt of a rock overhang.

Courtesy of BHP

FIGURE 10.23 A large slope failure at a surface mine in eastern Nevada

MONITORING

With any artificial excavation in rock, some sort of slope instability can be expected, varying from sloughing and raveling to large-scale mass movement. Horizontal displacements of pit slopes may be divided into four stages (Sullivan 1993):

1. Elastic movements
2. Creep movements
3. Cracking and dislocation
4. Collapse

Elastic movements are basically the reaction of the rock mass to unloading and are a function of stress and rock mass moduli. Creep movements are the relatively slow time- and stress-dependent movements that occur in some soil and rock masses. Cracking and dislocation are often taken to mean failure, as is collapse; however, they usually represent one of the first stages of collapse (i.e., the opening of tension cracks). Collapse can range from the sliding of an individual joint-bound block of rock to large-scale failure of the overall slope. An example of a large-scale slope failure (approximately 22.7 million t [25 million st]) is shown in Figure 10.23. This failure occurred at the intersection of a dominant fault structure (vertical scarp at top of photo) and a series of drag faults going off to the right side of the photo.

Major displacement may or may not cause difficulties for mining, depending on several key factors, which include (Sullivan 1993)

- The nature of the material involved in the instability,
- The type of instability,

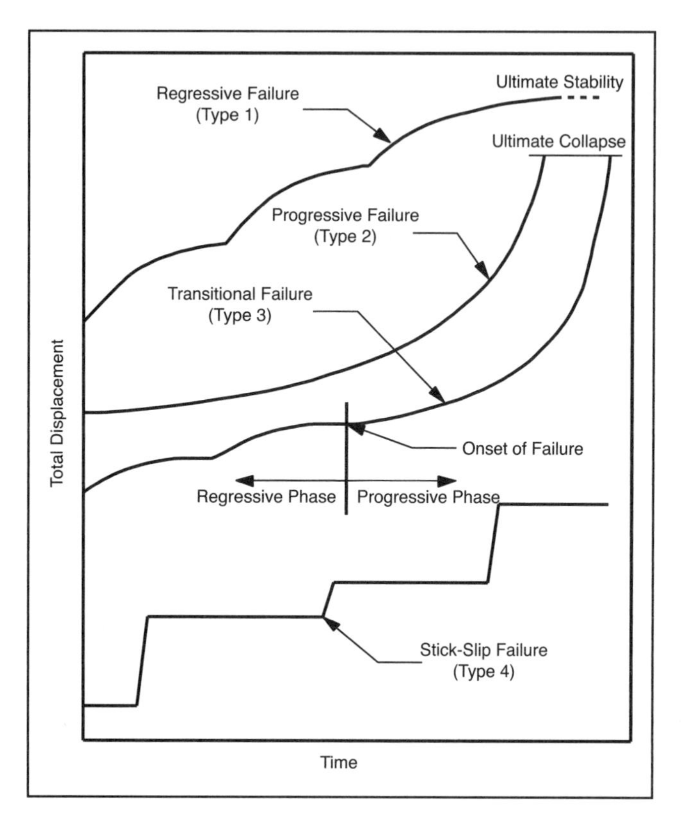

Adapted from T. Szwedzicki (ed.), Geotechnical Instrumentation and Monitoring in Open Pit and Underground Mining. In *Proceedings of the Australian Conference*, Kalgoorlie, June 21–23, 1993.

FIGURE 10.24 Typical displacement–time curves for pit slope failures

- The rate of movement,
- The type of mining system employed, and
- The relationship of the instability to the mining operation.

The preceding factors provide an economic basis for continuation of mining, support, excavation, and so forth. The protection of personnel and equipment is, however, another key consideration.

Displacement–Time Characteristics of Large-Scale Slope Failures

Broadbent and Zavodni (1978, 1982) analyzed many rock slope failures and determined that all existing or potential failures must be either regressive or progressive, depending on the tendency of the instability to become more stable or less stable with time. Regressive failure (type 1) is characterized by a series of short-term decelerating displacement cycles, ultimately leading to stability. A progressive failure (type 2), on the other hand, is one that will displace at an accelerating rate, usually an algebraically predictable one, and will ultimately culminate in overall failure.

Two other types of failure displacement–time curves have been defined: the transitional type (type 3), by Broadbent and Zavodni (1978, 1982); and the stick-slip type (type 4), by Sullivan (1993). A transitional-type failure starts as regressive and ends as a progressive type. The change in failure from regressive to progressive usually occurs as a result of a change in external conditions, such as a rise in groundwater level or a change in shear strength. The stick-slip type failure is characterized by sudden movements followed by periods of little or no movement. The movement phases often occur as a result of a change in groundwater level, rainfall events, or seismic events such as blasting. Typical displacement–time curves for the four failure types are illustrated in Figure 10.24.

Prediction of Time to Failure

A pit slope monitoring system has several functions (Call 1982):

- To establish a surveillance system to detect initial stages of slope instability
- To provide a detailed movement history in terms of displacement directions and rates in unstable areas
- To define the extent of the failure areas

Once an unstable area has been detected and delineated, another important reason for the monitoring system is to attempt to predict the time to failure. Kennedy and Niermeyer (1970) have presented techniques to predict the time to failure for the progressive type of failure system (type 2); Broadbent and Zavodni (1978, 1982) have done so for the transitional type of failure system (type 3).

Prediction of time to failure for the progressive system involves extrapolating the displacement–time curve for the fastest-moving monitoring point(s) outward in time to the point where the curve becomes vertical or near vertical (Figure 10.25). Mathematical exponential or power curve fitting techniques can be used to model the curve and then extrapolate into the future to estimate the failure date. A problem with this technique is identifying and monitoring the fastest-moving point(s), so multiple monitoring locations are often required. Another problem, discussed below, is that the shape of the plot of the displacement–time curve is scale-dependent.

Broadbent and Zavodni (1978, 1982) employed a similar technique to estimate the number of days prior to collapse for transitional (type 3) failures. Figure 10.26 shows the history of a failure prediction by Broadbent and Zavodni (1978, 1982) in terms of a displacement rate versus days prior to collapse. This figure shows the basis for the technique of predicting failure collapse times.

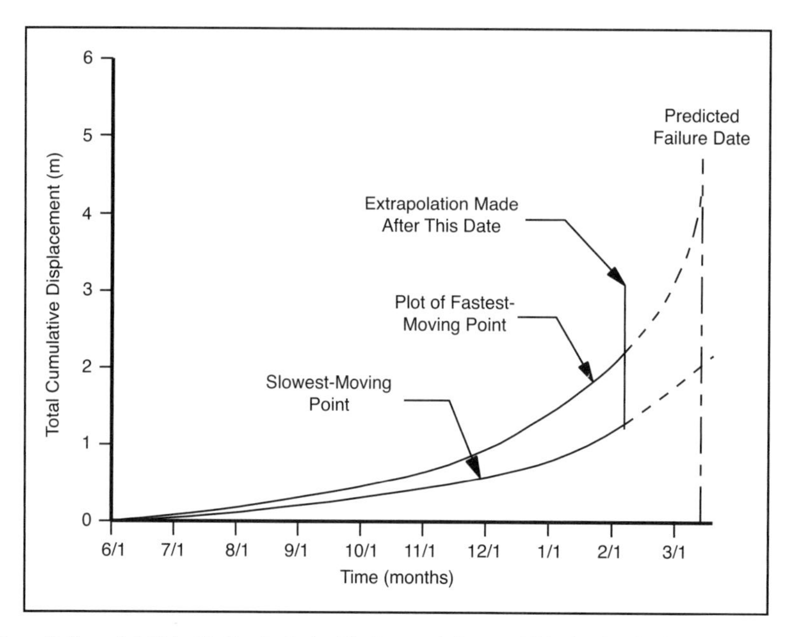

Adapted from T. Szwedzicki (ed.), Geotechnical Instrumentation and Monitoring in Open Pit and Underground Mining. In *Proceedings of the Australian Conference*, Kalgoorlie, June 21–23, 1993.

FIGURE 10.25 Plot of cumulative displacement versus time of the slowest moving point and the fastest moving point for a pit slope failure

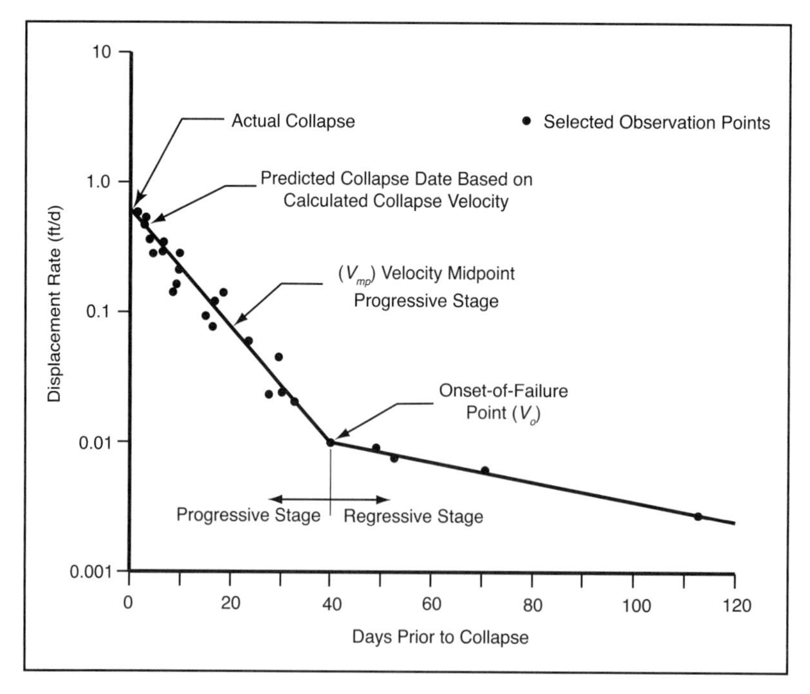

Adapted from Broadbent and Zavodni 1982

FIGURE 10.26 Failure prediction for an open-pit mine in eastern Nevada

On the basis of velocity–displacement data for nine transitional failures that progressed to total collapse, Broadbent and Zavodni developed a technique to estimate the velocity at the collapse point (V_{col}). They observed that

$$\frac{V_{mp}}{V_o} = K \quad \text{(EQ 10.1)}$$

where

V_{mp} = velocity at midpoint in the progressive failure stage (see Figure 10.26)
V_o = velocity at the onset-of-failure point (see Figure 10.26)
K = a constant (average = 7.21; standard deviation = 2.1; range = 4.6 to 10.4 for the nine transitional failures studied)

Given that the equation for a semilog straight-line fit has the form

$$V = Ce^{St} \quad \text{(EQ 10.2)}$$

where

V = velocity, in ft/d, as per Broadbent and Zavodni (1978, 1982)
C = constant
e = base of natural logarithm
S = slope of line, in days^{-1}
t = time, in days

and assuming that $t = 0$ at the collapse onset point, Equation 10.2 takes the following form for the progressive failure stage:

$$V = V_0 e^{St} \quad \text{(EQ 10.3)}$$

From this equation, as well as the empirical relationship of Equation 10.1, the velocity at the collapse point (V_{col}) can be determined as

$$V_{col} = K^2 V_0 \quad \text{(EQ 10.4)}$$

Equation 10.4 and a plot of displacement rate versus days prior to collapse can be used to estimate the number of days until collapse once the failure onset point is reached and the progressive-type failure pattern is established. To use this technique, one must remember that at the onset-of-failure point (V_0), $t = 0$. Then a plot of the displacement-rate-versus-time data of the monitor records starting at the onset-of-failure point will establish the progressive stage line on the plot (see Figure 10.26). Projecting this line to the calculated collapse velocity will give an estimate of the time to collapse. A difficulty with this technique is in recognizing the point in time when the failure onset point is reached

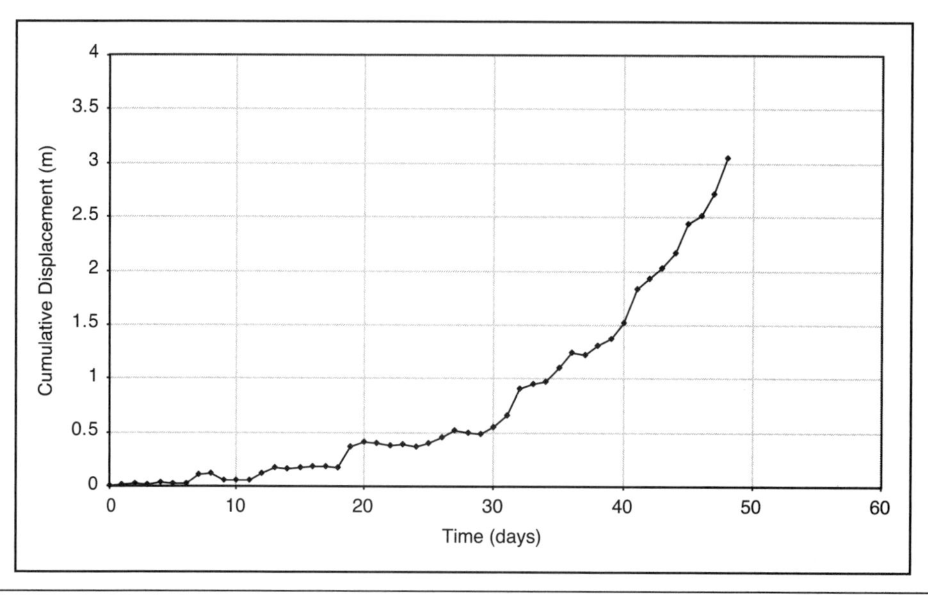

FIGURE 10.27 Typical cumulative displacement versus time progressive failure–type curve for a prism in an unstable slope

—that is, in establishing the progressive stage failure displacement rate pattern from the monitoring records.

Displacement–time plots, similar to those in Figure 10.25, have been used to predict the time (number of days) to failure. Figure 10.27 shows a typical cumulative displacement versus time progressive failure–type curve for a prism in an unstable slope.

One can estimate the time to failure using the plot of Figure 10.27 by (1) mathematically fitting a curve to the displacement–time curve and then projecting the curve forward in displacement/time space until it becomes vertical; or (2) projecting the curve forward "by hand," as is shown on Figures 10.28 and 10.29, then estimating the time until failure.

When the projected curve becomes vertical, it means that the prism is undergoing continual displacement with nearly zero change in time. Or

$$\Delta\text{displacement} \rightarrow \infty; \text{ as}$$
$$\Delta\text{time} \rightarrow 0$$

Since velocity is equal to (Δ displacement)/(Δ time), then the point from which the curve was generated is exhibiting high velocity movement at failure.

Using this technique for Figure 10.28, we get an estimated *time to failure* of 2 to 3 days.

However, there is a problem with this technique. It is scale-dependent. That is, if a second person plotted the same data but used a vertical (y-axis) scale as shown in Figure 10.29, that person would estimate the time to failure to be around 10 days hence.

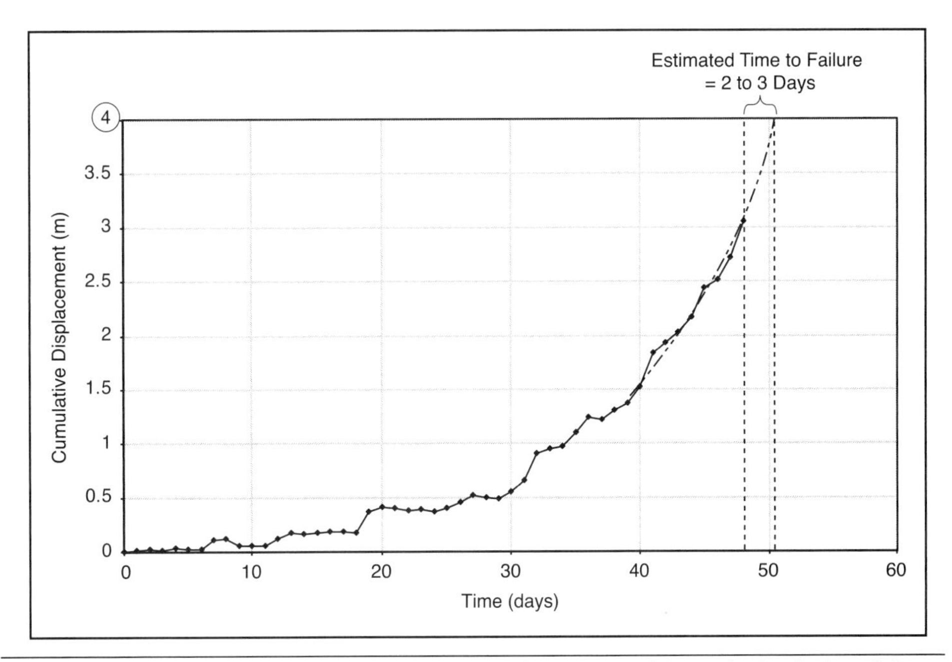

FIGURE 10.28 Cumulative displacement versus time curve with vertical axis (y-axis) ranging from 0 to 4 m and estimated time to failure

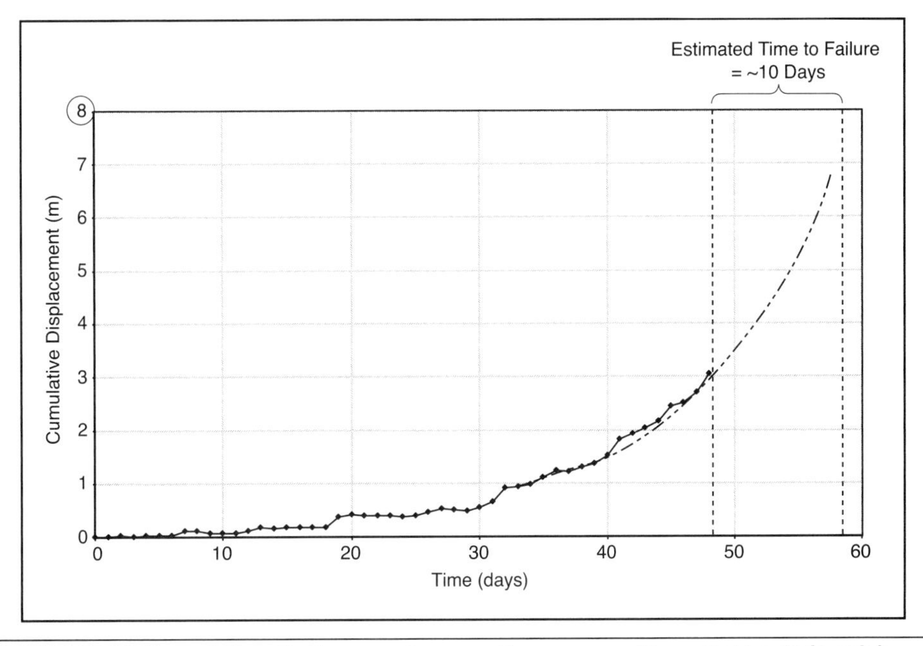

FIGURE 10.29 Cumulative displacement versus time curve with vertical axis (y-axis) ranging from 0 to 8 m and estimated time to failure

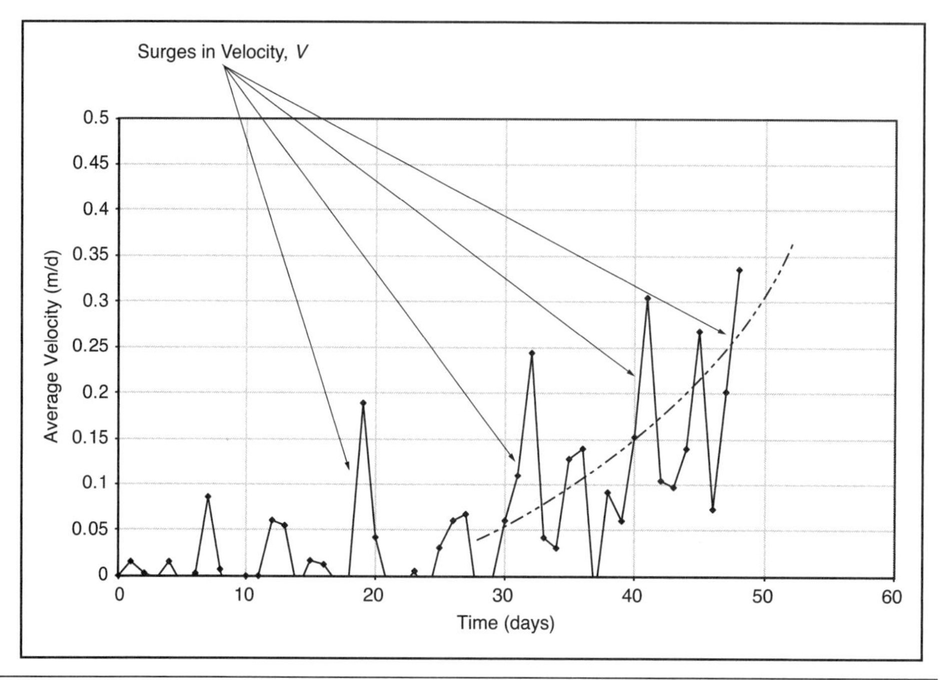

FIGURE 10.30 Average velocity per time period for the prism of Figure 10.27

Instead of looking at displacement with respect to time, if the velocity were plotted, one would get something looking like Figure 10.30. The average velocity per time period would be determined by

$$\text{average velocity per period} = (\text{displacement}_2 - \text{displacement}_1)/(\text{time}_2 - \text{time}_1)$$

$$= \Delta D/\Delta T\,(\text{m/d})$$

Again, one could look at the time (date) when the velocity approaches infinity ($V \rightarrow \infty$). But looking at Figure 10.30, one can see that there were several surges in velocity ($T \cong 18$–19, $T \cong 29$–32, $T \cong 39$–41, and $T \cong 46$–48), all of which could be mistaken for $V \rightarrow \infty$, but all of which settled back down for a significant period of time. One can see, though, that there is quite definitely an increase in velocity trend (dashed line) since about day 28.

A better estimate of the time to failure can be obtained by using time versus inverse velocity ($1/V$) as developed by Fukuzono (1985) and used for case study 2 below by Rose and Hungr (2006), and noting that as the velocity increases to infinite ($V \rightarrow \infty$), the inverse velocity approaches zero ($1/V \rightarrow 0$). Plotting inverse velocity versus time will often initially, in the early time periods, result in a great deal of scatter as the velocity fluctuates (see fluctuations on Figure 10.30). But as the velocity tends to stabilize and increase over time (accelerate), a regression trend line can be generated through the last portion of the T versus $1/V$ data points (Figure 10.31) and can be projected to the x-axis ($V = 0$) to determine

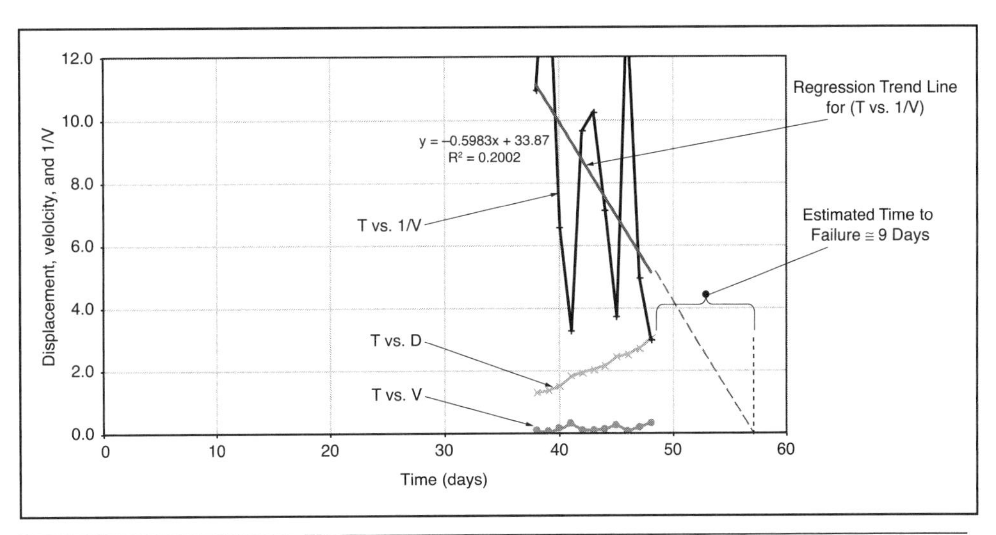

FIGURE 10.31 Plots of time versus velocity, time versus displacement, and time versus inverse velocity for the data plotted on Figure 10.27 for days 38 through 48

the estimated number of days until failure. As more data come in, the trend line and the estimated time to failure can be reestimated and readjusted. Also, the forecast line can be adjusted to accommodate potential trend changes (Rose and Hungr 2006; Rose 2011).

Monitoring Case Studies

This section presents two brief examples of the current use of instrumentation and monitoring in mining applications. Information used in both of the case examples was obtained by personal communication of the author with personnel from the respective mine operations.

Case study 1: Santa Fe Pacific Gold's Mega Pit slope failure. On December 26, 1994, at 1:00 a.m., Santa Fe Pacific Gold's (acquired by Newmont Mining Corporation in 1997 through the Newmont/Santa Fe Pacific Gold merger) Twin Creeks mine Mega Pit, located near Winnemucca, Nevada, experienced a major highwall failure (Figure 10.32) involving more than 2.27 million t (2.5 million st). In late August 1994, tension cracks were first noticed in the highwall of the west wall of the Mega Pit (Gray and Bachmann 1996). By early September 1994, 9 rebar hubs, 17 survey prisms, a permanent survey control point, and an automated slope monitoring extensometer (see Figures 10.13 and 10.14) were installed to monitor movement of the slide. The unstable area was located at a promontory, or "nose," of the pit, directly above an area being developed as the primary ore source for the first half of the next year. Additionally, tension cracks extended into the highwalls above the area currently being mined and above the area being stripped to supply ore for the second half of the next year. Significant site conditions that contributed to the instability included the following (Gray and Bachmann 1996):

Photo by Jim Bachmann, courtesy of Newmont Mining Company

FIGURE 10.32 Twin Creeks mine Mega Pit slope failure

- A high water table behind the highwall.
- A weak shale member dipping into the pit and overstressed by the weight of the 183-m (600-ft) highwall.
- A thrust fault near the toe of the failure zone.
- The promontory, or nose, configuration in the pit design (refer to Figure 10.4).
- Preexisting northeast-trending faults in the alluvium that influenced tension crack formation.
- The blasting and removal of a large catch bench, or step out, at the toe of the slope. This area was being developed for mining during the first half of the following year.
- Above-average precipitation during May and July. Additional substantial precipitation events occurred during October, November, and December, which likely contributed to the acceleration of movement.

Development work in the area directly below the slide area continued through the summer and fall of 1994 along with constant monitoring of the slide. As development work continued, movement of the slide increased. Figure 10.33 shows a plot of the total easterly displacement of prism 10 within the Mega Pit. (A generalized plan view map of the Mega Pit failure zone with prism locations is shown in Figure 10.4. On Figure 10.4, prism 10 is designated "SM10.")

In October, mining in the development area was limited to the day shift only. In November, increased precipitation resulted in greater acceleration of movement. In mid-December, as the movement curves showed even greater acceleration, all mining in the development area below the slide was halted, all people and equipment were pulled out, and access to the area was blocked off. Failure of the wall finally occurred at 1:00 a.m. on December 26 (see

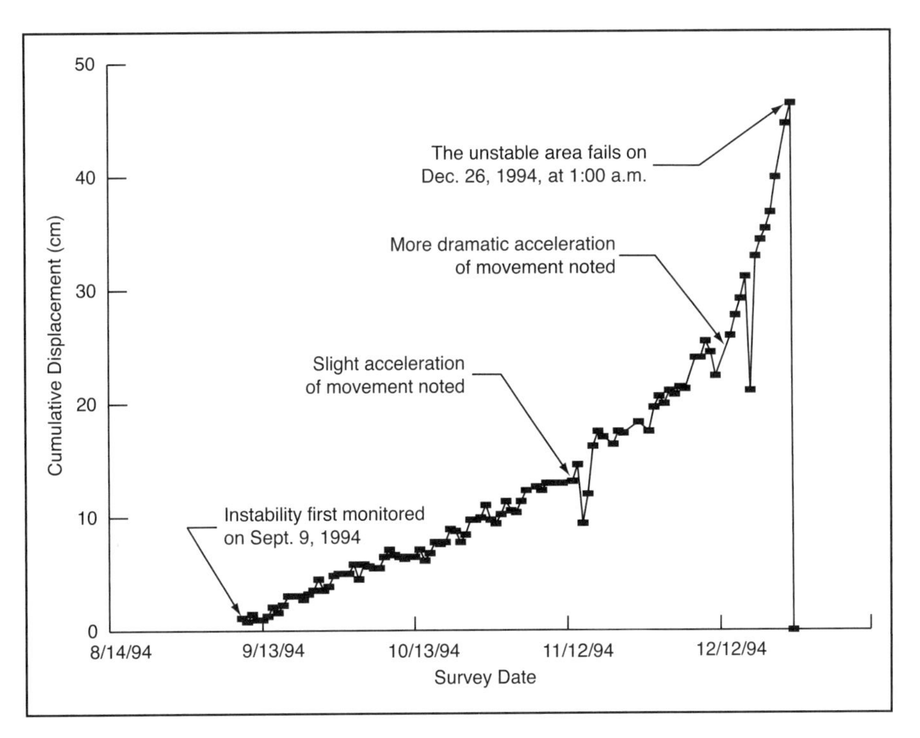

Courtesy of Newmont Mining Company

FIGURE 10.33 Mega Pit highway monitoring: Total easterly displacement of prism 10

Figures 10.32 and 10.33) while the mine was shut down for the Christmas holiday (Gray and Bachmann 1996).

After the failure, there was a 100-m (330-ft) vertical failure scarp remaining along with numerous overhangs. A plan was formulated to drill and blast behind the failure area to produce a stable, acceptable highwall angle. The blast plan was to relieve the toe area of the blast behind the slide scarp, thus causing the material above to cave upon itself because of its weight. A laser profiler was used to determine the exact topography of the highwall after failure. Following work with cross sections drawn by using the laser profiler data, mine personnel determined that a 55° drill-hole angle would be appropriate. This angle was largely based on the minimum required distance from the bottom of the holes to the free face. However, two other important factors were considered. First, the drill needed to remain on uncracked, stable ground. By projecting the holes inclined at 55° to the surface, this angle was found to be the maximum angle to give the required safety. Second, a hole angle of 55° was considered the maximum angle at which explosives could be loaded down the hole (Gray and Bachmann 1996).

The final blast pattern was drilled with a reverse circulation drill rig. Holes were 165 mm (6.5 in.) in diameter with an average depth of 76.2 m (250 ft). Spacing between adjacent holes averaged 5.5 m (18 ft), the burden ranged from 3.66 to 24.38 m (12 to 80 ft), and the

Photo by John Morkeh, courtesy of Barrick Nevada—Goldstrike Mine

FIGURE 10.34 Robotic theodolite for monitoring movement of survey prisms at a mine in north-central Nevada

powder factor averaged 0.055 kg/t (0.11 lb/st). On average, each hole contained 1,000 kg (2,200 lb) of explosives (Gray and Bachmann 1996).

The blast occurred in mid-March of 1995. Approximately 900,000 t (1,000,000 st) was successfully brought down at the planned angle. Following the blast, the material was removed and benches were reestablished on the highwall.

Case study 2: Barrick Goldstrike slope monitoring and failure S-01-A. At the Barrick Goldstrike mine's Betze-Post open pit, located near Carlin, Nevada, a computerized, automatic theodolite is used to monitor movement at approximately 100 survey prism stations. The robotic theodolite (Figure 10.34) is housed in a small shed on the south side of the pit (Figure 10.35) at a vantage point in direct line of sight with all 100 prism stations.

A computer system (Figure 10.36) is also housed in the robotic slope-monitoring shed. This system uses a specialized software package to control the theodolite. A joystick (shown at the center of the figure) is used to adjust the theodolite to sight in each prism. Once a prism is sighted in, pressing the "Enter" key on the computer accepts the location. This step is taken for each prism location. Once all the prisms are located initially, the software package takes control and the entire array of prisms is shot automatically on a cycle that takes

Photo by Dave Pierce, courtesy of Barrick Nevada–Goldstrike Mine

FIGURE 10.35 Shed for housing the robotic theodolite and remote computer system

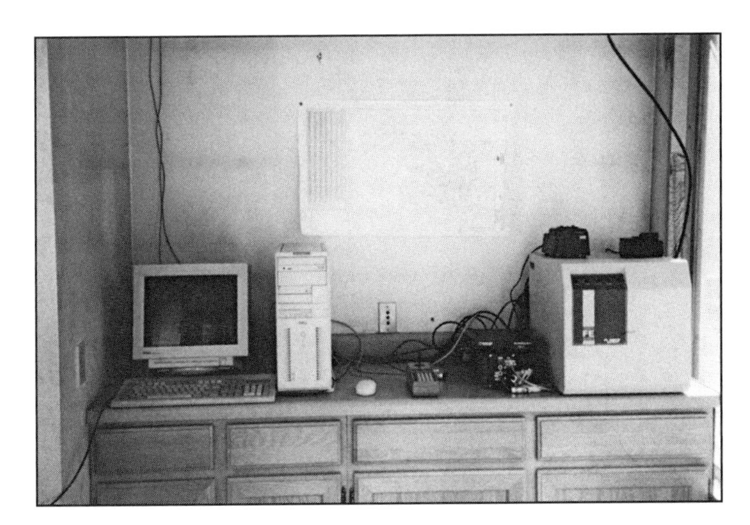

Photo by Dave Pierce, courtesy of Barrick Nevada–Goldstrike Mine

FIGURE 10.36 The remote computer system for controlling the robotic theodolite

approximately 3 hours to complete. After completion of a cycle, the system starts another cycle of shooting the set of prisms. The software package offers the option of programming the system so that subsets of prisms within the major set can be shot periodically to monitor specific areas. Each reading is transferred via modem link to a base computer in the main mine office.

The prism locations and movement data can be plotted on a mine plan map (Figure 10.37) to illustrate movement within both stable and unstable areas of the pit. Movement within

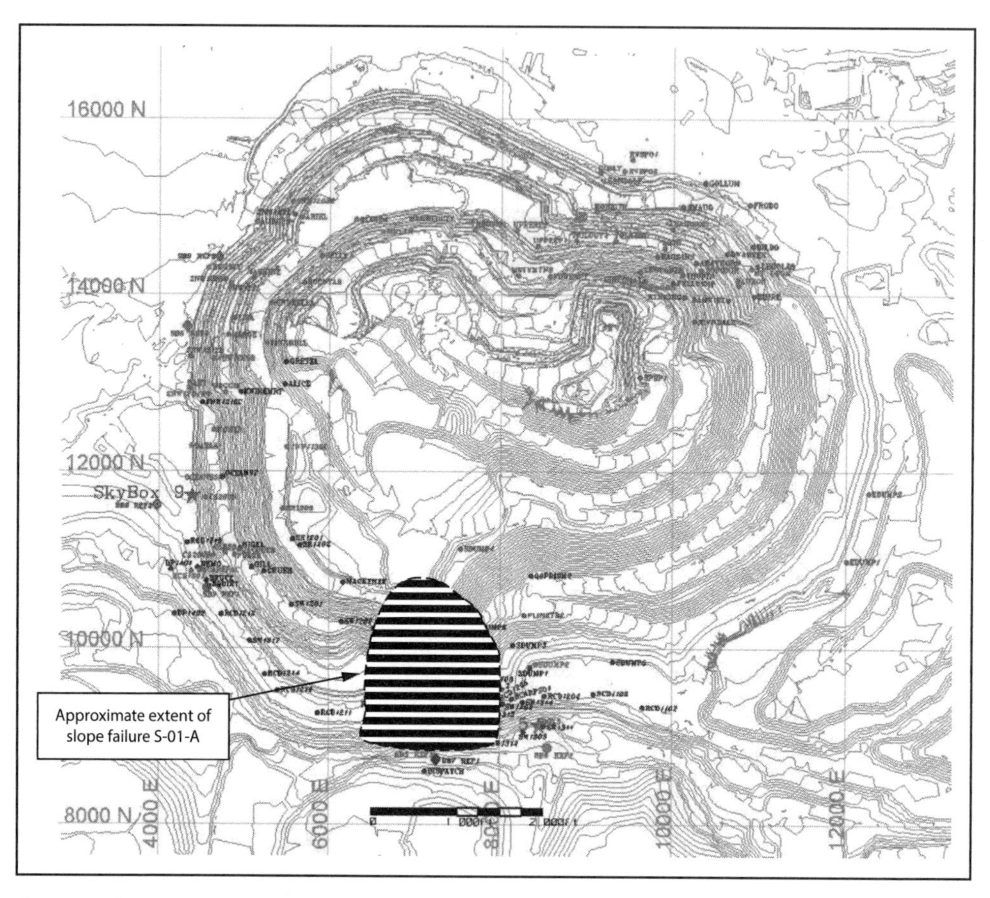

Courtesy of Barrick Nevada–Goldstrike Mine

FIGURE 10.37 Barrick Goldstrike's Betze-Post pit prism locations, skybox location, pit dimensions, and the S-01-A failure of August 2001

stable areas of the pit can occur as a result of ground rebound due to the removal of material from the pit. Plots can also be generated for each prism (Figure 10.38), showing total movement, elevation change, change in northing, change in easting, and slope distance change over any period of time. Real-time information for each prism can also be viewed on the computer system in the control shed or in the mine office, or it can be plotted as a screen dump.

On August 29, 2001, an approximately 550-m (1,800-ft) failure occurred on the southeast wall of Barrick's Betze-Post open pit (Figures 10.39–10.42). The southeast wall is situated in Jurassic granodiorite rocks that are argillically (clay) altered along major gouge-filled faults and shear zones. As a result, groundwater is highly compartmentalized and has a significant influence on slope stability (Rose and Hungr 2006).

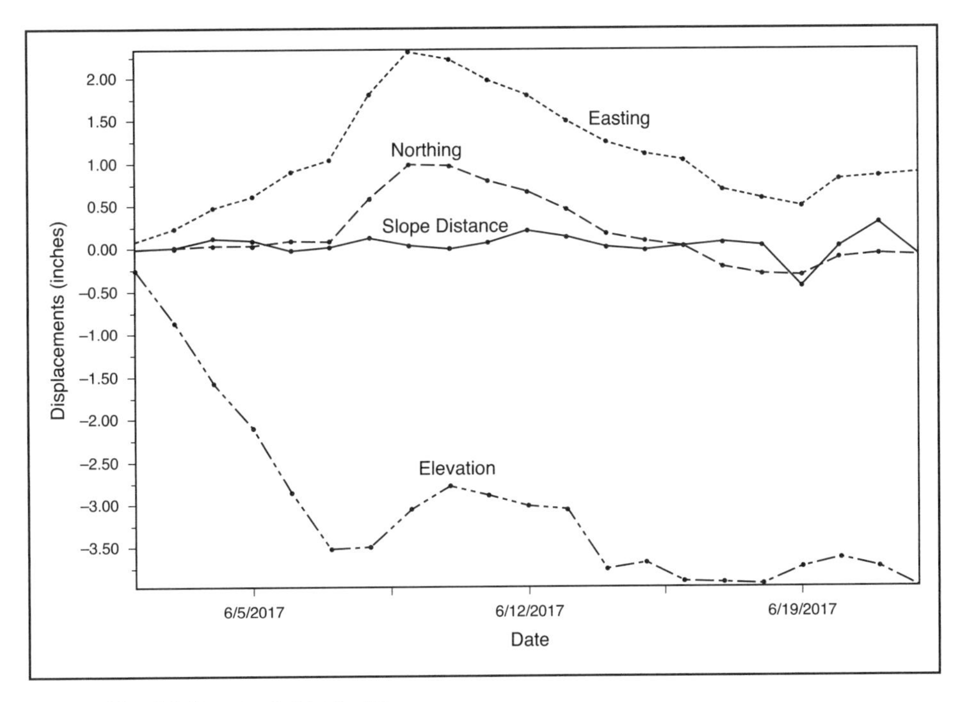

Courtesy of Barrick Nevada—Goldstrike Mine

FIGURE 10.38 Prism displacements plotted using Barrick's prism monitoring software, Canary

It was observed that complex wedge deformations on the upper southeast wall of the pit began as early as 1993. The failure mechanism involved shearing along deep wedge intersections that plunged moderately toward the open pit, causing upward heaving along shallow in-slope dipping faults that were oriented obliquely to the slope face (Rose and Hungr 2006).

Figure 10.39 shows an expanded view of the portion of the southeast wall of the Betze-Post pit of Figure 10.37 with the S-01-A slide delineated (Rose and Hungr 2006).

Figure 10.40 is a plot of inverse velocity versus time showing the trends of nine survey prisms located at various elevations on the southeast wall of the Barrick Goldstrike Betze-Post open pit over the last six weeks preceding the S-01-A slope failure in August 2001. Targets were monitored using a robotic total station at 2-hour intervals (Rose and Hungr 2006). Filtering (data smoothing) of two-hour robotic total station prism monitoring measurement data was achieved by calculating six-day average (incremental) slope distance velocities to reduce the effects of instrument error in low-level velocity values. As displacement rates increased, a clear inverse-velocity trend developed and began to converge on a failure time of August 29, 2001 (Figure 10.40). Linear regression was applied to the inverse velocity values, which defined a coefficient of determination (R^2) of 99% for

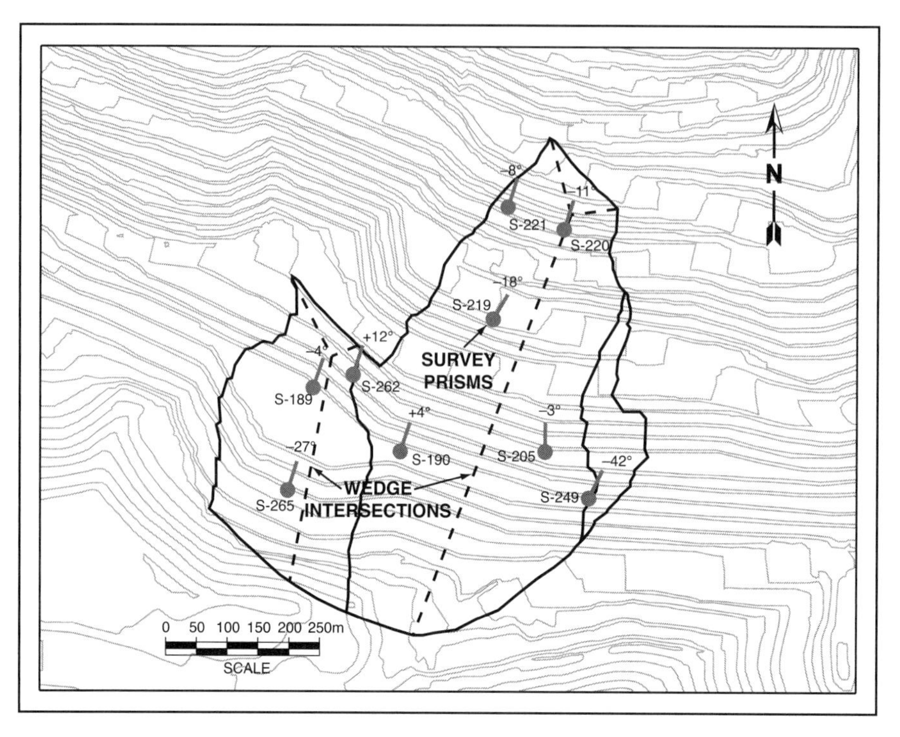

Source: Rose and Hungr 2006

FIGURE 10.39 Map of upper southeast wall showing complex wedge geometries, prism locations (solid dots), mean azimuth and plunge angles of displacement vectors (short solid lines), and plane intersections of the interpreted multi-planar complex wedge rupture surface (dashed lines)

all nine prisms. The failure (Figure 10.41) occurred on the predicted date encompassing an overall slope height of 550 m (1,804 ft) and an estimated 47 million t (51.8 million st). The instability occurred over several hours as a series of nested wedge failures and rock avalanches (Rose 2011).

In addition to the robotic theodolite, Barrick uses SSR units (Figure 10.6) as well as a dispatch-linked extensometer system (see Figures 10.13 and 10.14) to monitor stress crack movement. Real-time radar monitoring information can be displayed on the big screen in Barrick's Betze-Post geotechnical engineering command center (Figure 10.43).

For rough measurements of crack movement, Barrick (and other operators) also uses simple wooden lath crack gauges. These gauges are simply wooden survey stakes laid flat on the ground, one on either side of a crack, and within close proximity head to head. The bottoms of the laths are anchored, usually with loose rock. Measurements are manually taken periodically between the head ends of the laths (Figure 10.44).

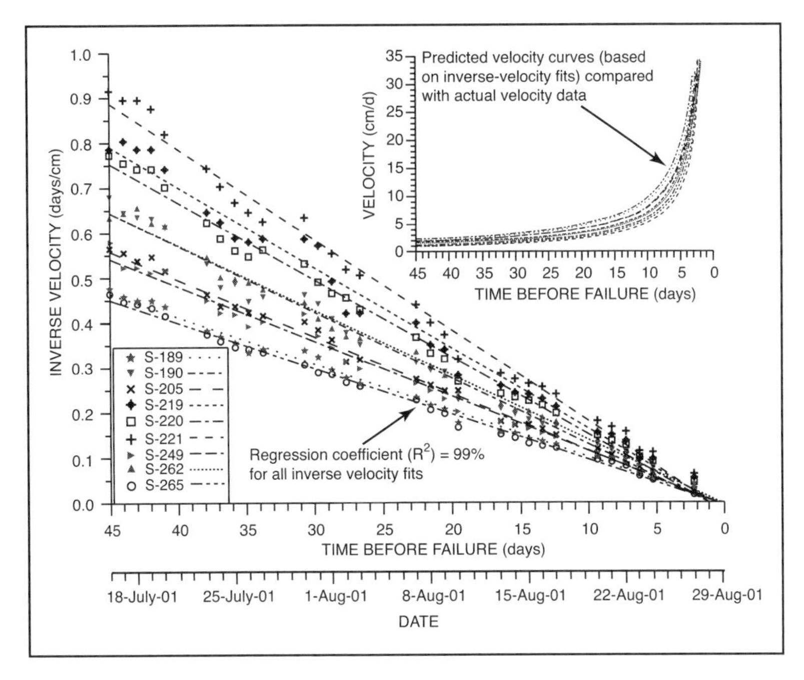

Source: Rose and Hungr 2006

FIGURE 10.40 Plot of 6-day average inverse-velocity (predicted curves versus actual values on inset graph) versus time for nine prisms (time 0 was the observed time of failure). Prism numbers correspond to Figure 10.39.

Photo by John Morkeh, courtesy of Barrick Nevada–Goldstrike Mine

FIGURE 10.41 Slope failure S-01-A, which occurred in Barrick's Betze-Post pit southeast wall

Photo by John Morkeh, courtesy of Barrick Nevada–Goldstrike Mine

FIGURE 10.42 The head of slope failure S-01-A, which occurred in Barrick's Betze-Post pit southeast wall on August 29, 2001

FIGURE 10.43 Screenshots of radar monitoring data at the Barrick's Betze-Post pit

REFERENCES

Boisen, B.P., and R.B. Monroe. 1993. Three decades in instrumentation. In *Geotechnical Instrumentation and Monitoring in Open Pit and Underground Mining*. Edited by T. Szwedzicki. Rotterdam: Balkema.

Broadbent, C.D., and Z.M. Zavodni. 1978. Slope failure kinematics. In *Proceedings of the 19th U.S. Symposium on Rock Mechanics*. Reno, NV: University of Nevada.

Broadbent, C.D., and Z.M. Zavodni. 1982. Influence of rock strength on stability. In *Stability in Surface Mining*, Vol. 3. Edited by C.O. Brawner. New York: SME-AIME.

Brown, E.T. 1993. Geotechnical monitoring in surface and underground mining: An overview. In Geotechnical *Instrumentation and Monitoring in Open Pit and Underground Mining*. Edited by T. Szwedzicki. Rotterdam: Balkema.

Photo by David Pierce, courtesy of Barrick Nevada—Goldstrike Mine

FIGURE 10.44 Simple crack extension monitoring gauge made from wooden survey laths laid head to head across a stress crack

Call, R.D. 1982. Monitoring pit slope behavior. In *Stability in Surface Mining*, Vol. 3. Edited by C.O. Brawner. New York: SME-AIME.

Collier, P.A. 1993. Deformation monitoring using the Global Positioning System. In *Geotechnical Instrumentation and Monitoring in Open Pit and Underground Mining*. Edited by T. Szwedzicki. Rotterdam: Balkema.

Fukuzono, T. 1985. A new method for predicting the failure time of a slope. In *Proceedings of the Fourth International Conference and Field Workshop on Landslides*. Tokyo: Japan Landslide Society. pp. 145–150.

Gray, C.T., and J.A. Bachmann. 1996. Blasting of the Twin Creeks' highwall failure. In *Proceedings of the 22nd Annual Conference on Explosives and Blasting Technique*, Vol. I. Cleveland, OH: International Society of Explosives Engineers.

Herring, T.A. 1996. The Global Positioning System. *Scientific American*, 274(2):44–50.

Kennedy, B.A., and K.E. Niermeyer. 1970. Slope monitoring systems used in the prediction of a major slope failure at the Chuquicamata mine, Chile. In *Proceedings of the Symposium on Planning Open Pit Mines*. Amsterdam: Balkema.

Longstaff, I.D. n.d. Comparing real beam and synthetic aperture radar techniques for slope stability radar. White Paper C1001. Brisbane: GroundProbe Pty Ltd.

Martin, A.G. 1992. Automated slope monitoring in open pit mines. In *MINExpo International '92 Session Papers*. Washington, DC: American Mining Congress.

McHugh, E.L, J. Dwyer, D.G. Long, and C. Sabine. 2006. *Applications of ground-Based Radar to Mine Slope Monitoring*. Report of Investigations 9666. DHHS (NIOSH) Publication No. 2006-116. Cincinnati: National Institute for Occupational Safety and Health.

Morkeh, J. 2017. Personal communication on June 6, 2017.

Noon, D., and N. Harries. 2007. Slope stability radar for managing rock fall risks in open cut mines. Presented at Large Open Pit Mines Conference, Perth, Western Australia, September 2007.

Rose, N.D. 2011. Investigating the effects of mining-induced strain in open pit slopes. Presented at Slope Stability 2011: International Symposium on Rock Slope Stability in Open Pit Mining and Civil Engineering, Vancouver, Canada, September 18–21, 2011.

Rose, N.D., and O. Hungr. 2006. Forecasting potential rock slope failure in open pit mines using the inverse-velocity method. *International Journal of Rock Mechanics and Mining Sciences* 44(2007):308–320.

Slope Indicator Company. 1994. *Applications Guide*, 2nd ed. Bothell, WA: Slope Indicator Company.

Sullivan, T.D. 1993. Understanding pit slope movements. In *Geotechnical Instrumentation and Monitoring in Open Pit and Underground Mining*. Edited by T. Szwedzicki. Rotterdam: Balkema.

University of Queensland. 2002. Slope Stability Radar Goes Commercial. On-Line News. News release, 26 June (accessed June 14, 2017). https://www.uq.edu.au/news/article/2002/06/slope-stability-radar-goes-commercial

Wilson, S.D., and P.E. Mikkelsen. 1978. Field instrumentation. In *Landslides Analysis and Control*. Special Report 176. Edited by R.L. Schuster and R.J. Krizek. Washington, DC: Transportation Research Board, Commission on Sociotechnical Systems, National Research Council, National Academy of Sciences.

Windsor, C.R. 1993. Measuring stress and deformation in rock masses. In *Geotechnical Instrumentation and Monitoring in Open Pit and Underground Mining*. Edited by T. Szwedzicki. Rotterdam: Balkema.

Glossary*

active rock-bolt force A rock-bolt force that loads the rock masses in which the bolt is placed immediately upon installation (e.g., point anchor or tensioned roof bolt).

active unsupported span The largest unsupported span in the tunnel section between the face and the supports.

angle of repose The maximum slope at which a heap of any loose or fragmented solid material will stand without sliding or come to rest when poured or dumped in a pile or on a slope.

apparent dip The angle at which a bed, stratum, vein, or borehole is inclined from the horizontal, measured non-normal to the strike and in the vertical plane.

aquiclude A body of relatively impermeable rock that is capable of absorbing water slowly but does not transmit it rapidly enough to supply a well or spring.

aquifer A stratum or zone below the surface of the earth capable of producing water, as from a well.

aquitard A low-permeability bed, in a stratigraphic sequence, of sufficient permeability to allow movement of contaminants and to be relevant to regional groundwater flow, but of insufficient permeability for the economic production of water.

asperity An instance of joint wall roughness that affects either the movement characteristics or strength properties of the joint. Two types of asperities are recognized. Waviness of a mean joint surface is regarded as a major, or first-order, asperity. Minor, or second-order, asperities are designated as roughness.

average vehicle risk The risk associated with the percentage of time a vehicle is present in a rockfall hazard zone.

backbreak Rock broken beyond the limits of the last row of holes in a blast.

bank (slope) stability *See* stability *and* slope stability.

bench A ledge that, in open-pit mines and quarries, forms a single level of operation above which minerals or waste materials are excavated from a contiguous bank or bench face. The mineral or waste is removed in successive layers, each of which is a bench, several of which may be in operation simultaneously in different parts of—and at different elevations in—an open-pit mine or quarry.

---

* Numerous terms in this glossary are reprinted with permission from the American Geological Institute © 1997.

bench angle The angle of inclination of a bench face measured from the horizontal.

berm A horizontal shelf or ledge built into the embankment or sloping wall of an open pit, quarry, or highway cut to break the continuity of an otherwise long slope and to strengthen the slope's stability or to catch and arrest loose, falling rock.

biased sample A sample taken in a manner resulting in a greater possibility of some members being selected or observed than others.

burden In rock blasting, the distance from the borehole to the nearest free face or distance between boreholes measured perpendicular to the spacing (usually perpendicular to the free face).

buttress A structure of massive dead weight used to support the toe of a slope.

cable bolt A rock anchor composed of one or more steel cables and used for ground support.

catch bench A bench designed to provide a sufficient width to catch loose, fallen rock.

chain-link mesh An interwoven wire mesh with small openings that is draped over a rock slope to control rockfall.

Clar-type stratum compass A portable, handheld instrument for geological, geophysical, mining, tectonic, and rock mechanical applications for measuring the azimuth and dip of a rock face in one single operation. Also called COCLA. Designed by Professor Clar of the University of Vienna.

coefficient of continuity (K_c) The ratio of the sum of the unattached area of the discontinuity to the whole area of the surface. $K_c = 1$ for major features such as faults, shear zones, and so forth; $K_c < 1$ for minor structural features.

coefficient of permeability *See* hydraulic conductivity.

coefficient of variation (CV) The normalized measure of dispersion of a probability distribution. CV is defined as the ratio of the standard deviation s to the mean $\bar{x}$.

cohesion A property of like mineral grains that enables them to cling together in opposition to forces tending to separate them. Cohesion is the portion of the shear strength, S, indicated by the term c in the Mohr–Coulomb equation. *See also* Mohr–Coulomb criterion.

cone of depression The depression, approximately conical in shape, that is produced in a water table or in the piezometric surface by pumping or artesian flow. The shape of the depression is a reflection of the fact that the water must flow through progressively smaller cross sections as it nears the well; hence, the hydraulic gradient must be steeper.

continuum model (in slope stability) Modeling of the continuum is suitable for the analysis of soil slopes, massive intact rock, or heavily jointed rock masses. This approach includes the finite difference and finite element methods that discretize the whole mass to a finite number of elements with the help of generated mesh.

crack gauge An instrument used to measure displacement across a crack in the rock mass. A crack gauge may also be known as a fissurometer.

creep An imperceptibly slow, more or less continuous downward and outward movement of soil-forming soil or rock. The movement is essentially viscous, under shear stresses sufficient to produce permanent deformation but too small to produce shear failure, as in a landslide.

crest The top of an excavated slope.

critical circle (critical surface) The sliding surface for which the factor of safety is a minimum in an analysis of a slope in ductile ground where average stresses can be used.

critical dip The minimum dip of a discontinuity striking parallel to a slope face on which sliding can occur, taking into account the frictional resistance of the discontinuity, seepage, and earthquake forces.

critical surface *See* critical circle.

daylight A slang term that means to intersect a slope face between the crest of the slope and the toe of the slope. This term applies to a plane (discontinuity, joint, fault, etc.) or line (line of intersection of two planes, drill hole, etc.).

decision sight distance (DSD) The length of roadway, in feet, a driver must have to make a complex or instantaneous decision.

decking In rock blasting, the act of separating charges of explosives by inert material and placing a primer explosive in each charge.

delay A blasting cap that does not fire instantaneously but has a predetermined built-in lag or delay.

dewatering well A well used to drain or draw down an aquifer adjacent to an excavation.

digital photogrammetry The science of determining the three-dimensional locations of objects from two-dimensional digital photographs.

dilatancy An increase in the bulk volume of a granular mass during deformation, caused by a change from closed-packed structure to open-packed structure, accompanied by an increase in the pore volume.

dip angle The angle at which a bed, stratum, vein, or borehole is inclined from the horizontal, measured normal to the strike and in the vertical plane. The inclination of a line, such as a borehole, is more accurately known as plunge.

dip direction The bearing of the dip of a slope, vein, rock stratum, or borehole measured normal to the direction of strike.

discharge Loss of groundwater (e.g., that which escapes through springs, seeps, and evaporation).

discontinuity A structural feature that separates intact rock blocks within a rock mass; a structural weakness plane upon which movement can take place. A discontinuity is also referred to as a weakness plane.

discontinuum model A discontinuum approach is useful for rock slopes controlled by discontinuity behavior. Rock mass is considered as an aggregation of distinct, interacting blocks subjected to external loads and assumed to undergo motion with time. This methodology is collectively called the discrete element method (DEM). Discontinuum modeling allows for sliding between the blocks or particles. The DEM is based on solution of dynamic equation of equilibrium for each block repeatedly until the boundary conditions and laws of contact and motion are satisfied.

drain hole A borehole drilled into a water-bearing formation or mine workings through which water can be withdrawn or drained.

drainage adit An adit driven beneath an excavation or proposed excavation for the purpose of dewatering the area.

drainage gallery A gallery driven beneath an excavation or proposed excavation for the purpose of dewatering the area.

drawdown The difference, measured in meters or feet, between the original pre-pumping water table level or potentiometric surface and the pumping water level at some time, *t*.

DSD *See* decision sight distance.

dynamic (rock) strength The amount of stress that a rock can withstand without failing while under dynamic or impulsive load conditions.

extensometer An instrument used to measure axial displacement between a number of reference points on the same measurement axis. When more than two reference points are used, the instruments are referred to as multiple-position or multipoint extensometers. A typical fixed borehole extensometer consists of a reference anchor, a linkage, and a displacement sensor.

fabric The orientation in space of the elements of a rock mass. Fabric encompasses such characteristics as texture, structure, and preferred orientation.

face The more or less vertical surface of rock exposed by excavation.

factor of safety (FS) The ratio of the ultimate resisting force (or stress, moment) of the material to the force (or stress, moment) exerted against it at fracture or yield.

fault A fracture or zone of fractures along which displacement of the two sides relative to one another has occurred parallel to the fracture.

friction angle *See* internal angle of friction.

FS *See* factor of safety.

gabion wall A retaining wall constructed of wire baskets that are filled with cobbles.

generalized Hoek–Brown strength criterion A method for obtaining estimates of the strength of jointed rock masses as a means to estimate rock mass strength by scaling the relationship derived according to the geologic conditions present.

geologic compass An instrument used to determine directions, consisting of a magnetized needle that is suspended at the middle so that it is free to point to the magnetic north pole. A geologic compass can also be used to measure vertical angles.

geostatic stress Earth stress.

geotextile Any permeable textile material used with foundation, soil, rock, or earth as an integral part of a constructed project, structure, or system. In addition, other types of geotechnical engineering-related materials are sometimes informally referred to as geotextiles.

Global Positioning System (GPS) A satellite-based navigation and timing facility deployed by the U.S. Department of Defense. It employs numerous satellites deployed in geosynchronous orbits around the earth to determine one's position on the earth to a high degree of precision.

gouge The clay or clayey material along a fault or shear zone.

GPS *See* Global Positioning System.

groundwater Water below the water table (i.e., in the zone of saturation).

groundwater discharge The return of groundwater to the surface.

gunite A mixture of portland cement, sand, and water applied by pneumatic pressure through a specially adapted hose and used as a fireproofing agent and as a sealing agent to prevent weathering of mine timbers and roadways.

histogram A histogram or frequency histogram consists of a set of rectangles having (a) bases on a horizontal axis (the x-axis) with centers at the class marks and lengths equal to the class interval sizes, and (b) areas proportional to class frequencies.

horizontal drains Perforated plastic or metal pipe installed in near-horizontal drill holes to provide underground drainage.

hydraulic conductivity A measure of the tendency of a medium to transmit water. A medium has a unit hydraulic conductivity if it will transmit in unit time a unit volume of groundwater at the prevailing kinematic viscosity through a cross section of unit area, measured at right angles to the direction of flow, under a unit hydraulic gradient.

hydraulic gradient (i) The change in hydraulic head, or potential, with distance measured in the direction of flow.

hydromulching The application to a disturbed area of a slurry spray of a mixture of seeds and other substances in suspension in water.

inclinometer An instrument used to measure the curvatures induced in initially straight boreholes. The sensors may be fixed into position, or they may be portable. A typical inclinometer system is composed of an inclinometer casing, the inclinometer probe, a control cable, and the readout unit.

infilling materials Materials that are created, deposited, or washed into rock discontinuities.

intact rock The primary unbroken rock as determined from a piece of core cut for compression testing. The term *rock substance* has also been used for the unbroken or intact rock.

internal angle of friction The angle between the perpendicular to a surface and the resultant force acting on a body resting on the surface, at which the body begins to slide.

interramp slope angle The inclination of the line joining the toe of the slope above one ramp to the outermost crest point of the ramp above.

inverse velocity method A method to predict time to failure of an unstable slope using the velocity of movement of key monitoring points across the unstable ground. The velocity of movement of the key points is determined for each increment of time (Δt).

Then, the inverse of the velocity ($1/V$) is plotted versus time (t). As the inverse velocity approaches zero ($1/V \rightarrow 0$), V approaches infinity ($V \rightarrow \infty$). That is, the velocity becomes very large, very real, and very noticeable and, therefore, failure is in process.

joint A divisional plane or surface that divides a rock and along which there has been no visible movement parallel to the plane or surface.

joint conductivity The hydraulic conductivity of a rock mass for which the joints, rather than the intact rock, are the controlling factor.

joint intensity The number of joints per unit distance normal to the strike of the set.

joint roughness coefficient (JRC) A dimensionless value used in mathematically estimating joint shear strength. The joint roughness coefficient varies from 0 for perfectly smooth rock to 20 in very rough rock.

joint (wall) compressive strength (JCS) A term in the joint roughness coefficient for the determination of the strength of rough joints, often estimated using some rapid testing apparatus such as the Schmidt hammer.

kinematic stability analysis A purely geometric means of examining which modes of slope failure are potentially possible in a rock mass with respect to an existing or proposed rock slope.

kinetic stability analysis A factor-of-safety analysis for the stability of a rock slope.

laser profiler An electronic distance measuring unit that uses a timed-pulse infrared laser distance-measuring device coupled with an instrument to measure vertical and, in most cases, horizontal angles. The infrared laser is of sufficient power not to require a reflector; that is, the laser light signal can be bounced off the rock. The profiler instrument is used to determine the rock face profile of a slope.

layer A bed or stratum of rock separated from the adjacent rock by a plane of weakness.

LiDAR (light detection and ranging) A remote sensing technology that measures distance by illuminating a target with a laser and analyzing the reflected light.

limit equilibrium analysis Limit equilibrium methods investigate the equilibrium of a soil or rock mass tending to slide down under the influence of gravity. Transitional or rotational movement is considered on an assumed or known potential slip surface below the soil or rock mass. Limit equilibrium methods are based on the comparison of forces, moments, or stresses resisting movement of the mass with those that can cause unstable motion (disturbing forces). The output of the analysis is a factor of safety, defined as the ratio of the shear strength (or, alternatively, an equivalent measure of shear resistance or capacity) to the shear stress (or other equivalent measure) required for equilibrium. If the value of this factor of safety is less than 1.0, the slope is unstable.

line drilling A perimeter control technique that uses a single row of unloaded, closely spaced drill holes at the perimeter of the excavation.

linear variable differential transformer (LVDT) An electromechanical device that produces an electrical output directly proportional to the displacement of a separate movable core.

major (geological) structures Geologic features such as faults or shear zones along which displacement has occurred and that are large enough to be mapped and located as individual structures.

Markland test A visual method that uses a stereographic projection of the great circle representing a slope face together with a circle representing the friction angle of a discontinuity to represent a "critical zone" within which plane failure or wedge failure is kinematically possible.

matrix The natural material in which a fossil, metal, gem, crystal, or pebble is embedded.

method of slices A method for analyzing the stability of a slope in two dimensions. The sliding mass above the failure surface is divided into a number of slices. The forces acting on each slice are obtained by considering the mechanical equilibrium for the slices.

minor (geological) structures These include fractures, joints, bedding planes, foliation planes, and other defects in the rock mass that are generally of lesser importance with respect to rock slope stability than are major structures.

Mohr–Coulomb criterion A rock failure criterion that assumes there is a functional relationship between the normal and shear stresses acting on a potential failure surface. The relationship takes the form $\tau = c + \sigma \tan\phi$, where τ is the shear stress, c is the cohesion, σ is the normal stress, and ϕ is the internal angle of friction. When shear strength, S, is used instead of shear stress, τ, the equation becomes the "shear strength criterion."

Mohr envelope The envelope of a series of Mohr circles representing stress conditions at failure for a given material. According to Mohr's rupture hypothesis, a failure envelope is the locus of points such that the coordinates represent the combination of normal and shearing stresses that will cause a given material to fail.

overall slope angle The angle measured from the horizontal to the line joining the toe of a wall and the crest of the wall.

passive rock-bolt force A rock-bolt force that is placed in the rock mass and does not play a role in ground support until the mass moves and subsequently loads the fixture (e.g., untensioned, grouted resin bolt or split set).

piezometer Either a device used to measure liquid pressure in a system by measuring the height to which a column of the liquid rises against gravity or a device that measures the pressure (more precisely, the piezometric head) of groundwater at a specific point.

pitch angle The angle that a line in a plane makes with a horizontal line in that plane.

plane shear failure Failure resulting from shear stresses whereby the failure plane can be approximated by a single plane or by a series of interconnected planes. A rotational component is normally not present.

plunge *See* dip angle.

presplit blasting A blasting method that uses lightly loaded, closely spaced drill holes fired before the production blast to form a fracture plane across which the radial cracking from the production blast cannot travel. As a secondary benefit, the fracture plane formed may be aesthetically appealing.

principle of superposition A principle of water drawdown that states that if the transmissivity and storage coefficient of an ideal aquifer and the yield and duration of discharge or recharge of two or more wells are known, then the combined drawdown or buildup at any point within their interfering area of influence may be estimated by adding algebraically the component of drawdown of each well.

probability The relative frequency with which one event can be expected to occur in an infinitely large population of events. Probability is concerned with events that individually are not predictable but that in large numbers are predictable.

probability density function (statistics) A function representing the relative distribution of frequency of a continuous random variable from which parameters such as its mean and variance can be derived and having the property that its integral from a to b is the probability that the variable lies in this interval. Its graph is the limiting case of a histogram as the amount of data increases and the class intervals decrease in size. Also called density function.

progressive-type slope failure A slope failure that will displace at an accelerating rate and eventually result in overall failure. *See also* regressive-type slope failure; stick-slip type slope failure; transitional-type slope failure.

random sample A sample taken in such a way that there is equal chance of every member of the target population being selected or observed.

raveling Rock that falls from a face as a generally ongoing occurrence.

recharge Replenishment of groundwater, its fundamental source being precipitation.

regressive-type slope failure Slope failure characterized by a series of short-term decelerating displacement cycles, ultimately leading to failure. *See also* progressive-type slope failure; stick-slip type slope failure; transitional-type slope failure.

retaining wall A structure built to provide stability for earth, soil banks, coal, ore or waste stockpiles, water, or other material where conditions do not allow the material to assume its natural slope.

rock bolt A bar, usually constructed of steel, that is inserted into predrilled holes in rock and secured for the purpose of ground control.

rock dowel A structure that consists essentially of steel reinforcing bars (rebars) that are cemented into boreholes. Rock dowels may or may not be subjected to post-tensioning.

rock mass The in situ rock made up of the rock substance plus the structural discontinuities: matrix plus fabric.

rockfall The relatively free-falling movement of a newly detached segment of bedrock or a boulder.

rockfall section Any uninterrupted slope along a highway where the level and occurring mode of rockfall are the same.

rotational shear failure Failure resulting from shear stresses whereby the failure plane can be approximated by a curved surface, a circular arc, or a curvilinear surface. A rotational component is normally present or assumed.

safety factor *See* factor of safety.

sector The length of wall, pie-slice-shaped portion of wall, or portion of an excavation that can be considered sufficiently homogeneous to allow use of a single set of structural data, strength data, and orientation data. A *subsector* may be used when the orientation of a face or excavation changes within a constant sector.

shear strength The stress or load at which a material fails in shear.

shear stress The stress component tangential to a given plane. Also called shearing stress; tangential stress.

shotcrete Gunite that commonly includes coarse aggregate (up to 2 cm [0.8 in.]).

slenderness ratio The width-to-height ratio, t/h, of a rock block or column.

slide A mass movement of descent resulting from failure of earth, snow, or rock under shear stress along one or several surfaces that are either visible or may reasonably be inferred. The moving mass may or may not be greatly deformed, and movement may be rotational or planar.

slope The degree of inclination to the horizontal. Usually expressed as a ratio, such as 1½:1, 2:1, 3:1 (indicating 1½*H* to 1*V*, 2*H* to 1*V*, 3*H* to 1*V*); or in decimal fraction (0.04); or in degrees (2° 18'); or a percentage (4%). It is sometimes expressed as steep, moderate, gentle, mild, flat, etc. Also called gradient (or grade).

slope monitoring The surveillance of engineering structures either visually or with the aid of instruments. It includes observational techniques and instrumentation techniques.

slope stability The resistance of any inclined surface, as the wall of an open pit or cut, to failure by sliding or collapsing.

slope stability radar (SSR) State-of-the-art technology that uses radar (radio detection and ranging) interferometry for real-time monitoring of slope movement in mine walls and general slopes.

spacing In rock blasting, the distance between adjacent boreholes, usually in the same row. It is measured perpendicular to the burden.

specific capacity (of a well) The yield per unit of drawdown of a well, usually expressed as cubic meters per day per meter ($m^3/d/m$) or gallons of water per minute per foot (gpm/ft) of drawdown, after a given time has elapsed, usually 24 hours. Dividing the yield of a well by the drawdown, when each is measured at the same time, gives the specific capacity.

specific gravity The ratio of the mass of a body to the mass of an equal volume of water at a specified temperature.

stability The resistance of a structure, slope, or embankment to failure by sliding or collapsing under normal conditions for which it was designed, for example, bank stability and slope stability.

stand-up time The length of time which an underground opening will stand unsupported after excavation and barring down.

stereographic projection The projection of the orientation of a plane or the normal to that plane on a half sphere.

stick-slip type slope failure A slope failure characterized by sudden movements followed by periods of little or no movement. *See also* progressive-type slope failure; regressive-type slope failure; transitional-type slope failure.

storage coefficient The volume of water that an aquifer releases from or takes into storage per unit surface area of aquifer per unit change in the component of head normal to that surface. The storage coefficient is also known as the storativity.

stress The force per unit area, when the area approaches zero, acting within a body.

strike The course or bearing of the outcrop of an inclined, bed, vein, joint, fault plane, or borehole on a level surface; the direction of a horizontal line perpendicular to the dip direction. The course of a line (i.e., a borehole) is more accurately called the "trend."

structural feature In geology, a feature representing a discontinuity of mechanical properties, such as a joint, fault, or bedding plane. *See also* discontinuity.

surcharge An additional force applied at the exposed upper surface of a restrained soil. May be added to a slope by natural actions or by the actions of humans.

tension crack A crack that developed in rock as a result of tensile stress.

threadbar A bolt that has been threaded.

tiltmeter An instrument used to measure the horizontal and/or vertical tilt of structures, soil, or rock masses.

toe The bottom of a slope or cliff.

toppling A type of failure in which steeply dipping slabs tip toward an excavation.

transitional-type slope failure A slope failure that starts as a regressive type of failure and ends as a progressive type. *See also* progressive-type slope failure; regressive-type slope failure; stick-slip type slope failure.

transmissivity The rate at which water of prevailing kinematic viscosity is transmitted through a unit width of aquifer under unit hydraulic gradient.

trend *See* strike.

trim (cushion) blasting A control blasting technique in which the cushion blast is shot on a later delay than the production blast or at a much later date.

true dip The dip angle, in degrees, measured in the dip direction.

undercutting discontinuity A discontinuity dipping at an angle into the slope face.

unit weight Weight per unit volume; previously called density.

weakness plane *See* discontinuity.

wedge failure A type of failure involving rock bounded on two sides by discontinuities that form a tetrahedral wedge configuration.

well yield The volume of water per unit of time discharged from a well, either by pumping or free flow. Commonly measured in units of cubic meters per day (m^3/d) or gallons per minute (gpm).

wire line extensometer An extensometer employing wire from a spool located on stable ground strung to a stake anchored on unstable ground. Movement of the unstable ground extends the wire proportionally. Two general types are in use: the portable wire line extensometer, which must be read manually, and the automated dispatch linked extensometer system, which is computerized.

working face (or slope) The series of benches or berms making up a typical working pattern.

Index

NOTE: *f.* indicates figure, *t.* indicates table.